FEEDLOT EMPIRE:

Beef Cattle Feeding in Illinois and Iowa, 1840-1900

★

JAMES W. WHITAKER

A Replica Edition

The Iowa State University Press
Ames, Iowa
1975

For Mother and Father
who have helped their sons beyond measure,
and the rest of the clan
who have been interested and interesting observers

JAMES W. WHITAKER is Associate Professor of History, Iowa State University.

Composed and printed by
The Iowa State University Press
First edition, 1975
International Standard Book Number: 0-8138-0625-9
Library of Congress Catalog Card Number: 74-14276

The Iowa State University Press REPLICA EDITIONS, reproduced from typescript, are specialized studies selected for their significance and scholarly appeal.

Feedlot Empire: Beef Cattle Feeding in Illinois and Iowa, 1840-1900, is based on a thesis submitted for the doctor of philosophy degree in history at the University of Wisconsin, Madison.

COVER: A turn of the century photograph illustrating the ope lot method of cattle feeding. (Courtesy: State Histor ical Society of Iowa)

CONTENTS

Preface ix
1. Illinois and Iowa: Physical Setting and Settlement . 3
2. Illinois-Iowa Agriculture in the Nineteenth Century 9
3. Illinois-Iowa Beef Production to 1860 18
4. Chicago and the Rise of the Packing Industry 35
5. Dressed Beef and the Expansion of the Packing Industry 45
6. Structural Change in Illinois-Iowa Cattle Feeding to 1900 55
7. Regional Concentration of Illinois-Iowa Cattle Feeders 74
8. Age of Improvement I: State and Private Interests . 85
9. Age of Improvement II: Federal Influence 106
10. Age of Improvement III: Culmination 115
11. Illinois-Iowa Cattle Feeding in the Nineteenth Century 122

Appendix A: Maps
- Map 1. Iowa and Illinois Counties 127
- Map 2. Area of Wisconsin Glaciation 128
- Map 3. Illinois-Iowa General Soil Types 129
- Map 4. Illinois-Iowa Land in Farms 130
- Map 5. Railroads, 1861 131
- Map 6. Railroads, 1900 132
- Map 7. Beef Cattle per Hundred Acres, 1850 133
- Map 8. Beef Cattle per Hundred Acres, 1860 134
- Map 9. Beef Cattle per Hundred Acres, 1870 135
- Map 10. Beef Cattle per Hundred Acres, 1880 136
- Map 11. Beef Cattle per Hundred Acres, 1890 137
- Map 12. Beef Cattle per Hundred Acres, 1900 138
- Map 13. Registered Cattle Calved or Imported, 1870 . 139
- Map 14. Presumed Areas Influenced by Improved Breeds of Cattle, 1870 140

Appendix B: Tables and Figures
Table B.1. Population Growth 141
Table B.2. Illinois-Iowa Position as a Percentage of Total U.S. Amount, 1900 141
Table B.3. Illinois-Iowa Agriculture: National Position, 1850 142
Table B.4. Illinois-Iowa Agriculture: National Position, 1870 142
Table B.5. Illinois-Iowa Agriculture: National Position, 1900 142
Table B.6. Selected Chicago Beef Prices, 1840-60 . 143
Table B.7. Illinois-Iowa Railroad Development: Square Miles for Each Mile of Track . . 144
Table B.8. Receipts and Shipments of Cattle at Chicago Union Stock Yards, 1865-1900 . . 145
Table B.9. Dressed Beef Shipments from Chicago . . 146
Table B.10. Live Cattle and Dressed Beef Shipments . 146
Table B.11. U.S. Exports: Fiscal Year Ending June 30, 1900 147
Fig. B.1. Value of U.S. Exports of Pork, Lard, Live Cattle, and Beef, 1866-1900 by Five-Year Averages 148
Table B.12. Chicago Cattle Receipts, Selected Years 149
Fig. B.2. Average Monthly Prices of Fat Cattle and Feeder Cattle, 1874-1900 150
Table B.13. Beef Cattle on Farms 151
Notes 153
Bibliography 181
Index 193

P R E F A C E

This book is concerned with the development of beef cattle feeding as a major agricultural industry during the latter half of the nineteenth century. The study is limited to Illinois and Iowa because the industry reached its height within these two political divisions. This is not to say that man's political boundaries limited agricultural activity but rather that, because agricultural data in the census are gathered by and expressed in terms of political units, the historian must take counties and states into account. Though the industry described here went beyond Illinois and Iowa boundaries into southern Wisconsin, western Indiana, and northern Missouri, these areas are not discussed because it is believed they would not significantly alter the patterns that were developed.

The chronological limits of 1840 and 1900 are at logical dividing points in the development of the industry. By 1840 men were transferring the pattern of cattle feeding in Ohio, Indiana, and Kentucky to the prairies of Illinois and Iowa. This movement was not new. From earliest Colonial times, men had raised cattle on the fringes of settlement and driven them to town and city markets. Carolina cowpens, backcountry Pennsylvania, and western Massachusetts had provided cattle for eastern markets; and as settlement moved westward, cattlemen moved further west. Soon after the American Revolution, cattlemen crossed the mountains into the Ohio River valley to raise and drive cattle to Baltimore, Philadelphia, New York, and even Boston markets. Pennsylvania farmers also fattened Ohio cattle on corn before sending them to market.[1]

Once settlements crossed the mountains, the spread of population forced cattlemen to move westward again to open new grazing land. Ohio drovers appeared in Illinois in the 1820s and 1830s, and by 1840 cattlemen were well established on Illinois prairies, raising cattle to be driven east. However, when settlers reached the corn country of Illinois and Iowa, instead of moving on in a few years they remained to raise and feed cattle through the end of the century and to the present day. Though men continued to carry the raising of cattle westward to the plains, the center of prime fat-cattle production remained in Illinois and Iowa.

By the end of the 1850s, cattlemen of central Illinois and eastern Iowa had shifted from raising cattle to be driven east for fattening in Ohio and Pennsylvania to fattening their own on corn and grass for the Chicago and New York markets. At this early date, the commercial outlines of Corn Belt agriculture were visible. This important combination of corn and livestock was present from the early nineteenth century. How to market the combination was the problem for the farmers. Although the Corn Belt is often thought of as an area of corn and hogs, it was and is just as much an area of corn and cattle.

Cattle feeding in Illinois and Iowa started as a commercial enterprise; and contrary to the experience in earlier areas, the business in these two states survived the technological changes and geographical shifts in areas of supply and markets that took place during the nineteenth century. By 1900, compared to the 1840s and 1850s, the industry had changed in types of feeding routines, in quality of beef produced, in age and weight of cattle at market time, and in principal markets; it had evolved a new approach that remained well into the twentieth century. Through all this change, Illinois and Iowa cattle feeders quickly moved from the fringes of the fat-cattle trade in the 1840s to be the leading prime-beef producers in the last half of the century.

This book is the story of the early development and the subsequent growth of this distinctive aspect of Corn Belt agriculture. External forces such as the development of the railroad network, the rise of the urban market, and the development of refrigeration by the Chicago meat packers, which completely changed the processing and marketing system for fresh meat, had made changes possible and necessary in the way cattle feeders operated. Illinois and Iowa men who responded to the outside market influences by changing their feeding systems to produce better quality beef at a younger age led the way in the adjustment to new circumstances, which kept these states the center of prime-cattle production.

This study could not have been completed without the

patient encouragement and help of Vernon Carstensen. Grateful acknowledgment is also made to Morton Rothstein; to Allan Bogue for sharing some statistics; to Iowa State University Sciences and Humanities Research Institute, Graduate College, and Department of History for financial aid; to the library staffs of the State Historical Society of Wisconsin, the University of Illinois, and Iowa State University; to the Illinois Historical Survey, especially its director, Robert Sutton; and to Jan Miller for double-checking notes.

FEEDLOT EMPIRE: Beef Cattle Feeding in Illinois and Iowa, 1840-1900

CHAPTER 1. ILLINOIS AND IOWA: PHYSICAL SETTING AND SETTLEMENT

The major portions of Illinois and Iowa, the core of that great agricultural area of the United States known as the Corn Belt, possessed an excellent combination of natural resources in soil, climate, topography, potential transportation routes, and lack of serious barriers to agriculture. Farmers quickly exploited this rich land and produced millions of bushels of corn as their major crop. Corn—a bulky, low-value grain—cost too much to transport far to market; but when fed to beef cattle and swine, it became a very profitable crop. This combination of prairie grass, corn, cattle, and swine early became the commercial base of Corn Belt agriculture; and though changes occurred during the nineteenth century, Illinois and Iowa farmers remained leading producers of fat cattle.

Corn grew on a wide variety of soils, but highest yields came from dark, well-drained, silt loam with an abundance of humus or organic matter. Humus was an important ingredient because it held moisture and prevented rain from running off or leaching the mineral elements too rapidly. Humus directly supplied nitrogen, the soil element needed in greatest amounts for corn.[1]

Geologic forces had endowed Illinois and Iowa with large areas of fertile soil ideal for producing corn and rich pasture. A series of glacial advances across the two states left them mantled with glacial till, crushed sedimentary rock (mainly limestone), wind-blown loess, and lacustrine deposits from glacial lake beds—all excellent soil-building materials. A good balance of acid and alkali in the soil south and west of Lake Michigan provided an excellent area for tall prairie grass. The deep roots and growth processes of this grass incorporated much organic material (humus) into the soil. This stored plant food, held moisture, fostered biotic life, and gave structure to the black prairie soils. Favored with a humid continental climate and warm summers, this land could produce in abundance for years with proper farming practices.[2]

At least four different glaciations combined to help create the bountiful soil of Illinois and Iowa. As each glacier advanced, it scraped off the high points and filled in the low spots. When the Wisconsin glacier (the last) receded, wind and water erosion shaped and modified the morainic and drift cover that remained. These elements created a terrain in the area of Illinois that was a gently undulating plain in the north, west, and south; nearly flat in the east-central part; and quite hilly and broken in the extreme northwest and in the south where the glaciers did not scrape off the tops of the Ozark hills. The Mississippi and Illinois rivers cut their channels below the level of the plain and in some areas created wide lowlands of rich alluvial or terrace soils. The area of Wisconsin glacial drift in the northern and eastern part of Illinois was the last of the state to be disturbed by glaciers and thus was exposed for a shorter period to the actions of wind and water. (See Map A.2. Map A.1 containing county outlines and names is the first of the maps in the Appendix.) Fewer mineral elements had been leached out of the soil by rain; therefore, the soil was more fertile than in other parts of the state. There also had been less time to develop a good drainage pattern for the area, with the result that successful cultivation awaited drainage projects. This was especially true of the flattest portion of the state in the east-central counties of Champaign, Ford, Iroquois, Kankakee, Livingston, and Vermilion.

During the last glacial recession wind-borne silt (loess), blown from the west, covered parts of the state. Thickest in the western part, it thinned out to the east and south. Variations in loess cover, age of the glacial drift, and degree of weathering accounted for important variations in Illinois soils and consequent agricultural productivity.

Most of western Illinois was covered with loess four feet or more thick. Exceptions were the alluvial area of the lower Illinois River valley; sandy, glacial drift and wind deposits in Mason County; and a basin in the north, covering parts of Whiteside, Lee, Bureau, and Henry counties. The loess and alluvium areas were very fertile, but the sandy areas were less so and subject to poor drainage. Wind deposited very little loess upon the fertile drift cover left by the last glacier in the northeast quarter of Illinois. In the eastern counties of Kankakee, Iroquois, and Ford were some less fertile sandy deposits. Just south and west of this area loess existed from two to four feet thick in the east-central part of the state; the southern part south of Coles County had only small deposits. Distribution of the fertile loess duplicated many of the other fertility producing factors, and its absence from most of southern Illinois decreased that area's prospects for raising corn. This was one of the reasons that southern Illinois farmers did not concentrate on beef cattle production.[3]

The vegetation that developed on top of this soil material modified it. By the time of American settlement prairie grass covered much of Illinois north of Adams County west of the Illinois River and north of Christian County east of the river except for deciduous tree cover along the watercourses and in the extreme northwest. South of Adams and Christian counties tree cover was the rule and occasional prairies the exception. The dark prairie soils of northern and central Illinois, further enriched by generations of grasses, were more productive than the lighter forest soils of southern Illinois and the more broken country. Much of southern Illinois had a layer of hard, relatively impervious subsoil known as claypan, which prevented good drainage, reduced absorption of rainfall, restricted the rise of subsurface water, and favored development of shallow root systems. The presence of claypan became critical during wet and dry spells by either keeping the topsoil too wet or letting it dry out too quickly. (See Map A.3.) Forest soils (other than that of the claypan area), though less fertile than prairie soils, were nevertheless suitable for a wide variety of crops.[4]

Iowa topography was a gently rolling plain with the highest elevations in the northwest. The roughest, most broken terrain with the highest percentage of land not suited for cornfields was in the northeast corner and comprised Allamakee County and parts of five nearby counties. After the action of the same four glaciers that affected Illinois, Iowa had no area of the claypan soil that hindered southern Illinois agriculture. During the retreat of the Wisconsin glacier winds covered over two-thirds of present Iowa with four feet or more of rich soil-building loess. North-central Iowa from Dickinson County south to Dallas County and north east to Franklin County received no loess cover during the last glaciation (as did east-central Illinois) and was the flattest portion of the state. Thus it had the most drainage problems, making it suitable first for grazing country rather than grain crops. With the exception of the counties adjoining the Missouri River in the west and the river valleys in the eastern part of the state, most of Iowa was prairie. Combined with loess and the favorable glacial drift, centuries of biotic action by the prairie grass created the brunizem prairie soils over most of Iowa, perhaps the "most productive soils known to man."[5]

Corn Belt climate was as favorable as the soil resource. All Illinois and Iowa had an average growing season of over 145 days from the last killing frost in spring to the first in fall. Varieties of dent corn dominant in Illinois and Iowa in the last half of the nineteenth century needed from 90 to 125 days of tolerable temperature and rainfall conditions to mature. Corn grew best with hot wet weather in July and August and in this, too, climate favored the Corn Belt. Most of the yearly precipitation for Illinois and Iowa came in the warm months with an average of 20 to 25 inches of rain between April 1 and October 31. On the average at least half of this rain came in the three crucial months of June, July, and August. Periodic droughts and variations in the average dates of frosts occurred, causing variations in the crop yield; but natural conditions usually favored agricultural productivity.[6]

In addition to soil fertility and climate, natural transportation routes for movement of people and produce favored agricultural development in Illinois and Iowa. The drainage systems created by glacial runoff and subsequent centuries of weathering formed the Great Lakes and the Ohio, Illinois, Mississippi, and Missouri rivers, which with their tributaries made it possible to till much of the land without artificial drainage and also provided a low-cost transportation outlet to markets to the south and east. The low elevation of the divide between Lake Michigan and the Illinois River permitted construction of a canal linking the two, and the relatively level terrain made it easy soon after to connect lake ports by railroad to the productive agricultural interior.

Physical factors made most of the rich land of Illinois and Iowa fully able to produce an abundance of food for men and animals if farmers put it to use. These factors also differentiated the area and in part accounted for different types of agricultural uses. Well-drained fertile prairie soil was good for grains, especially corn; poorly drained prairie served best in early years as pasture for cattle. Claypan soils inhibited corn production. Rough terrain was better suited to pasture crops than grain and encouraged dairying. Nevertheless, the past experience, current predilections, and ability of settlers on this rich soil accounted in an unmeasurable but important way for the use they made of available resources.[7]

Westering Americans (and immigrants) did not take long to learn of, covet, and possess Illinois and Iowa land. People came by ones and twos, by families and towns, by foot and horseback, by wagon and canal, by lake and riverboat. By the early 1830s the Erie Canal and the Ohio River had opened the gates to a flood of humanity that surged west from Lake Michigan and up the valleys of the Wabash, Big Muddy, Kaskaskia, Illinois, Sangamon, and Rock rivers in Illinois and started up the Des Moines, Skunk, Iowa, Cedar, Wapsipinicon, Maquoketa, and Turkey river valleys in Iowa.

Enticing reports of the fertile and beautiful country had been appearing for years. Colonel George Rogers Clark, reporting on his military campaigns during the Revolution, in 1779 described southern Illinois as "more beautiful than any idea I could have formed of a country almost in a state of nature." He predicted that someday "meadows, extending beyond the reach of your eye," would "excel in cattle."[8]

Scottish farmer, Patrick Shirreff, after his travels through Illinois in 1833, pronounced it "not greatly surpassed by any portion of the earth" for agricultural quality and potential and predicted that it would have "sufficient room for all who are likely to desire a settlement." Two years later Albert Miller Lea writing of his reconnaissance with the United States Dragoons heralded yet another "Eden" in the Iowa River valley, which "presents to the imagination the finest picture on earth of a country prepared by Providence for the habitation of man." Many other travelers and residents testified to the agricultural potential of the region. These early estimates of the productivity of the area received a back-handed seconding from Horace Bushnell, Connecticut minister, lecturer, and essayist, who in 1846 labeled the incipient Corn Belt, not as a beautiful, hilly paradise such as New England, but rather the visually unappealing "great American Corn-field, the Poland of the United States."[9]

By 1846 Iowa had garnered enough people to become a

state even though Indians still held part of it; and Illinois, though a state since 1818, showed a remarkable ability to absorb all those who desired to settle. By 1850 only the Grand Prairie area of Champaign, Ford, Iroquois, Kankakee, Livingston, and Piatt counties and the wet bottomland of Bureau and Lee counties had less than six people per square mile.[10]

The transition from Indian country to statehood went quickly once settlement had begun. Beginning with the first Indian land cessions (1795 in Illinois and 1824 in Iowa), an evolution of events followed: federal land survey and sales, increasing settlement, and the development of political structures. Indians ceded land in Illinois from 1795 to 1833 and in Iowa from 1824 to 1851. Federal land surveyors had finished their work in Illinois, for all practical purposes, by 1844 and in Iowa by 1861.[11]

Early settlers urged their friends and relatives to come to "the flower of the World, the Garden of Eden, a land flowing with milk and honey." From these passionate urgings, from general reports about soil fertility, and from an assumed salubrity of climate as well as the many common motivations for human migration, settlers flocked to Illinois after 1815 and to Iowa after 1833. They moved up the river valleys first where there was a transportation route and timber for fuel and shelter and then pushed on to the prairies, and rapid population growth continued. By 1900, 25.6% of Iowans and 53.9% of Illinoians were classified as urban dwellers in towns of 2,500 population or more. (See Table B.1.) This meant that an increasingly smaller percentage of the population accounted for the growing agricultural production. Illinois had 264,151 farms in 1900 and Iowa had 228,622, but from these farms came an abundance making the two states leaders in farm production.[12] (See Table B.2.)

As people moved into this region, they quickly shifted farming from a pioneering way of life to a business. Corn and livestock thrived on the rich land and favorable climate from the beginning of settlement; and although distractions appeared from time to time, corn and cattle remained major products. By the end of the century, they accounted for the largest percentage of the value of farm products of the two states.[13] (See Table B.2.)

CHAPTER 2. ILLINOIS-IOWA AGRICULTURE IN THE NINETEENTH CENTURY

As farmers settled in Illinois and Iowa in the 1840s and 1850s, they sought to develop a commercial agriculture. Rather than attempting to be self-sufficient by producing a wide variety of crops and livestock, frontier farmers quickly concentrated on a few income-producing activities, which they changed as market conditions changed. No exception to American agricultural experience, these men made little attempt to conform to a literary ideal of the self-sufficient pioneer. They simply were not interested; the long hours, versatile talents required, and lack of opportunity for the capital accumulation necessary for an attempt at self-sufficiency did not appeal. The productivity of the land, the options before them, and the developing transportation system encouraged farmers to think in terms of commercial rather than subsistence agriculture.[1]

Men turned from the sod-corn and wheat of the pioneer years to cash-grain and livestock production. This was a quick transition in central and northern Illinois and eastern Iowa with less experimentation with a variety of crops than took place in Wisconsin, for example. There farmers spent two decades growing wheat, sugar beets, sorghum, hops, flax, tobacco, and sheep before they found dairying to be the best combination of resources for commercial agriculture.[2] No need existed for this long period of searching for commercial combinations in Illinois and Iowa where the grass and abundance of corn made livestock feeding a logical choice.

Distractions at times drew some farmers away from cattle feeding, but these did not last. In the 1840s high prices for wool encouraged many Illinois farmers to raise sheep, with a resulting increase in wool production from 1,500 pounds in 1842, to 246,610 pounds in 1846, to an estimated 1,200,000 pounds in 1849. Sheep raising declined in popularity during the 1850s but increased again in Illinois during the Civil War because of the cotton shortage. After the war, however, wool prices dropped and farmers turned to more rewarding enterprises. Iowa, too, experienced the sheep boom and the disillusionment in the late 1860s. In Cedar County, for example, "the sheep mania . . . proved a bad speculation." As an alternative to sheep or the unrewarding wheat, men called for more "capital and labor engaged in dairy, and stock raising."[3]

Chinese sugarcane (sorghum) and sugar beets were two other minor enterprises tried in the 1850s and 1860s. By 1860 Iowa accounted for 18% and Illinois for 12% of gallons of sorghum produced in the United States.[4] The first successful Illinois sugar beet refining operation began at Chatsworth in Livingston County in 1867 and processed 40 tons a day in 1868. Two years later a Shelby County farmers' club issued an "Invitation to Capitalists" to establish a sugar beet factory there. The Chatsworth factory moved to Freeport in 1871 to be closer to land better suited for growing beets.[5] But neither sorghum nor sugar beets was ever a major crop in acreage or value of product.

A surprisingly short time elapsed after settlement until Illinois and Iowa farmers turned to commercial production of grain and livestock. Little effort was required to plant corn and even less to harvest it because cattle could be turned into the fields and allowed to harvest the crop themselves. What the cattle knocked down and wasted, swine could salvage. Corn was the basic survival crop for both men and animals, so that corn and cattle were present from the time of settlement and only gave settlers the problem of finding their most profitable use. Henry R. Schoolcraft, United States Indian Agent and member of several reconnaissance expeditions in the upper Mississippi valley, remarked on this production and the potential of an outside market in Sangamon County, Illinois, as early as 1821. Schoolcraft thought the region was prepared to compete with Michigan and Ohio in provisioning government posts on the Great Lakes. Timothy Flint, missionary to the Mississippi valley and writer, reported in 1828 that eventually prairies would be habitable after people planted trees on them, but in the interim the "inexhaustible" summer range for cattle gave Illinois an advantage over eastern and southern states in the cattle trade. Bluegrass, said Flint, had appeared in areas where settlers had killed prairie grass, and the state "sent great numbers of fine cattle and horses to New Orleans." Five years later in 1833 Illinois farmers had plowed under sufficient prairie for grain production to warrant Patrick Shirreff's observation that forest agriculture "is seldom followed, the cultivation of the prairie being so much more simple and profitable." Farming seemed easy to Shirreff in terms of labor and capital investment, and some settlers attested to this in noting that sickness was no "great misfortune" because it was so easy to make a living. An Iowan reported the land so rich he could raise as much with half the labor as in Ohio. Others disagreed, at least to the extent of saying that it was a hard country for women and cattle. Two universal comments, complaints that applied well into the latter half of the century, were that farmers did not provide adequate shelter for livestock and farms were too large to be worked efficiently.[6]

From 1830 to 1860 wheat was a major money crop, but after the 1830s, increasing amounts of corn, cattle, and swine were produced and exported from the area in some form. With the exception of southern Illinois, the two-state area justified the name Corn Belt before the Civil War. In 1859 a Knox County, Illinois, farmer exclaimed: "Corn, the country is one immense corn field."[7]

By the time of the 1850 census one-third of the land in Illinois, with less than 900,000 inhabitants, was already in farms. (See Map A.4.) But the land was so productive that Illinois farmers (with only 5.2% of the farms and 4.4% of the improved farmland in the nation) produced an abundance of corn and wheat disproportionate to their numbers and a slightly greater proportion of oats, "other" or beef cattle (i.e., not dairy cows or oxen), and swine (which followed the cattle in the feedlot during corn feeding at the approximate rate of two swine for each steer). Iowa lagged about 10 to 15 years behind Illinois in the growth of agricultural production, and in 1850 only 7.7% of the land in the state was in farms. While the less than 200,000 inhabitants of Iowa held 1% of the nation's farms and only 0.73% of the nation's improved farmland, production of wheat and corn exceeded a proportional share of the national total.[8] (See Table B.3.) Corn and livestock were already a major part of Corn Belt agriculture.

Production totals for the two states do not tell the whole story, for in 1850 concentrations of the various crops existed within each. Illinois farmers in northern counties close to lake ports or central counties along the Illinois River and Illinois-Michigan Canal could raise wheat as a cash crop after 1848. East-central and southern Illinois farmers concentrated more on corn for the southern market or on local livestock. Swine were distributed rather evenly over the state, but there was a concentration in the corn counties (except De Witt and Vermilion-Ford); in the southeastern counties from Crawford to Williamson; and in the west-central counties of Knox, Warren, McDonough, and Cass. Counties north of Marshall had the least number of swine. "Other" cattle were most plentiful in the central counties of Cass, Greene, Macoupin, Menard, Morgan, Sangamon, and Scott with lesser concentrations in a belt from Edgar through Vermilion-Ford, McLean, Tazewell, Fulton, McDonough, and Adams and in Madison and Randolph counties east of St. Louis. The corn-cattle-hog pattern of Corn Belt agriculture already existed though it was not yet highly developed.[9]

The decade 1850 to 1860 saw many changes. After several years of good crops, some improvement in transportation, and good prices because of the Crimean War, wheat became less reliable in northern Illinois. Crop failures because of spring rains, late frosts, and insects after 1855 caused many farmers to take to heart the calls from the agricultural societies and periodicals to give up wheat for more stock raising. By 1860 northern Illinois farmers placed considerably less emphasis on wheat, a condition that lasted through 1870. Farther south, wheat cost more to produce than it was worth in some inland areas; some even considered it a "nuisance" that just bred chinch bugs. More often than not, standing water killed winter wheat in central and eastern Illinois in areas of poor drainage. The drill, which provided a better method of planting wheat; a shift to winter wheat, which could be harvested earlier in the season; and, most significant, the coming of railroads to southern Illinois in the 1850s caused an increase in wheat production in the south. In 1870 Commissioner of Agriculture Horace Capron felt it necessary to devote a major paragraph in his speech at the Illinois State Fair to urge farmers to depend less on wheat. The total wheat acreage fluctuated between 1.7 and 2.5 million acres from 1859 to 1874; but as total acres in farms increased, wheat became proportionally less important.[10]

Because of soil and terrain and, to some degree, because of the type of farmers who stayed there, southern Illinois did not share in the development of the corn-livestock farm economy of the rest of the area. Though the state was rather backward in Corn Belt farming techniques and attitudes (a reaper salesman called southern Illinois farmers the "coon dog and butcher knife tribe"), an effort to excel in horticulture was successful. As early as 1836 Gershom Flagg had 500 apple trees on his Madison County farm, and a major part of his income came from fruit sales; but interest in large-scale production did not come until the 1850s and later. When production increased, the Illinois Central Railroad operated an express train to carry peaches, apples, and berries into Chicago. By 1868 one station in Shelby County claimed it had shipped 50,000 boxes of peaches in one season.[11]

From 1860 on, in response to the growing urban market, northeastern Illinois included more dairying and truck garden production. Gail Borden opened his first milk condensing factory at Elgin in 1865; and by 1870 Kane, McHenry, De Kalb, Cook, Lake, Du Page, Kankakee, Boone, and Winnebago counties were important cheese-producing centers.[12]

The rise of dairying was only one of the signs of growing regional specialization in farming. In the 1850s east-central Illinois turned to corn and livestock when the Illinois Central Railroad opened the area by bringing in settlers and providing transportation to market. Farmers in counties close to the Mississippi, Illinois, and Ohio rivers had shipped corn, beef, and pork to the south; but the Civil War blockades interrupted this, forcing men to seek markets in the east by way of Chicago. The higher cost of freight (in proportion to the value of grain) by wagon and railroad to Chicago caused many farmers to concentrate on feeding corn to cattle and swine. Livestock could walk to the nearest railroad, and freight charges on livestock were a lower percentage of the value of the product than charges on grain shipments.[13]

In the meantime Iowa farming went through changes similar to those in central and northern Illinois. Though by 1860 not nearly as much of Iowa as Illinois had over 50% of the land divided into counties in use as farms, the grain-livestock economy was well developed in the eastern part of the state. (See Map A.4.)

Early settlers had planted wheat and corn in the Cedar, Iowa, Des Moines, and other river valleys. No area corresponded to southern Illinois and its problems. The river system channeled trade to the south until the railroads extending west from Chicago crossed the Mississippi. By 1860 four Iowa railroad lines met Illinois railroads from Chicago at the river. In 1850, for example, Wapello County, about 70 miles up the Des Moines River, had over 51% of its land in farms and over 35% of the land was improved. Of the 978 householders for the 1850 census, 178 came from Illinois, 369 from Indiana, and 242 from Ohio, so that most were familiar with farming as practiced in the old Northwest.[14]

Some areas of agricultural concentration existed in Iowa in the 1850s. Eastern Iowa counties north of the Cedar River and Benton, Iowa, Johnson, Mahaska, Louisa, Des Moines, and Lee counties concentrated on wheat. (See Map A.1.) Muscatine, Taylor, and Page counties and the area from Keokuk to Madison county and south embraced the higher corn-producing areas. The swine concentrations centered in the corn counties and in Marshall, Fremont, Washington, and Des Moines counties. Cattle concentrations occurred in widely scattered counties such as Allamakee and Delaware in the northeast; in Pottawattamie, Fremont, Page, Taylor, and Decatur in the southwest; and in Clinton, Muscatine, Louisa, Des Moines, Lee, and Washington in the southeast. Many Iowa farmers also had turned to corn, cattle, and hogs.[15]

But as the corn-cattle pattern appeared, changes in agricultural technology encouraged men to continue to return to grain production. The periodic lure of corn or wheat as a cash crop continued to attract farmers away from cattle feeding during the 1850s, 1860s, and 1870s. Through a combination of availability of railroads, type of land tenure, cost of drainage, and price of beef in the 1880s and early 1890s, east-central Illinois was fully and finally turned toward producing cash grains rather than fat cattle. Most important in encouraging grain production were the increased invention and use of agricultural machinery in both states from 1840. Perhaps most dramatic in their impact on agriculture were the plow and the reaper. Prairie plows opened up more new land, and the reaper made larger wheat and oat fields per farm possible. By 1851 the McCormick factory produced over 1,000 reapers a year at Chicago, and John Deere turned out 10,000 plows a year by 1857. In 1859 one man in Piatt County, Illinois, had a portable steam engine, four threshing machines for steam or horse power, a corn sheller, and more work for them than he could handle. During the last half of the century, mechanical innovations continued as inventors refined tillage machinery, such as Oliver's introduction of the chilled-iron plow in 1870, and harvesting equipment to make each man-hour more productive.[16]

Of great importance in aiding the expansion of farms across the two states was the growth of the transportation network. Railroads played the most important part in marketing the agricultural produce of the interior counties and providing an all-weather outlet to the east. From only 788 miles of track in Illinois and none in Iowa in 1854 to 4,823 miles in Illinois and 2,683 miles in Iowa in 1870, the railroad networks spread out from Chicago. By 1900, Illinois had 11,002 miles and Iowa had 9,185 miles of single-track line.[17]

Innovations in machinery, development of the transportation network, and growth of population caused the number of farms in Iowa and Illinois to increase from 1850 to 1860 by 313% and 88% respectively and from 1860 to 1870 by 90% and 41%, while the United States increase in the two decades had been only 41% and 30%. (See Map A.4.) The amount of improved land in Illinois increased by 159% from 1850 to 1860 and by 47% from 1860 to 1870. In Iowa improved land increased by 359% and 147% for the same periods. Thus by 1870 farmers had turned 72% of the land in Illinois into farms and 43.7% of the land in Iowa. Production figures for 1870 compared to 1850 indicate the same quick growth in a rich area and reveal the dominance of feed grains over wheat in Illinois. (See Table B.4.) The feed-grain dominance was less pronounced in Iowa than in Illinois in 1870 than it would be later, but the trend was established.[18]

From 1870 to 1900 further regional shifts occurred in agricultural production within Illinois and Iowa, but the result was more corn and less wheat. As settlement moved westward in Iowa and the transportation network developed, Iowa agricultural regions changed. The center of corn concentration moved toward the southwest, the size of the area where sheep were important became smaller, and farmers in the northeast counties moved toward more dairy production. East-central Iowa farmers shifted toward meat production of swine and beef fed on corn, and those in the north-central counties began raising more oats in addition to the dairy and beef cattle. Wheat production centered in the northeast. Iowa was moving toward the early twentieth-century divisions of northeastern dairy area; north-central cash-grain area; western corn, beef, and swine area; south-central and southeastern pasture areas; and east-central livestock area, with swine predominating.[19]

After 1870 Illinois farmers in the northeastern counties

around Chicago concentrated increasingly on dairy and truck produce. Those in the northwest counties, having found a good combination, continued the trend toward cattle and hogs. Farmers north and west of the Illinois River continued in livestock and grain, but men in the east-central region turned increasingly to cash-grain operations in place of cattle feeding. Farmers in the Sangamon River counties, who had raised both wheat and corn, turned more to corn. From Shelby County south, operations could be called "general farming" where there was a combination of corn, wheat, hay, cattle, and swine. Around East St. Louis some concentration of dairy cattle and produce existed and there were areas of fruit specialization in the southern counties; but on the whole the southern fifth of the state was out of the main trend of Corn Belt agriculture.[20]

By 1900 the two-state area, which had only 3.7% of the land surface of the United States, contained 13.9% of the improved farmland of the nation. Most farmers had given up raising wheat and turned to feed grains and livestock, which had been indicated as the major commercial combination as early as 1850. (See Table B.5.) Corn was the most important income producer of the grains, accounting for 54.2% of the value of all farm crops produced in 1899 in Illinois and 50.6% in Iowa. Oats accounted for just over 17% of cereal crop values in each state, and wheat contributed less than 6%. Of the value of domestic animals on farms in June 1900, "other" or beef cattle (i.e., not dairy cows 2 years old or over) accounted for 25.6% in Illinois and 35.3% in Iowa. Swine, though in larger numbers on farms, accounted for only 12.6% of the value of domestic animals in Illinois and 16.0% of that value in Iowa. Sheep in each state accounted for less than 2% of domestic animal values. The cattle and swine figures do not take into account the number of animals marketed annually, so they are not completely representative of the two-state share of cattle raised and marketed during the year. Because many of the hogs marketed came as an adjunct to cattle feeding, one can discuss cattle as the major livestock interest of farmers feeding corn without excluding swine. Surely the man prophesied rightly who had described this area as "the great bank from which our sustenance is discounted."[21]

This concentration of the Corn Belt on feed grains and livestock was taking place at a time when the population of the United States continued to grow at a fast rate. More significantly, an increasingly greater proportion of the population was urban and did not produce its own food. The growing urban areas provided Corn Belt farmers with the opportunity as well as the need to change their agriculture from simply a way of life to a highly organized commercial operation. By 1900, 37% of the United States population lived in places of 4,000 or more residents, and the trend was toward greater urbanization. The 1900 census figures represented a 20% increase over 1890 in total population, and a 36% increase in urban population over 1890. For the same period the population of Illinois increased by 26%, but its urban population increased by 53% for a total of 51% of the state living in places of 4,000 or more. Iowa, much less urban, gained 16% in total population and 36% in urban dwellers from 1890 to 1900, when 20% of the state lived in places of 4,000 or more.[22]

The land of the Corn Belt was suited to produce for the growing urban market by geography and the type of farmers who settled it. Nowhere was this response to that market more noticeable than in the changes that took place in the cattle feeding industry in Illinois and Iowa. Feeding corn to cattle was well established by 1850 in the Corn Belt; but by 1870, in response to the expanding urban market, men devised refrigeration methods to ship fresh beef to any part of the country. In turn the demand for better quality meat for the refrigerated dressed beef trade brought changes in methods of feeding and in the quality of beef cattle in Illinois and Iowa.

CHAPTER 3. ILLINOIS-IOWA BEEF PRODUCTION TO 1860

Corn Belt farming is perhaps most often thought of as a concentration on corn and hogs, but equally important in the nineteenth century was the combination of corn and cattle. Feeding corn to cattle and then using hogs to salvage the waste was a successful commercial agricultural pattern in central and northern Illinois and eastern Iowa.

In the years before the changes of the 1860s and 1870s, Illinois and Iowa cattlemen developed a pattern of operations similar to that of areas in Ohio, Indiana, and Kentucky. From Colonial times, cattlemen operated on the fringes of the settled area where land was cheaper and animals could graze over large areas. A range cattle business developed in western Massachusetts, in central Pennsylvania, and in the backcountry of the South. Men collected their cattle periodically and drove them to city or town markets. As the area of settlement expanded in the late eighteenth century, Pennsylvania and Virginia cattlemen moved their operations over the mountains into the Ohio River valley. There they found the combination of corn and grass in central Ohio and Kentucky ideal for raising and feeding cattle. Livestock feeding was also a way to market the frontier corn crop, which was not profitable to ship any distance as grain because of high transportation costs. However, when distilled into whiskey or fed to cattle and hogs, corn became profitable. Ohio and Kentucky farmers drove corn-fattened cattle over the mountains to market. As early as 1805 George Renick had driven cattle from southern Ohio to Baltimore, and by 1820 cattle driven over the Wilderness Road, Cumberland Road, and Mohawk-Erie Trail supplied the eastern markets. The rich farming counties in eastern Pennsylvania specialized in finishing Ohio cattle for the Philadelphia and New York butchers. In the middle Scioto River valley in Ohio, bluegrass country in Kentucky, and upper Wabash River valley in Indiana, farmers increasingly specialized, first on raising and driving cattle east, then on fattening those obtained farther west.[1]

By 1819 Ohio cattle feeders had traveled as far west as Missouri and Illinois in search of young animals to return to Indiana, Ohio, and Kentucky for fattening on corn in preparation for the overland drive to eastern markets. Some of the cattlemen soon took advantage of the rich prairies of Illinois and Iowa as places to raise cattle and corn more cheaply than in Ohio or Kentucky. By the 1820s and 1830s, even before the complete removal of the Indians, travelers commented upon the lush prairie grass that made Illinois a natural grazing area. They also noted that Illinois men already drove cattle to eastern markets or sent them to New Orleans. By 1831 in Madison County "cattle raising was carried on at a large scale"; in Morgan County settlers remembered that in the "early years," the great profits came from raising and feeding cattle; and by 1833 in the area around Springfield, Illinois, there were "a great many cattle . . . reared on the prairies."[2]

In addition to supplying stock to Ohio corn raisers, Illinois man fattened cattle on prairie grass, bluegrass pasture, and corn. In the early 1830s, Patrick Shirreff reported that farmers in central Illinois complained that cattle did not "thrive well" on cured prairie grass. He reported that the more experienced farmers used prairie grass only for summer grazing and planted timothy for cutting because it yielded a ton more per acre than either bluegrass or red clover. Farther north in Putnam County, by 1849 farmers replaced prairie grass with timothy because the wild grass was not ready for grazing until late spring and failed earlier than timothy in the fall. Also, increased use of fencing prevented men from driving cattle to open prairie and back every day.[3]

Drovers brought the cattle feeding pattern to Illinois when they returned to settle after seeing the advantages of cheap prairie grass and corn there. They changed from being itinerants in the west to settled landowners and made the adjustment to a new kind of operation. Such men as John T. Alexander and Jacob Strawn of Morgan County, Isaac Funk of McLean County, and Benjamin Harris of Champaign County set the example for Corn Belt agriculture as they made the cattle feeding operation work in Illinois. In a few years they began to compete with Ohio cattlemen in driving fat cattle from Illinois to eastern markets. Benjamin F. Harris, for example, was born in Virginia in 1811 and moved with his parents to Ohio in 1833 where he worked as a drover taking stock from Illinois, Indiana, and Ohio to the eastern markets. He came to Illinois as early as 1835, and by 1841 he grazed large numbers of cattle on the prairies of east-central Illinois. When he found little market for the grain he also raised, he developed a major corn feeding operation.[4]

Harris was just one of the many who moved west and brought familiar cattle operations with them. Agricultural developments in Illinois and Iowa had begun to follow the pattern of transition from range country to a grain growing and cattle feeding region. By the 1850s eastern Iowa men fattened cattle for Chicago and eastern markets, and southwestern Iowa around Council Bluffs and areas in northern Iowa supplied feeder cattle for eastern Iowa and Illinois. Increased land values in Ohio, rising prices for corn, and lower costs of cattle production in the states to the west gradually aided the shift in the emphasis on feeding from Ohio to Illinois and Iowa.[5]

Agricultural writers such as Solon Robinson (Indiana farmer, frequent correspondent to the Prairie Farmer, and occasional agricultural editor of the New York Tribune) felt that not enough farmers took advantage of natural conditions in Illinois and Iowa favoring cattle feeding. In 1850 Robinson wrote: "Now, my friends, is it not time for you to begin to think that wheat is not the most natural and profitable staple crop of this part of Uncle Sam's big pasture? Does any land in the world produce better beef than the prairies of Indiana, Illinois, Wisconsin, and Iowa? Grass, either wild or cultivated, is ever growing luxuriantly upon an inexhaustible soil. Indian corn, the best crop in the world for beef, rarely, if ever fails." Not alone nor the first to say so, Robinson, one of the more eloquent and important agricultural writers of the day, saw the possibilities for the prairie states. His 1850 letter to the Prairie Farmer came after another wheat failure; he then argued that the Middle West should concentrate on producing cattle because they were more profitable than wheat. Maryland and Virginia, he argued, could raise wheat successfully but could not raise cattle as cheaply as the Middle West with its prairies for grass and corn. Robinson hoped the wheat failure would "induce the people of all this great grass growing region . . . to sit down and count the cost" of growing poor wheat crops against raising cattle.[6]

Charles Mason (United States Commissioner of Patents and an Iowa farmer and stock raiser) heartily seconded the idea of switching from wheat to livestock, claiming that "the great defect in our agriculture is the failure to rear the proper number and quality of animals." In addition, men argued that use of cattle manure would increase the total grain crop by continuing a high state of fertility on cropped fields. Cattle also could walk to market and could consume surplus grain in a profitable way, and prices for good animals were more stable than fluctuating grain prices. Comparing all this to wheat failures and the cost of transporting grain to markets, many saw cattle feeding as the logical enterprise.[7] It was only about twenty years since Ohio drovers had settled in Illinois, but by 1856 enough farmers marketed fat cattle to have Illinois beef frequently make up the largest percentage of reported western cattle offered in the New York markets, thus surpassing Ohio. This rising Illinois business even caused problems in Ohio where cattle fattening declined in some areas that were formerly dependent upon feeder cattle from Illinois.[8]

By the time of the Civil War, cattle feeding operations of a similar type were well established in central and northern Illinois and eastern and southern Iowa. Abundant prairie beyond settlements provided range and corn land. Range grazing of cattle moved west and south, though some remnants remained on the wet prairies in east-central Illinois and in north-central and western Iowa until the 1880s. Before 1860 men had difficulty finding enough feeder cattle. Pioneers kept a cow and a few work animals, often turned loose on the prairie, but few were willing or able to sell more than a calf or two or a dried-up cow to a drover. Men rode far and wide in search of feeder cattle.

Isaac Funk, after learning the business in Ohio, moved to central Illinois in 1826 in order to have sufficient rangeland, but he found his operation limited by the small number of cattle in McLean County. This forced him to travel into Sangamon County to find settlers willing to sell. As settlement increased, so did the number of available feeder cattle and the magnitude of Funk's business. Eli Strawn had a system of subsidiary buyers in counties around his Ottawa, Illinois, headquarters who bought feeder cattle and collected fat-cattle herds for him. Others went further afield than Funk and Strawn.[9] Large-scale operators gathered their herds from farmers who felt that feeding cattle was not profitable unless they had two dozen head or so.[10] These feeders usually did not raise their own calves, so of necessity they developed a discriminating eye for estimating the current weight and fattening potential of calves they inspected.

In search of larger supplies of relatively low-cost feeder cattle, cattlemen reached out to Cherokee country (Oklahoma), Texas, and even New Mexico. This movement of western range cattle into the Corn Belt for fattening was a repetition of the drover pattern that had been connected with cattle feeding in Ohio. Here men repeated the three stages of the cattle feeding business: a progression from range country to feedlot to market. The range had moved from the Colonial backcountry to Ohio, Illinois, and Iowa, and then onto the plains. In the early 1850s, Illinois and Iowa cattlemen became aware of the possibilities of feeding cattle obtained from the plains. Men from Piatt County, Illinois, fed Texas cattle in 1852; and in 1853, Tom C. Ponting drove a herd to Illinois and sold them in New York in 1854, supposedly the first such cattle entered in the New York market. The next year cattle from Cherokee country and Texas passed through Springfield on their way to eastern Illinois feeders. Also in 1855 two Illinois men drove a herd from Texas to Iowa; a Chicago newspaper reported that a man named McCoy, operating from La Salle, was driving a herd of Texas cattle to Illinois. The next year, an Illinois man reported a profit of $24 a head on Texas cattle sold in New York. Encouraged by these prospects, Corn Belt cattlemen shipped to the Allerton Stock Yards in New York City each year from 1856 to 1859 an estimated 1,000 to 2,000 head of Texas cattle fattened on grass or corn in Illinois and Iowa.[11]

After the Civil War, feeders returned again to Texas and the plains for cattle. The increasing use of the western range as a source brought significant changes to the cattle feeding industry in the Corn Belt. Cattlemen had to decide which of the two available methods to follow in feeding. They could buy cheap western cattle, feed them for a year, then sell them for a small profit per animal and expect to make money on the large volume of business; or they could hope to profit on a low-volume business of better quality cattle that brought a larger profit per head. Not all who chose the latter practice, however, followed it completely, Many did not switch from common cattle to improved beef breeds. Common cattle, though of better quality than those from Texas, were not as efficient in converting corn into beef as stock improved by cross-breeding to purebred bulls. Improved or graded cattle also had better quality meat than the common variety. From the 1840s through the 1870s the Prairie Farmer complained that farmers showed too little interest in improving their beef cattle. However, many saw the problem and tried to do something about it. A Peoria County, Illinois, farmer wanted to know where he "could obtain some cattle that are of better blood than those that have been bellowing over the prairies from time immemorial. Such as a man could take some satisfaction in looking at and feeling of—such as will repay a man for feeding and taking care of."[12]

James N. Brown of Sangamon County, Illinois, provided one answer. The first man of record to bring purebred Shorthorns into the state for breeding purposes, Brown came from the Kentucky bluegrass area in 1834 to found a stock farm; he and his sons were important beef cattle breeders and feeders during the rest of the century. A graduate of Transylvania University and nephew of Benjamin and Elisha Warfield, well-known Shorthorn breeders in Kentucky, Brown was one of the "first to recognize that the best way to get the most profit out of good grass and good corn without robbing the land of its fertility was to stock it with good cattle." In 1855 he had 500 steers, 400 "stall-fed" beeves, and 200 to 400 hogs on his 2,250-acre farm. ("Stall feeding" in contemporary usage did not mean confining cattle to a building but merely to a feedlot.) The farm had 480 acres of corn; 120 acres of wheat, oats, and grass; and almost 1,400 acres of pasture, but Brown's enterprise was the exception before the 1870s. Though several herds like Brown's were in Illinois from the 1840s and in Iowa from the 1850s, only a small number of the common cattle of either state showed much improvement from crossing with purebreds.[13]

Despite the efforts of cattle breeders and agricultural publicists, through local and state fairs, to have "the good blood of our stock . . . widely diffused and encouraged in every condition, thereby elevating the stock of the country generally above mediocrity," most of the fairs reported few improved cattle entries before the middle 1850s. Newspaper accounts of fairs and agricultural society transactions generally praised the cattle exhibited but also noted the relative scarcity and less than desired quality. Except for places such as the Morgan-Sangamon County area in Illinois and Clinton, Scott, Lee, and Story counties in Iowa where purebreds were raised, there was little competition among purebreds for prizes.[14]

Regardless of which procedure a man followed in his

feeding business, he needed capital. The financing of early cattle graziers and feeders was not well documented. It has been argued that the great scarcity of capital resources or their involvement in land purchase prevented the optimum use of much of Illinois and Iowa prairie for cattle feeding. Supposedly, men just did not have the cash with which to finance feeding operations. This lack of capital might have prevented men from starting purebred herds or from buying purebred bulls to improve their stock. Some farmers felt they could not pay "all we have got" just to buy a purebred bull.[15] But feeders such as Funk, Strawn, Alexander, Harris, and Ponting succeeded one way or another in obtaining local cattle on credit or having sufficient cash to buy them in distant areas. Isaac Funk, in effect, borrowed from the neighboring farmers who supplied his feeder cattle by deferring payment until he sold them. After Funk became known to Chicago packers, he occasionally fed fat cattle on contract from a packer who supplied the money for the initial purchase.[16]

Cattle drovers and feeders could also borrow large sums from some local persons of affluence or small amounts from many different people in the neighborhood. In March 1834 Shaw Smith of Paris, Ill., tried to borrow enough cash for a cattle buying trip. He borrowed $60 from one neighbor (but for only a "few weeks"), which would be due before he sold the cattle. He therefore wrote to another man seeking a larger, longer term loan to cover the $60 due and perhaps an additional $60. When some notes came due before the stock had been sold, one man even offered to pay in cattle. Jacob Strawn at first raised wheat and operated a flour mill in order to generate capital with which to buy cattle and land. After that, he frequently borrowed from local merchants and then paid the merchants' bills in St. Louis when he sold the cattle.[17]

Because few men accepted bank notes at face value very far from the place of origin, cattle buyers carried or shipped sizable sums of specie. Alexander and Strawn reportedly carried as much as $20,000 in gold in their saddlebags at one time. Benjamin Harris had some close escapes, but was never robbed of the gold he carried on buying trips. In later years men shipped money by stage or railroad express.[18]

Once a man had surmounted the problem of acquiring cattle, he was faced with how best to feed them for market. Corn and grass were the basic feeds, but cattlemen differed on how to utilize them in the most profitable manner. During the 1840s some insisted that grain was too expensive for winter feeding and, in addition, argued that once cattle had tasted grain they would not want to eat grass. These men believed that grass would be just as effective for growth as grain. If any grain were used, it should have been as a supplement to hay rather than the exclusive feed. Other farmers sought substitutes for hay in root crops such as carrots, rutabagas, mangel-wurzels, and corn fodder. A writer for the U.S. Patent Office estimated that root crops could save from one-quarter to one-half the cost of winter feeding. The Patent Office publications also offered other advice mainly by reporting on English experiments and reprinting articles from the _New Genesee Farmer_ and the _Cultivator_ which, for example, listed the percentage of "flesh" and "fat-forming principles" in certain foods in an attempt to show that foods with the most oil fattened animals the quickest. Thus, farmers should feed such things as sunflower seeds and "all food should be cooked, if possible, and fermented."[19]

Despite the preferences of Patent Office clerks or correspondents, few Illinois and Iowa feeders prepared cooked food for their cattle. The _Prairie Farmer_ contended that availability of corn voided the need for cooked food or root crops. Root crops and cooked food were practical in Europe where little corn was available, but, asked the _Prairie Farmer_, just because Europeans fed root crops, "is it advisable for us to throw away our corn, and run after turnips and oil cake too?" Rather than follow the suggested variations, Illinois and Iowa feeders stayed with the classic pattern of "stall feeding" developed relatively early in Ohio and Kentucky and brought west. In the stall feeding system men grazed cattle during the summer and fall, then put them in lots during the winter and fed them shocked corn for grain and roughage. As Chicago replaced distant eastern markets and railroad shipping replaced driving cattle to the east, farmers increased the corn ration as market time approached. The proximity of the market and improved transport made it possible to profit from feeding cattle to a better finish than when the long and difficult trip to eastern markets negated the efforts to have a high-quality finish at market time.[20]

Regardless of what periodicals advocated for cattle feeding systems, we do know how some individuals fed cattle before 1850. John T. Alexander gathered local and Missouri cattle for fattening on corn. In the spring and summer he concentrated on raising a large corn crop. In periods of drought, such as 1854, he drove his herd west to the Illinois river bottom to be close to enough corn for winter feeding. As the corn matured, it was cut and shocked in the field where it remained until fed. A few weeks before the pasture gave out in the fall, Alexander gradually introduced a corn ration for the cattle on grass. Then he moved them into feedlots and fed them daily on corn stalks and ears for four to six months. The stalks were thrown from a wagon that followed the same circular path around the lot, so that by the end of the season the "ground was literally floored or paved with cornstalks." The ration averaged about one-half bushel of corn per head per day. Alexander claimed cattle usually gained 200 to 300 pounds in four to six months and doubled in value. He used two feed yards, feeding in alternate lots each day. Swine were placed in the lot not used by the cattle to salvage corn scattered and trampled into the dirt and to scavenge through droppings for kernels that had escaped digestion. Most men raised hogs in conjunction with cattle feeding in order to get full value for the corn fed.[21]

James N. Brown lived several miles east of Alexander in Sangamon County. He specialized in raising and feeding cattle on bluegrass, adding corn only during the last few months before market. On his 2,250-acre farm in January 1856, he had 250 acres of timber, 1,400 acres of cultivated bluegrass and timothy, 480 acres of corn, and 120 acres of wheat and oats. At that time, he had 500 steers of which 300 were going to market in February while the rest remained on hay and grass for the winter. He tried to have enough grass for grazing the whole year. This necessitated having two summer pastures, a winter pasture not used in summer, and harvested hay for lot feeding in March and April when he kept cattle off the pastures. Brown figured it cost him $20 to raise a steer to three years on pasture and then about $18 more to prepare it for the butcher ($4 for grazing and $14 for corn for the last six months). He favored stock improved with Shorthorn blood; he claimed these should yield a profit of $9.50 a head more than average native cattle.[22]

By 1856 Brown advocated marketing finished steers at three years of age instead of four, something that was not common for a good many years. In this system it was best to use calves born in March and April so they would have a full summer on grass. After a winter feed supplement of crushed corn and cob at the rate of two quarts per head per day, the calves came "to grass at one year old fat and sleek, with cost of two dollars per head in corn and cob meal." The calves spent the second summer on grass until mid-December when Brown added a ration of crushed corn at the rate of a peck per day. After eating about twenty bushels of corn in addition to the grass and fodder, the young steers weighed close to 1,000 pounds in May. Grass the third summer in addition to stall feeding one-quarter bushel of corn per day with grass from October to December brought the steers to the feedlot for all the corn they would eat until May. When sold in the late spring, the cattle would average about 1,450 pounds at a cost of $40, which yielded a profit of about $10 per head at the current price of $3.50 per hundredweight.[23]

Benjamin Harris had consulted with Jacob Strawn in 1835 before raising cattle. By December 1836 he had over 100 head feeding on shocked corn at an estimated cost of 10 cents a bushel. About the following April 1, he combined those with others locally purchased for April delivery and turned the whole herd out to graze until July when he started them toward Philadelphia. In the 1850s Harris raised few cattle on his 4,000-acre farm, preferring instead to buy two-year-olds in August and September and stall feed them for the spring market. In the winter of 1853-54 he bought over 65,000 bushels of corn from Sangamon and Mason county farmers in addition to that he raised himself. Cattle were herded on the prairie without shelter in the day and kept in a pen at night; each herd of 100 steers was tended by one hired man. Harris aimed at adding 450 to 600 pounds per animal and expected to make a profit of about $20 per head. In 1856, however, in a contest with a neighbor he produced one lot of cattle averaging 2,377 pounds gross weight per head for a group of 100. In the early 1840s he drove herds to the East and then supplied the St. Louis market, but after 1854 he shipped into Chicago on the Illinois Central Railroad.[24]

Jacob Strawn was one of the first to put corn in shocks for cattle feed, and he was given credit for "setting the example of introducing the present mode of farming," that is, the switch from grazing to feedlot. Strawn, whose great passion in life was "steers—cattle; to drive them, feed them, buy them, sell them," had farms in both Sangamon and Morgan counties, totaling over 16,000 acres in 1854. On much of this land hired hands raised corn. In 1854 he had 2,900 acres in corn and 1,600 head of cattle on the Morgan County farm. Some of the corn land yielded 80 bushels to the acre, but the average was closer to 40. Like Harris, he did not raise calves but preferred to buy cattle for finishing. For many years before the Civil War, he reportedly dominated the St. Louis cattle market, several times allegedly having a corner on it.[25]

All these men had swine that scavenged in the lots either with or after the cattle. Most men believed swine represented a large part of the profit in the operation, and for small-scale feeders swine were assumed to be a decisive element in making the system pay. The ratio of hogs to steers varied from 2 to 1 to 3 to 2.[26]

The above four examples represent common methods of operation for cattlemen, although on a scale larger than typical. Some Illinois and Iowa men fed cattle raised on the farm, though many claimed this was less profitable than buying two- and three-year-olds. It was more profitable, however, than selling the calves to others to feed. The *Prairie Farmer* waged a vigorous campaign to persuade dairy farmers to raise their calves for beef rather than selling them as young veal or feeders. Some farmers fed more corn, some less; a few tried stable feeding, but most used open lots with a minimum of shelter. Northern Illinois wet areas, east-central Illinois, and frontier Iowa had more grazing on prairie than corn feeding. Some men insisted that rushes in the wet areas were a very succulent forage. Others fed cattle on husked or snapped corn dumped in troughs, rather than corn on the stalk, as a way to save on grain; but they then had to supplement this with hay or straw as a roughage.[27]

Not all men were successful as cattle feeders. The *Prairie Farmer* through editorials and printed correspondence tried to encourage the inexperienced or less ambitious to develop better practices. Farmers were constantly reminded that they lost money by careless handling of cattle, by a false sense of economy in refusing to spend money on salt or full winter feed, and by not trying to improve quality by crosses with purebred stock.[28]

One much discussed problem of the careless and inexperienced feeder was the "mad itch." To supplement roughage, men turned cattle into picked cornfields or, if they could not afford the labor to shock corn, into fields of standing corn, which were often rented by the acre. Unless plenty of water was available and the cattle had been gradually introduced to dry fodder, they might get dry husks wadded in their third stomach and die of the "mad itch." Men reported the problem fairly often before 1860 and sometimes after. Generally, the advice offered by Prairie Farmer editors and correspondents was to permit cattle in fields only for short periods at first and to have sufficient water available.[29]

Much of the space in livestock columns of the Prairie Farmer during the early years dealt with the specific questions of whether or not it was necessary to provide shelter for cattle on feed. The main argument followed two views: those of humanity to animals and economy of feed. Some argued that leaving cattle in open fields during winter was cruel and sheltering them in barns reduced food costs because they ate less when warmer. Central Illinois men rarely sheltered animals in a barn. To inquiries such as one in 1844 wanting to know "whether the mode in use of yard-feeding cattle . . . is found to combine more advantages than the feeding under shelter," the Prairie Farmer invariably replied that the only right and proper thing to do was to provide shelter, at best in a barn and at worst in an open shed built on the north and west sides of the feedlot. It was "barbarous" and expensive to leave cattle out in storms. The printed correspondence on the subject indicated that many thought feeding cattle in standing corn, timber lots, or on the south and east sides of slopes gave sufficient natural protection. Others constructed hayracks to form enclosures for wind protection. Relatively few mentioned that they stabled cattle. The writers often claimed that cattle sheltered in some way came through the winter in better condition and sold for more in the spring or did not need any extra feeding to bring them back to their prime weight of the previous fall. On the other side of the question, men who did not shelter stock claimed that if the cattle got too comfortable they would not forsake their "nests" to get water or eat at their best pace for gaining weight. Also, it seemed too expensive (to many who did not have the ready capital) to build barns or hire help for the extra work of feeding stabled cattle. Poorer farmers felt that the stock was not worth the expense of shelter or full feed during the winter, but their cattle often sustained a net loss in weight. When these thin animals sold at low prices in the spring, the owners complained that the cattle business did not pay. What cattle needed most was adequate food rather than shelter, and the large-scale operators recognized this in their disregard of more elaborate shelter other than a grove of timber. Despite all efforts to the contrary, barns or even open sheds for stock did not appear to be common before the Civil War.[30]

In addition to feed rations, a cattleman had to cope successfully with market conditions and price fluctuations if he were to operate at a profit. Thus he needed to produce a type and quality of animal suited to the seasonal demands of the particular market open to him. Investing money in better quality beef than the available buyers sought was as likely to be unprofitable as offering poor quality beef for the Christmas trade. The whole industry early shaped itself to the demands of the outside markets in Ohio, New York, and finally Chicago, as well as to local outlets. Illinois cattlemen had first been concerned with Ohio as a marketplace, but limited local markets quickly developed also. O. W. Kellogg pioneered the Peoria-Galena cattle route in 1827 when he took three wagons and a herd from Peoria to the lead mining area. In the early 1830s, Dudley Willits recalled, about once a year young men from Mercer County, Illinois, "would get around fifteen or twenty fat cattle and scare them down" to the St. Louis market for sale at about $8 to $12 a head. By 1851 Morgan and Sangamon counties sent as many as 4,000 head of cattle a week by river to St. Louis.[31]

Southern and central Illinois operators sent herds of cattle weighing from 600 to 1,000 pounds to the East or to New Orleans. Rapidly increasing settlement in northern Illinois and southern Wisconsin created such a demand in the 1840s that many "very sorry animals [that] could never find sale but for want of better" came to the Chicago market from downstate.[32] By 1850 McLean County, Illinois, claimed that it supplied one-fifth of the cattle sold in Chicago. Isaac Funk made Chicago his main outlet after 1835, though he still sold some cattle to New York drovers as late as 1850. Funk even sent his brother Absalom to Chicago to act as his agent over a period of several years. Funk often sold on contract to packers a year in advance at a fixed price and then bought and fed to fill his orders. When driving cattle to Chicago, he moved just enough in one herd to keep a packer busy for one day and spaced his herds one day apart on the road to Chicago.[33]

For several years Iowa cattlemen supplied people heading for California. By the middle 1850s Iowa and Illinois feeders were able to ship over railroads to Chicago and New York, though some men along the rail lines continued to walk cattle into Chicago as late as 1860.[34] New York market reports began listing the number of Illinois cattle offered each week. By 1856, Solon Robinson reported in the New York Tribune that almost the whole New York supply in one week came from Ohio and Illinois and that they were generally three- to five-year-old steers that gave "good evidence of having been where corn is plenty and cheap." Jacob Strawn even sent one of his sons to New York for a short time as his agent for selling the family cattle in the eastern market.[35]

That seemingly endless flow of beef to the East led Robinson to predict that the price was too high in New York and would soon fall. His predictions proved sound but enraged at least one feeder frustrated by a bad year; he replied in common farmer fashion to the workings of the market: "What can we do to stop the mouth or pen of that swaggering and blatant agricultural editor of the New York Tribune . . . who is calling upon us farmers to make up our minds for the low prices? Please tell Solon that we consider him an ass and defy him. . . . We venture to assure all New Yorkers who scribble that we shall sell our cattle to the drovers only at paying prices. Grass fed beef in this quarter is a failure, and we must stall feed what few cattle we turn off, or else sell half fatted stock."[36] Unfortunately, many feeders did the usual thing during a bad crop year when they did not have enough corn and pasture. They sent poorly fattened stock to market and prices fell as predicted.

The representative prices shown in Table B.6 fluctuated from $5 per hundred pounds in 1851 to $1.50 in 1859, and prices were lower on the average toward the end of the period. Prime beef animals for the local fresh meat trade might sell above the top figure given in the newspaper tables. A 1,500-pound average steer would have been worth anywhere from $22.50 to $75 in the Chicago or New York market in the years 1840 to 1860. During the 1850s in Chicago the net weight of the four quarters of beef (dressed weight) from a slaughtered steer averaged between 500 and 700 pounds. Dressed weight probably equaled from 40% to 50% of liveweight.[37] Average gross (live) weights for cattle in the 1840s and 1850s varied from 800 to 1,600 pounds, with weights over 1,200 pounds more frequent in the 1850s. Occasionally, very fat cattle were sent to market, such as the famous herd of 100 cattle averaging 2,377 pounds that Benjamin Harris marketed in 1856. They were four years old and sold at $7 per hundred pounds. A dozen of the herd kept back from the 1856 sale and fed until February 1857 averaged 2,786 pounds when sold. These phenomenal tributes to the art of cattle feeding were reportedly decorated with ribbons and paraded through Chicago streets preceded by a band and followed by one hundred butchers uniformed and mounted. The *Prairie Farmer* noted that fairs might be giving too much attention to big animals. All things being equal, an 1,800-pound steer was better than an 800-pound animal; but, the *Prairie Farmer* warned, men were in danger of sacrificing quality for large size.[38]

Feeders generally sent their fat cattle to market some time after three years of age. In 1847, four- to six-year-old beef was being sold at 600 to 800 pounds net weight after a winter on cheap inferior corn. In the 1840s it cost from $7 to $9 a head to raise cattle to three years of age, while steers could be sold for $15 to $35 a head as feeders. In the 1850s the cost to produce three-year-old cattle increased to an average of $15 to $20 a head, while the market value was about twice the cost. In 1855 an Adams County, Illinois, man paid $30 a head for 1,200-pound steers in October for winter feeding. He anticipated selling them in the spring for about $60 a head after a gain of 300 pounds.[39]

Bloodline was an important factor bearing on the weight, maturing age, and profitability of cattle. Native cattle matured later, were smaller, and were less economical to feed than purebreds or good grade cattle sired by purebred or improved bulls. James N. Brown figured that he could make a profit of at least $12 a head on 1,800-pound three-year-old Shorthorn grade steers. Grade steers fed the same rations weighed more than natives of the same age; and because there was more and better quality beef on the carcass, they sold for more. However, because grade steers cost more to buy as feeder cattle, many men with limited capital tended to continue raising natives.[40]

Generous feeding to produce a good fat finished animal was important for profit, but the feed had to be wisely used, for as Solon Robinson reported of a poor quality shipment of Iowa cattle in the New York market, "lots of corn" went "to very little purpose" with bad management.[41] Cattlemen had to learn how to get the most gain in flesh on a steer with the least quantity of corn. Two factors were necessary for such a profitable operation: first, men must select steers that had the frame and disposition to readily convert corn into beef, and second, they must provide the proper proportions of corn and grass to keep them gaining weight without wasting feed.

Contract feeding such as that practiced by Funk provided no sure profit. At times contract deliveries upset the market and affected prices and profits. If the price of corn went up because of a poor crop or the grass dried early from drought, drovers were forced to market cattle they had contracted for even if they were in poor condition. This happened in June 1857 when drovers had to take the remainder of their spring contract cattle to market before they were properly fattened, and prices declined. Financial crises, such as that in the fall of 1857, and the timing of the packing season affected Chicago prices too. Packing generally began in October and continued through December, but it varied with the demand, the general state of the economy, and the duration of cold weather.[42]

By the end of the 1850s central Illinois and eastern Iowa cattlemen had shifted their type of operation from one of raising animals to be driven east and fattened in Ohio and Pennsylvania to one of fattening cattle on their own corn and grass for the Chicago and New York markets. The business expanded in volume as the number of Illinois and Iowa cattle sent to market increased, yet the pattern of operations within the region remained the same. Three- to five-year-olds fattened on corn and grass were prepared for either of two types of market. For the butcher market throughout the year cattle were driven or shipped to the city where they were slaughtered and consumed. There was also the seasonal demand of packinghouses that prepared salted or pickled beef during the winter, which could then be shipped to distant places regardless of the season. The increasing number of Illinois and Iowa cattle appearing in both of these markets brought some changes in feeding practices further east. In order to compete with Illinois and Iowa cattle, Ohio feeders began to send animals to market at three and one-half years of age instead of four and one-half.[43] Though Illinois and Iowa men could raise and fatten stock at less cost than Ohio feeders, the Ohio men concentrated on producing fat cattle of better quality in a shorter time, thus eliminating the cost of an extra year in the process. This tended to equalize the difference in production costs. Farther east where costs for fat cattle were higher yet, competition from midwestern stock shipped by railroad caused some New England feeders to give up fattening cattle for the Brighton market serving Boston.[44]

Despite problems in procuring, feeding, and marketing, cattle fattening in Illinois and Iowa was a successful and profitable business by the late 1850s. Events were occurring that would change the patterns of cattle feeding in the next three decades. Until the late 1850s there was insufficient outlet at a good price for large quantities of quality beef cattle. In Chicago, beef not consumed locally was pickled and sold in barrels, a process that reduced all barreled beef to one quality. Differentiation in quality began to be rewarded when men shipped live cattle to New York for the city butcher market, but the greatest incentive for quality beef production came with the development of refrigeration with ice. This made possible the dressed beef trade that offered fresh meat slaughtered in Chicago and shipped in iced refrigerator railroad cars to the urban markets of the nation. In fresh meat sales, better quality brought a higher price. People paid more for fresh meat than for pickled beef, which in turn raised the market price for quality cattle. Farmers could profit by improving their breeds and changing their feeding practices as they did from the 1860s through the 1880s. The entire industry changed. Up to 1860 many Illinois and Iowa cattlemen were ready and able to feed better quality animals, but changes in the market structure were necessary to make this profitable. By the 1860s buyers for the eastern metropolitan markets began to look to the Middle West for their supply of beef; and cattlemen, shippers, and butchers began to look for faster and cheaper ways to get better beef to the cities. This set the stage for the great changes in distribution and marketing that encouraged beef cattle production in Illinois and Iowa.

CHAPTER 4. CHICAGO AND THE RISE OF THE PACKING INDUSTRY

In the years after 1865 changes and technological advances contributed to continued growth and expansion of the beef industry and made Illinois and Iowa leading producers of corn-fattened cattle and Chicago the world's leading cattle market. The organization and expansion of a central market in Chicago both illustrated and embodied many of these changes. Several elements combined to bring this market into being: the new railroad network, the concentration of meat packers in Chicago, the development of refrigeration facilities, the reorganization of retail meat marketing, and growth of the export trade in live and dressed beef. As a result demand increased, and further development of the Illinois and Iowa farm economy ensued with its advantages for combining the production of cattle and corn.

The necessary first stage in the expansion involved centralization of marketing facilities in Chicago. Lake transportation and railroads made this possible by bringing increasing quantities of agricultural produce to Chicago for redistribution to eastern markets. As a consequence, a whole range of commodity handling facilities developed—grain elevators, stockyards, banks, and a Board of Trade.

Located advantageously for lake transportation by the Illinois-Michigan Canal Commissioners in 1827, Chicago grew slowly at first. The canal, which was to bring many benefits, did not open for 20 years; but militia service during the Black Hawk War acquainted many men with northern Illinois, and they returned after the war to establish farms. These veterans and other new residents turned to Chicago as the supply center for the region. The federal government constructed harbor facilities and cleared the bar from the mouth of the Chicago River so that it could serve as a port. North into Wisconsin and as far south as Danville and Bloomington, Chicago men traded eastern manufactures for western produce. Chicago merchants in turn shipped wheat and pork to balance their accounts with eastern suppliers. In 1838 wheat became the major export factor in this balance of payments and remained so for a decade. By 1842 Chicago shippers sent 586,907 bushels of wheat east, and in five years this had increased more than three times to 1,974,304 bushels. In 1848, the year they organized the Board of Trade, grain dealers had storage space for 700,000 bushels of wheat.[1]

The population of the city grew steadily. In 1840 it had 4,470 residents; in 1850 there were 29,963 residents, and by 1860 the population had risen to 112,172. By 1870 the population was 298,977 and grew to 503,185 in 1880; by 1890 it was 1,099,850. The city itself offered a growing market for midwestern farm produce.[2]

The Great Lakes system provided easy access to the East, and in 1839 a company established regular steamship connections between Chicago and Buffalo. Inland transportation, however, met some serious obstacles. In 1848 completion of the long-promised canal to the Illinois River opened a water route to the area southwest of the city. In the same year newly constructed plank roads offered another way through the wet prairies of the low watershed between Lake Michigan and the Illinois River. Some of these extended almost 60 miles from the city, but they were relatively unimportant in comparison to the railroads and water routes.[3] Though both the canal and plank roads were important in transporting grain, they also aided the marketing of meat products, the other mainstay in agricultural commerce.

Pork packing was the first animal product business of any magnitude in Chicago and the Middle West. Cincinnati held the first large concentration of pork packers, but practically every river town in the Ohio and upper Mississippi river valleys had a packing establishment at some time. For example, during the six-month season ending in March 1848 six Chicago firms packed 19,666 hogs, and 14 towns on the Illinois River and 9 Illinois and Iowa towns on the Mississippi River packed from 1,500 to 35,000 hogs each. Though pork was their major product, some packers also slaughtered beef in Chicago and in the larger towns such as Alton, Peoria, Rock Island, and St. Louis.[4]

George W. Dole (pioneer merchant) started the first regular packing business in Chicago in 1832. In the first season he slaughtered and packed 152 head of cattle and 338 hogs.

Soon after that Gurdon Saltonstall Hubbard (fur trader and frontier merchant) entered the business, and by 1836 his firm packed 6,000 hogs in one season. The financial crisis of 1836 dealt a temporary blow to the economy by halting work on the Illinois-Michigan Canal. As local demand for pork and beef temporarily declined, men tried to market their barreled meat in the East. In this trade pork products maintained a considerable lead over beef in the 1830s, but by 1843 Chicago beef shipments totaled 10,380 barrels compared to 11,112 barrels of pork. The firm of Wadsworth, Dyer, and Chapin (the leading beef packers in the 1840s) at times slaughtered 100 head of cattle a day and packed beef for the U.S. Navy and the British government. Thomas Dyer, partner in the firm, so systematized the disassembly process that by 1847 the plant could handle 45 animals an hour. Coincident with this development of beef packing, tanners, coopers, and fat processors established businesses.[5] Most cattle driven to the Chicago market originally consisted of grass-fed animals, but by the late 1840s Chicago packinghouses began to offer an alternative market for Illinois corn-fed beef formerly driven overland to eastern markets.[6]

Before the development of artificial refrigeration, beef packing was a seasonal operation limited to the period of cold weather. In Chicago the packing season generally opened in late October or early November and could last until early March, depending upon demand and the state of the weather. The operations of Wadsworth, Dyer, and Chapin offer an example of large-scale beef packing in the late 1840s. During the 60-day season in the fall of 1848, the company planned to slaughter over 3,500 cattle. On peak days 75 to 80 men slaughtered and packed from 120 to 180 cattle a day. Men cut carcasses into eight-pound pieces, put the meat through a three-step pickling process, packed 38 pieces in salt into a tierce (a cask between barrel and hogshead size) and then shipped the tierces to New York where they sold at $18 each for the English market. After the better pieces of meat had been barreled, the workmen cut, salted, and dried the necks and shanks for sale the next spring in the West Indies as jerked beef. The refuse fat produced 40 barrels of tallow a day, which sold for 7.5 cents a pound. Wadsworth, Dyer, and Chapin sold tongues, hoofs, horns, and gall, but did not use the liver, lungs, bones, or blood.[7]

To supply their packinghouse, Wadsworth, Dyer, and Chapin bought cattle from as far south as De Witt County,

Illinois, and the upper Wabash valley. Isaac Funk was one of their major suppliers. He furnished 1,200 head in October 1848, 1,650 head in November 1849, and 1,000 head in November 1850. Most dressed out at between 600 and 750 pounds and were purchased for about $3 to $3.50 per hundred pounds liveweight. Sometimes the packers put up the capital for Funk to buy, feed, and drive cattle to Chicago on contract.[8]

As early as 1849 the Prairie Farmer declared that "Chicago is at present the first point in the United States in slaughtering and putting up of beef. There are some two or three establishments engaged in this business here, who have no rivals at any other place in the United States—so we are informed, and so we believe."[9] In 1850 the eight establishments listed by the Chicago Tribune expected to pack 27,500 cattle, which at a conservative estimate represented an investment of almost $0.5 million in animals alone. Despite the Prairie Farmer's boast Chicago's total quantity of packing was not that remarkable. Beardstown, Peoria, and Quincy did a large pork packing business and also packed some beef.[10]

The further development of Chicago as a meat packing center depended upon an improved transportation system. Canal and lake routes closed during the winter, and the vagaries of the weather made driving livestock to the city a risky venture. In the spring, roads were almost impassable. The railroads overcame the handicap by providing all-weather connections with the hinterland. With the Chicago market readily accessible by rail, prices there became the standard for other western markets. By the early 1850s centralization of packing facilities had taken place to the extent that for a century the history of the meat processing industry in Illinois was largely the history of meat packing in Chicago.

In the late 1840s and early 1850s men around the state began to revive earlier plans for internal improvements. This time there was a considerably larger economic base and immediate prospect of revenue. The Galena and Chicago Union Railroad began pushing out to the wheat country west of Chicago in 1847 as a direct result of agitation by the business community to surmount the transport difficulties in that direction. Petitions to the legislature asking the state to build the railroad had drawn little support. However, Chicago interests raised the capital and began work; by 1849 trains ran to a point 21 miles west of the city. Six years later the railroad ran to Freeport, which was 122 miles northwest, straight west to Fulton on the Mississippi, and to Beloit, Wisconsin.[11]

Other railroad projects followed. Construction of the Chicago, Alton, and St. Louis Railroad began in 1847, and by 1854 it offered through service to St. Louis. The great Chicago, Burlington, and Quincy system was launched in 1849; reached the Mississippi across from Burlington, Iowa, by 1856; and was extended to Ottumwa by 1859. By 1870 a direct line ran through southern Iowa from Council Bluffs to Burlington and on to Chicago. The Chicago and Northwestern won the railroad race across Iowa to Council Bluffs in 1867 by taking over the Galena and Chicago Union and consolidating and extending Iowa short lines. The Chicago, Rock Island, and Pacific linked Chicago to the Mississippi at Rock Island in 1854, and by 1869 it too had connections with Council Bluffs. The Illinois Central Railroad, started in 1851 as a north-south line down the center of the state, had 700 miles of track in operation between Chicago and Cairo and Cairo and Dunleith in the northwest corner of Illinois by 1858. It too then crossed the river by 1870 and had built westward to Sioux City in western Iowa.[12]

In 1860 six major rail systems extended south and west of Chicago, tying Illinois and Iowa closer to the city and encouraging settlement of the prairies. (See Map A.5.) Illinois had 2,790 miles of railroad in operation or 1 mile of railroad to every 19.8 square miles of land. Iowa, at the same time, had only 655 miles of railroad or 1 mile for each 84 square miles. By 1900 both states were thoroughly criss-crossed by railroads. (See Table B.7.) However, railroad building continued at a brisk pace, so that in 1880 Illinois had 7,851 miles of track and Iowa had 5,400 miles. By 1900 the railroad systems of Illinois and Iowa had grown to the point where they had 11,058 and 9,391 miles of track respectively. (See Map. A.6.) Railroad development tied Illinois and Iowa together as one economic tributary to Chicago where the railroads converged. By 1900 at least 19 railroads brought cattle into the Chicago market.[13]

This railroad building had immediate and far-reaching impact on Chicago and the cattle business. By making beef more readily available to Chicago and the eastern markets, the railroads stimulated both consumption and production. As one Iowa farmer wrote in 1856 after the Burlington Railroad had penetrated Wapello County, "many of us are turning our attention to feeding cattle." The coming of the railroad also led him to ask the *Prairie Farmer* to print wholesale cattle prices and to publish information about shipping rates and routes to New York.[14] It cost about $15 a head to ship live cattle from northern Illinois to New York and about $20 a head to ship cattle from Burlington, Iowa, to New York; over two weeks was required for the trip from Burlington.

The *Prairie Farmer* calculated that people in New York consumed 185,574 beef cattle in 1855, while in Philadelphia and Baltimore the population consumed at least 100,000 head. Half of the total beef consumed in these cities came from Ohio and the West. Illinois alone sent 17,000 cattle east by rail in 1855.[15] One prominent cattle raiser estimated that the coming of the railroad increased cattle prices in central Illinois by 10%. The railroad also encouraged rapid settlement of lands in eastern Illinois and in Iowa. Prairie-county populations in east-central Illinois increased from 150 to 600% from 1850 to 1860, and the acres under cultivation in the prairie region went from 74,000 acres in 1850 to 2,902,000 in 1860.[16] After the Illinois Central Railroad began operation, Effingham County, 200 miles south of Chicago, "increased rapidly in population, and many of the farmers are now driving a very respectable business." Similarly, in Iowa, farm products brought "a good price" in Cass County once the railroad arrived in 1869. Seeking the same benefits, Jefferson County Iowans organized to secure a railroad line through their area, which would make "things begin to look up." At that time, corn in Jefferson County (without a railroad) sold for 15 cents a bushel less than in Cass County, which was served by a railroad.[17]

Initially, however, railroads seemed to have an adverse effect upon the number of cattle packed in Chicago. Compared to the 27,500 packed in 1850, during the short season of fall 1852 packers processed a little less than 22,000 cattle. There had been a heavy summer demand in the East and the railroads permitted shipment of live animals to New York and other eastern markets. The shipment of beef animals to eastern markets increased from an estimated 2,657 in 1853 to 42,638 head in 1858. In 1858 Chicago firms packed 34,675 head of beef. By 1860, 97,474 animals were shipped east, enough to have supplied 43% of the number received in New York City that year. Of the remainder of the Chicago receipts of 177,101 cattle, 34,623 animals were packed for shipment and 42,074 were consumed within the city. Chicago Board of Trade figures indicated that about 20,000 cattle were driven to Chicago in 1860; the rest arrived on the railroads. By 1864, under the influence of the Civil War, receipts had climbed to over 300,000 with over 70,000 packed for markets outside Chicago.[18]

As railroads brought more cattle and hogs into Chicago, local boosters proclaimed its preeminence as the packing center of the nation. Chicago establishments packed half a million hogs in 1862, surpassing Cincinnati as a pork packing center. At the same time, receipts of cattle for slaughter and packing increased. A Chicagoan proclaimed the city the "great _meat manufactory_ of the world."[19]

But even as Chicago made this boast, in 1865 the number of cattle packed started to decline. Live cattle receipts and pork packing continued to grow, but the beef pack of 92,459 animals in 1865 was the high point for many years. In 1870 Chicago packers slaughtered only 11,963 cattle. The decline occurred because some packers had moved west to Kansas City to utilize cheaper range cattle for barreled beef where quality differentiation was not important. In addition, eastern buyers paid to ship numbers of live cattle to the East for slaughter there.[20]

The Civil War was a boon to Chicago as well as to Illinois and Iowa beef producers by increasing demand at a time when railroads could deliver beef to the city. The war and the railroads also brought centralization of market facilities as increasing receipts of livestock created chaos in handling transactions between several markets in the city. Relief came with the formation of the Union Stock Yards in 1865.

Before 1865 Chicago had several small stockyards located in various parts of the city. In the early days drovers immediately sold the few cattle they brought in. There was no need to pen cattle while the drover sought a buyer. But as the trade increased, there was need for a place to keep live animals before they were sold. The slaughter houses had some pens, but private yards soon superseded them. In the late 1830s Willard Myrick opened one of the first of these yards as an adjunct to his tavern where drovers stayed while seeking a sale. In 1848 the Bull's Head Yards opened near the southwestern plank road. Entrepreneurs opened similar yards around other taverns, so that a number of possible marketplaces were available. With the advent of the railroad, yards not advantageously located fell into disuse, so the duplication of facilities declined. In the late 1850s, however, the existence of several yards still required buyers and sellers alike to ride the circuit to close an advantageous deal. Cattle prices could vary as much as 50 cents per hundred pounds between yards on the same day.[21]

During the Civil War, representatives of the packers and the railroads began negotiations to end the inefficient proliferation of stockyards. Packers had to have buyers at several different places, and railroads had difficulty getting shipments from the yards of a western railroad to the yards of an eastern one. Buyers had attempted to meet the problem by designating Tuesday and Friday only as market days for cattle; and in 1856 the junction railroad linked all the railroads entering Chicago, but that was no solution to the problem. At a meeting in 1864 the Packers' Association called for a central stockyard. The railroad managers accepted the idea and applied for a charter from the state legislature.

In June 1865 the Union Stock Yard and Transit Company of Chicago began construction of the Union Stock Yards and opened the gates for business December 25. Four representatives of the packers and five for the railroads composed the first board of directors. The company had a capital stock of $1 million, all but $75,000 held by nine railroads. The yard covered 345 acres (only 100 acres of pens at first); had two artesian wells, a drovers' hotel, and a bank; and could care for 100,000 head of stock. Four separate unloading areas served the five western and four eastern railroads that entered the yards.[22] The railroads controlled the yards until 1894 when their representatives resigned after losing control to new interests. As a result of a complicated corporate shuffle typical of that decade, Armour and Company (one of the major meat packers) and Frederick Prince (Boston banker and broker) took control. The packers had wanted a share in the profits and management for some time. They finally threatened to move their operations and build new yards of their own, at which point some of the old stockholders and railroads gave in and allowed a corporate reorganization.[23]

Farmers complained in the 1870s and 1880s that the Stock Yards made an exorbitant profit from the yardage fee of 25 cents a head for cattle and feed charges of $1 a bushel for corn and $20 to $30 a ton for hay. Yard prices ranged from 50 to 300% above the market price outside the gates. In 1877 an Illinois legislative investigation found that a basket of corn called a bushel by the yards did not in fact weigh 70 pounds but only 55. The current market price for a 70-pound bushel was 43 cents, while the yards charged $1. A bill to have the state take over the management received little support in the legislature, but the State Railroad Commission obtained supervisory powers allowing it to set maximum limits on rates for feed and handling. A later bill in 1887 to regulate the commission and feed charges of the yards did not pass.[24]

Despite complaints the yards immediately proved their usefulness, and as years went by they were improved and enlarged. By 1887 the capacity was 20,000 cattle, 150,000 hogs, 10,000 sheep, and 1,500 horses. By 1900 the yards covered 500 acres and had 250 miles of railroad track, 130,000 pens, and a capacity for 75,000 cattle, 80,000 sheep, 300,000 hogs, and 6,000 horses. Over 200 commission firms had offices there.[25]

With minor fluctuations the number of cattle handled each year continued to increase until 1892 when it reached the high point for the nineteenth century: 3,571,796 cattle and 197,576 calves. In 1900 the yards handled only 2,729,046 cattle and 136,310 calves. (See Table B.8.)[26] The reduction in receipts after 1892 came partly because of the decentralization of the meat packers to Kansas City, Omaha, St. Paul, and Iowa cities and partly because of a series of years of low prices during which breeding herds had been reduced and some Illinois and Iowa men gave up cattle feeding.

In addition to the handling services the yards provided space for commission firms and financial institutions. The Union Stock Yards National Bank, established in 1869, became the National Live Stock Bank in 1888 and had assets of over $9 million in 1900. The Drovers National Bank, organized in 1883, had assets over $5 million by 1900. There was also the Chicago Cattle Loan Company, with much the same directorate as the National Live Stock Bank which had several Union Stock Yards men and packers' representatives on its board.[27]

By 1870 centralization of marketing and packing facilities tied terminal and transshipping railroads, commission houses, financial institutions, and packing companies together into a "packing town" built around the Union Stock Yards on the south side of Chicago. However, with all this change in handling cattle, there had been none in beef processing procedures for a growing market. Quality cattle for the eastern market were sent live, a costly and inefficient way of handling a large volume of business, which tended to make Chicago only a transshipping point.

The Chicago market structure and processors could handle more cattle than the barreled beef market demanded. Chicago was prepared to handle range cattle before they were readily available after the Civil War; and yet when shipments increased the volume of the market in Chicago, cattlemen found that the quality of range beef offered more meat for barreling than for the live trade to New York. Range cattle were thus no answer for the rising urban demand for quality fresh meat. However, the combination of the ability of Illinois and Iowa feeders to produce quality beef cattle and the centralization of marketing, processing, and shipping facilities of Chicago with a new technique for handling fresh beef brought renewed prosperity to the industry. Quality beef could be slaughtered in Chicago and sent to the East in refrigerated railroad cars, where it could be marketed as fresh beef without the expense of shipping all the offal, the liability in shipping live cattle. The development of this dressed beef trade accounted for much of the growth of the beef cattle industry in the last half of the nineteenth century.

CHAPTER 5. DRESSED BEEF AND THE EXPANSION OF THE PACKING INDUSTRY

The demands of the new urban market, created by the expansion of the railroads, gave rise to the dressed beef industry. Consumer demand for processed agricultural goods in urban areas made it possible for those supplying the products to build national marketing organizations and to concentrate the national trade of any line in the hands of a few firms. The processors of consumer goods such as flour, tobacco, and fresh meat were the first big businesses created by the rise of the urban market.[1] A rising demand existed for fresh meat in the East as the cities outgrew their local supplies after the Civil War. A supply of quality beef was potentially available in Illinois and Iowa feedlots if there were sufficient price rewards. Chicago market facilities could handle an increased volume of cattle. Both the producers and the processors were prepared for the change when the possibilities of the lucrative urban market encouraged men to combine the railroads with refrigeration to ship fresh beef slaughtered in Chicago to the eastern markets.

The introduction of effective means of refrigeration prevented the early spoilage of fresh meat and lengthened the period of time after slaughter and the distance from point of slaughter that fresh beef could be consumed. The basic principles of refrigeration had been understood and applied, with ice compartments in meat cooling rooms, before the Civil War; but the packing industry was slow to use cooling systems in marketing fresh meat in conjunction with railroads. There had been some experiments with ice refrigeration in railroad cars, however, and by 1869 George Hammond, a Detroit packer, sent fresh beef to Boston in a refrigerator car. Packers and shippers began to appreciate the possibilities of refrigeration in the early 1870s and installed chill rooms in the packinghouses. Most of this early refrigeration resulted from air circulating around ice or an ice-salt mixture. Not until the 1890s did mechanical refrigeration of the ammonia-compression type come into common use in the meat processing industry.[2]

Hammond continued his experimentation with refrigerator cars and started sending small shipments regularly to Boston in 1872. Nofsinger and Company, Kansas City packers, sent two carloads of dressed beef to Philadelphia in 1875 as well as in the next few years, but did not continue partly because railroads raised the rates to discriminate against the product. A Chicago group gave a hint of things to come when in 1875 they tried a scheme to undercut butchers supposedly planning to raise meat prices in Philadelphia during the centennial exhibition. They sent dressed beef to Philadelphia in ice-packed cars at a temperature below 35 F and sold directly to consumers at a reduction of 34% from the current butcher price. The _Prairie Farmer_ suggested that such direct retailing be done on a large scale in New York City.[3]

Gustavus Swift (a butcher and retailer from New England) led the way in employing the new technology of refrigeration to concentrate beef slaughtering and processing in Chicago and as a result changed the marketing system for fresh beef. Swift had moved west first to Albany, N.Y., and then in 1875 to Chicago in order to eliminate middlemen in his marketing operations. At first he shipped live cattle east to his Massachusetts partners; but when he saw some of the experiments in shipping dressed beef, he decided that the competitive advantage lay with the man who perfected the method. Swift set about creating the dressed beef processing and distribution system that became the standard of the packing industry well into the twentieth century. In the winter of 1876-77 Swift shipped dressed beef in boxcars without refrigeration; this was unsatisfactory, so he experimented with refrigerated cars. When the major railroads refused to build refrigerator cars, Swift ordered them from independent builders and by 1879 he owned 100.[4] Once Swift had created a demand for dressed beef by underselling local butchers through his own retail outlets, he borrowed all the money he could, as well as reinvesting his profits, for expansion to match demand. Swift erected a Chicago slaughterhouse in 1880 and opened houses in Kansas City in 1888; Omaha in 1890; East St. Louis in 1892; St. Joseph, Mo., in 1896; and St. Paul, Minn., in 1898. By 1899 Swift had 193 branch houses distributing his products to retailers across the country.

Incorporated with a capital of $300,000 in 1885, Swift and Company had a capital of $25 million and annual sales of $160 million by 1903.[5]

The dressed beef trade became important almost immediately. (See Table B.9.) By 1882 the Prairie Farmer reported that Swift's shipments had caused a "sensation" in eastern markets. The next year, realizing the potential of the dressed beef trade, Armour and Company and Nelson Morris, leading Chicago pork packers, joined Swift in shipping dressed beef. At first, New England was the major market; as it developed, there was an immediate decline in the number of live cattle shipped from Chicago as well as from St. Louis to New England. (See Table B.10.) But soon packers shipped to New York, Philadelphia, and Baltimore. Indeed, from January 1881 to November 1, 1884, almost 89% of the dressed beef received in those cities and all of New England came from Chicago. By 1883 domestic and export demand was so great that even with record cattle receipts during September prices did not decline.[6]

There was a bitter 15-year struggle before dressed beef was fully accepted by consumers, butchers, and railroads. Consumers distrusted the quality of meat shipped hundreds of miles after being slaughtered, but dressed beef was sufficiently cheaper than local butcher stock that people were usually willing to try it. Swift had a much harder time convincing butchers to sell Chicago-dressed beef. Local butchers controlled retail meat distribution in the nation. They sometimes obtained their meat from large city packinghouses or killed cattle on the farms where they bought them; in most cases they bought cattle at a stockyard and dressed them in a local slaughterhouse or their own shop. Butchers and owners of local slaughterhouses, who saw the threat to their business, fought back and had the support of railroads that had an investment in the live cattle trade.[7]

A stockcar of live cattle produced fewer pounds of salable meat than were carried in a refrigerator car of dressed beef. If railroads charged the same rate per pound for both live cattle and dressed beef, the processed meat could be shipped at considerably lower cost per salable pound. Retailers of dressed beef could then afford to undersell local butchers; if enough butchers sold dressed beef, there would be less demand for live cattle shipments. The shippers saw this clearly and prevailed upon the sympathetic railroads to set rates per car or ton-mile for dressed beef considerably higher than for live cattle. This reduced the price advantage for dressed beef to the benefit of cattle brokers, stockyards operators, and the railroads with their investment in cattle cars. New York butchers joined the group pushing for higher rates in order to reduce or eliminate competition from dressed beef. The Prairie Farmer thought the railroad interests had organized the butchers but assured Cornelius Vanderbilt (of the New York Central Railroad) that "manipulating stocks is child's play compared with this new venture" to ruin the dressed beef trade.[8]

The Vanderbilt railroads and the Pennsylvania system, with large investments in cattle cars and stockyards, refused to provide refrigerator cars or to haul them at advantageous rates. As a result, packers not only built their own but turned to the Grand Trunk and the Baltimore and Ohio railroads to haul them since those lines had very little live cattle business. Even so, the other railroads and live cattle shippers did succeed through rate pools in keeping published freight rates for dressed beef usually at least 50% above those for live cattle from 1879 to 1892, even during the rate wars. In 1883 Chicago firms could place choice cuts of beef in New York for 11 cents a pound. The New York butchers claimed they had to pay 12 cents a pound for live cattle and charge 25 to 35 cents for choice cuts in order to pay for labor, waste, and loss on less desirable cuts and still make a profit. If butchers bought from one of the packer branch houses, they could have the cuts they wanted at 12 to 13 cents a pound, sell them for 18 to 20 cents, and still make a profit. At that time live cattle freight rates to New York were 40 cents a hundred pounds and dressed beef rates were 64 cents.[9]

Despite shipper protests and resolutions by such groups as the National Cattle Growers' Association supporting lower rates for dressed beef, the railroads refused to reduce charges to the same level as for live cattle. Only the Baltimore and Ohio and the Grand Trunk railroads seemed inclined to bolt railroad rate pools and thus caused a rate war in 1887 and 1888. Over the last quarter of the century, the level of rates on all kinds of beef generally declined, but there was no equality for live and dressed beef. However, by 1890 the Chicago packers had sufficient organization and economic power to influence railroad rates to their advantage.[10]

Local butchers, especially wholesalers, were most vehement against dressed beef and combined to fight the packers. They organized the Butchers' National Protective Association in 1886 and attacked dressed beef on the grounds that it was unwholesome and that only the locally killed product was sure to be sanitary. In some areas they enlisted the support of local trade unions to boycott butchers who retailed Chicago dressed beef. The packers responded by establishing branch wholesale houses in cities, opening retail outlets, and selling it as low as 3 cents a pound. When local butchers agreed to sell dressed beef from the branch house, the packers closed their retail stores. Packers also established the peddler car, a refrigerator car filled at a branch outlet to deliver small amounts of dressed beef and other perishable goods to little towns along the railroad line. This technique created an almost limitless national market for fresh beef, mutton, and pork and seriously threatened local butchers in every town served by a railroad.[11] Some states adopted legislation forbidding price cutting and other packer retail practices, but U.S. district courts, the Supreme Court affirming, struck down these state laws as restraints of trade. By 1894 all the obstructionist efforts had failed and the _Butchers' Advocate_ acknowledged defeat.[12]

The hermetically sealed tin can provided another way that Illinois and Iowa corn-fattened cattle could be packed and shipped to domestic and foreign markets. The tin can had been known and used since the Napoleonic wars, but not until the Civil War and the use of canned condensed milk did it become popularly accepted in the United States. Soon after the war Chicago packers began to investigate canning as a replacement for packing pickled beef in barrels. William J. Wilson (a leading packer) began canning beef in the early 1870s; Libby, McNeil, and Libby began in 1875. After a court ruling invalidated patent claims on the canning process, all packers handled canned meat. A regular canner grade of cattle came to be recognized in the market. This was mostly western but also included some lightweight Corn Belt cattle averaging about 950 pounds each. Two canning processes developed. One consisted of placing boiled compressed meat in square or pyramidal cans, which were then reboiled and sealed after cooling. The other consisted of putting raw meat into round cans, which were cooked and sealed before cooling. By 1884 a Bureau of Animal Industry agent reported that half the canned meat of Chicago packers went to Great Britain.[13]

In addition to the domestic trade, exports of livestock,

dressed beef, and canned beef had an important influence on the industry. The foreign trade took up the slack as American beef production increased, and this had a definite influence on market conditions. There had always been some export of cattle and beef products from the United States, but it did not increase significantly until after the Civil War.[14]

Most exported live cattle went to England from the late 1870s to the late 1880s, while through the end of the century the majority of dressed beef went to England. The English were faced with a serious problem, as the urban population grew much faster than local agricultural production. At the same time, the United States had a growing surplus of cattle, and livestock reporters noted with favor the growing foreign demand that buoyed the market.[15]

Nelson Morris (the first of the major packers in Chicago to cater to the European trade in beef) shipped some live cattle to England in 1868, but not on a large scale. Next, a Scottish firm imported some American cattle in 1873 and 1874, but did not continue. Finally, Timothy Eastman (a New York cattle dealer) revived live shipments to England in 1876; and from then until 1885 he was the major exporter of live cattle. During July, August, and September 1878, for example, England imported 22,129 live cattle from the United States.[16] Continental cattle feeders even went to the extent of importing feeder cattle for fattening and resale in England. A Schleswig-Holstein group paid as high as $137 a head for 1,100- to 1,200-pound steers delivered in Hamburg to be fattened for German and English markets. This prompted the Prairie Farmer to predict the development of a whole new market for American feeder stock in Europe. However, in the late 1870s Britain restricted cattle imports from the Continent for fear of spreading cattle diseases, and pressure from English feeders and fear of pleuropneumonia in American cattle brought restrictions requiring them to be slaughtered on the wharf at the port of entry. The Prairie Farmer immediately called for federal laws to control contagious diseases so that beef cattle could be certified disease-free. Chicago prices fell in January 1879 when the new British restrictions went into effect, but these affected the live cattle trade only; dressed beef shipments increased.[17]

Despite a temporary decline, shipments of live cattle to England continued. The rule requiring immediate slaughter of imported live cattle remained in effect until the early twentieth century, even after American beef processors and feeders had succeeded in obtaining government action on disease control and inspection programs for livestock. By 1900 the United States exported 396,977 live cattle, the majority of which went to England and almost two-thirds of which passed through the Chicago market.[18]

Dressed beef had the same advantages over live cattle in the export trade as it had in the domestic market in that more salable pounds of dressed meat than live cattle could be shipped in the same space. The Prairie Farmer and other farm periodicals, as well as Illinois and Iowa fat-cattle producers, were interested in the foreign market. Articles reported opportunities in the English market, London prices, and problems posed by various types of embargoes by foreign governments for protection from American pleuropneumonia. Papers also followed reports from English newspapers on beef shipments originating in the Middle West. Illinois beef carcasses arrived in Glasgow, Scotland, as early as June 1876, having been slaughtered in New York just prior to shipment. A Scottish newspaper reported in the early fall that sales had averaged 150 carcasses a week.[19] During July, August, and September 1878 England imported 59,512 quarters of beef from the United States, while during the first six months of the year 165,422 quarters of beef entered that country. All this dressed beef came iced, but refrigerator-equipped ships were being planned to reduce the load of ice and increase the amount of beef carried. Liverpool received 2,120 live cattle and 4,280 quarters of beef in one week in October 1878. The magnitude of shipments grew so much from 1875 to 1878 that shipping lines converted passenger ships to beef carriers.[20]

By 1900 the United States had exported 329,078,609 pounds of dressed beef, mostly to England. Table B.11 compares 1900 exports in volume and value with selected other products. The value of live cattle slightly exceeded that of dressed beef, and the combined total value of beef made up 60% of the value of the total meat exported, the third largest item in value of American exports.[21] (See Fig. B.1.)

The great expansion of the market through technological and marketing changes in the dressed beef trade created large, vertically organized corporations based in Chicago. These corporations controlled, for the most part, both the wholesale and retail domestic markets for beef products. Known as the "Big Five," Armour and Company, Patrick Cudahy,

Nelson Morris, Swift and Company, and Wilson and Company frequently acted together in buying cattle in the Union Stock Yards and in apportioning retail markets among themselves.

Occasionally the Big Five met to set prices and attempt to eliminate some of the worst aspects of competition. With high capital demands and rapidly expanding business, they were pressed for cash and took every means to reduce overhead. The first pool organized in 1885 included Swift, Morris, Armour, Samuel Allerton (farmer, stockyard owner, and packer), and George Hammond and concerned only shipments of meats on a pro rata basis. The second known pool operated from 1893 to 1896 when Henry Veeder (a lawyer for the packers) acted as pool secretary and compiled reports of shipments to the major eastern market areas, apportioned them among the pool, and assessed fines for overshipments. Representatives of the packers met every Tuesday afternoon to discuss operations. The pool broke down in 1896 and 1897 when Schwarzchild and Sulzberger (a New York based wholesaler and packer) flooded the market; however, that firm joined the pool in 1898, and the packers set price margins and shipment quotas through the end of the century. The pool even set prices by ascertaining the net carcass cost of average operations for each packer and then establishing a uniform minimum cost system for the industry regardless of the price of livestock. By basing the wholesale price of meat on an artificially high cost, they all made a profit.[22] A Federal Trade Commission investigation revealed this only after the statute of limitations had expired; and when they were under investigation, the packers denied operating any pools in the early twentieth century. Contemporary rural newspapers and a Senate committee attacked the packers in the 1890s for being against the cattle feeders' interests. Packers were criticized for driving local butchers out of business by selling beef below cost. Once the local butchers closed their doors or agreed to be distributors for the packers, retail meat prices rose and wholesale prices to feeders supposedly declined. Packers were also accused of using range cattle as a "club with which to beat down the price of natives [cattle raised in the Middle West] and compel the farmer to compete with the range." Some farmers became discouraged and let the quality of their cattle deteriorate to range level. The only remedy offered by the press was for farmers to breed and feed prime cattle for the eastern or export market. Live shipments to the East had increased in 1890 and 1891 so newspapers urged farmers to "breed past Chicago."[23]

Some packers even went into cattle feeding themselves. As early as 1878 Nelson Morris kept cattle in Peoria, Illinois, which were fattened on distillery mash. Morris even helped finance construction of the Great Western Distillery, said to be the largest in the world at the time. Cattle chained to feed bunks received wet mash with hay for balance. In 1880, 16,000 cattle were fed on mash in Peoria. The number fluctuated and rose as high as 28,000 in 1893 but declined to less than half that by 1895. Morris also had western ranches and a large farm in Indiana where he fattened range cattle. He turned out as many as 75,000 head a year from his feedlots.[24]

By the early twentieth century when the federal government investigated the packers, they were large corporations. The Big Five killed 70% of all livestock and 82% of cattle slaughtered for interstate commerce. All were very much interested in meat substitutes such as dairy products, fish, poultry, eggs, and vegetable oils. Swift was the greatest butter distributor in the country by 1916, and the five controlled at least half the interstate commerce in poultry, eggs, and cheese through their wholesale branch houses and retail outlets. They refined 31% of American cottonseed oil in 1916, giving them a big stake in the oil-cake business as it became a more popular cattle feed supplement. They were also major factors in the hide and leather industry. The Big Five owned 91% of all the refrigerator cars equipped to ship dressed beef on the railroads, and they controlled the major stockyards, terminal railroads, market papers, and livestock banks. Packer men sat on the boards of 3 banks in Boston; 9 in New York; 25 in Chicago; 5 in St. Joseph, Mo.; 3 in Kansas City, Omaha, and Wichita; 2 in San Francisco and St. Paul, Minn.; and 1 each in East St. Louis, Denver, Fort Worth, Oklahoma City, Portland, Oregon, and Sioux City, Iowa.[25] Much of the magnitude of other interests of packers became apparent after 1900, but the initial basis for this growth came from the rise of the nineteenth-century midwestern beef and pork processing industry.

Chicago remained the meat marketing and processing center of the nation in 1900, though the packing industry began to decentralize. By the end of the century the leading packing centers and number of cattle slaughtered were: Chicago

(1,794,397), Kansas City (1,092,804), Omaha (516,669), East St. Louis (484,008), and St. Joseph (228,977). Lesser centers were Des Moines, Sioux City, St. Paul, and Fort Worth. The packers had so revolutionized the meat trade that butchers in some cities claimed that "not a retail butcher has made a fair profit and a living in the last ten years" because of packer branch houses.[26] The packers had also affected the raising and feeding of cattle in Illinois and Iowa.

CHAPTER 6. STRUCTURAL CHANGE IN ILLINOIS-IOWA CATTLE FEEDING TO 1900

Expansion of the packing industry went hand in hand with changes in beef production. Demand for beef increased because of growing population, improved transportation, and better preservation of meat. The vision of profits, touted by word of mouth, newspapers, and books, attracted men and capital to the Great Plains to establish or enlarge cattle ranches. The increasing range cattle production, in turn, encouraged changes in midwestern feeding operations. Illinois and Iowa producers who did not fatten range cattle were required to improve the quality of animals they raised and at the same time market these cattle at a younger age to compete with cheaper western stock. This brought changes in feeding practices, financing, and marketing. Men of tradition and experience shifted to intensive feeding of improved quality beef over a shorter period of time. The dressed beef trade, which made it pay to produce quality, helped Illinois and Iowa cattlemen who were willing to adjust from being caught in the same position as Ohio feeders in the 1850s who faced lower cost competition to the west. The range cattle boom also created problems for the industry because of overexpansion and the decline of prices in the 1880s and early 1890s, which caused some in the Middle West to turn to other farm products.[1]

With the end of the Civil War and the abnormal demand for wool and the continued poor results with wheat in older areas, many Illinois and Iowa farmers turned to growing feed grains which they could sell or feed to their own stock. New farms in northwestern Iowa, eastern Kansas and Nebraska, and southern Minnesota added to the abundance of wheat and corn. This contributed to a decline in grain prices, which gave a relative advantage to stockmen who could utilize feed grains. The agricultural press had been campaigning for some time (and continued through the next decade) for farmers to "raise corn always in preference to wheat. Learn to convert Corn into Pork [and] Beef," farmers were told. Inclusion of livestock in the farm system became the watchword of agricultural writers such as "Tama Jim" Wilson and Coker F. Clarkson in Iowa and editors of the Prairie Farmer from Chicago.[2]

Civil War demand had, however, reduced the number of cattle in the Middle West. At the end of 1863 the Commissioner of Agriculture estimated the number to be down by 30% from the previous year in Illinois, the leading cattle-fattening state. In 1865 the Prairie Farmer assured midwestern farmers they had no rival for beef production in the East at least. They were urged to replenish war-depleted cattle resources in preparation for anticipated meat demands from the South and the growing population of the North. "Self-interest demands that strenuous efforts be made to increase [cattle] numbers."[3]

Cattlemen obtained their supply in a variety of ways. The most common practice was to raise cattle from calves on the farm, but most large-scale feeders had neither the capital nor the desire to maintain the number of brood cows necessary. Even so, an Iowa farmer contended that there was less personal satisfaction in buying feeder cattle than raising them.[4] Occasionally, a large-scale feeder raised his own cattle, especially before the 1880s. John D. Gillett, one of the most successful fat-cattle feeders in Illinois, raised a strain of Shorthorn grade cattle on his 12,000-acre property in Logan County. In 1876 Gillett had 500 cows and fed 700 fat cattle a year.[5] To supplement his own calves, Gillett bought cattle from neighboring farmers, one of the easier ways to obtain them without the expense of a large cow-calf herd. County reports in the newspapers frequently mentioned the scarcity of feeder cattle or that buyers had been active in the area taking surplus stock from farmers who did not specialize in feeding. Such local buying reached its height in the late summer and early fall when cattle came off pasture in good condition.[6]

The immediate neighborhood seldom supplied enough, so men went further afield into unsettled portions of the state where large herds grazed; into neighboring states; and to the Chicago, St. Louis, or Kansas City markets to buy feeder cattle shipped from adjoining states and the western range. Even before range cattle became available, Champaign County, Illinois, feeders bought poorer offerings in the Chicago markets and fed them up to a better grade. In 1876 Sangamon County feeders bought three-year-olds from Wisconsin, Iowa, and Missouri in the Chicago, Kansas City, and St. Louis markets.[7] In the 1870s much of northwestern Iowa remained open range and a source of feeder cattle. Farmers in the northeastern Iowa wheat area sold their surplus at one and two years of age to farmers in the southern half of the state and in Illinois, where corn abounded. The Iowa range supplied feeder cattle until the middle 1880s in competition with Texas, Indian territory, Kansas, Nebraska, and the increasingly important northern range—Wyoming, Montana, and Dakota. In the 1870s and 1880s midwestern farmers aided development of western ranches to which they sold stocker cattle and from which they later bought feeder cattle. Eugene S. Ellsworth (Iowa farmer-banker) brought cattle in the late 1880s and early 1890s from his Dakota ranch to farms in Emmet, Hancock, and Wright counties for fattening. As the business emerged from a period of depression in the 1890s and demand for feeder cattle increased, range men bought Canadian cattle as well.[8]

The source of feeder cattle and the place of market depended in certain areas upon the available railroads. Many Iowa men bought stock at the Missouri River markets where western railroads converged—Kansas City, Omaha, Council Bluffs—because it was cheaper than buying in Chicago and paying the freight back to Iowa. On the other hand central Illinois men found Chicago, where railroads running south and west converged, a better market for western feeder cattle. Absence of railroads limited feeding in some areas such as O'Brien County, Iowa, where no one fed western cattle until the railroad came through in 1881. In parts of Franklin County, Iowa, this was so until the advent of the Chicago Great Western Railroad in 1903.[9]

There was some argument over which was the most profitable method of operation, to feed fewer quality cattle or to handle a larger number of average steers. The *Prairie Farmer* and the *Iowa Homestead* agreed that the farmer who raised his own cattle, if of good quality, made the most money; but both newspapers assumed it was "better" in noneconomic ways for farmers to raise their own. As the *Prairie Farmer* put it, "the feeder has no business to buy his stock; he should breed and raise it himself."[10]

The movement of Texas cattle to northern markets after the Civil War was in response to the demand for more beef and the cheap price there. Though Texas cattle were not of major significance through the rest of the century as a supply, they were the first feeder cattle to come from outside the Middle West and were heralds of the later western range industry which came to supply the majority. As the cattle feeding industry became larger, the trend toward regional specialization continued; the range country raised cattle and Corn Belt farmers fattened them.

The business of shipping Texas and Cherokee (Oklahoma) cattle north resumed, rather than began, after the Civil War. The influx of Texas stock initiated range competition for Illinois and Iowa feeder cattle; and though the center of this competition moved to more northern ranges during the last quarter of the century, the consequences remained the same. Illinois and Iowa men adjusted to the presence of cheap range cattle by upgrading the quality of their beef and marketing it at a younger age.

Texas cattle reappeared in the North in 1866. They were shipped up the Mississippi River to Cairo and the Illinois Central Railroad yards or driven overland through Missouri. Supposedly, 10,000 Texas cattle were shipped to Chicago in 1866. George Duffield (a farmer from Keosauqua, Iowa) went to Texas in 1866 and bought 1,000 cattle at $12 a head and drove them back to Iowa during the summer. Less than half of Duffield's herd made it to Iowa in October because of stampedes and night losses, but the low price encouraged others to follow suit.[11] In 1867 Chicago received almost 40,000 head of Texas and Cherokee cattle, and the railroads and feeders prepared for a bigger year in 1868. Because of the low amount of shrinkage during shipment, a valuable hide, and lower initial cost, demand grew for southwestern cattle. The Illinois Central Railroad built a stockyard with a capacity of 4,000 head at Tolono, in Champaign County, where 15,000 Texas cattle were distributed to central Illinois feeders in June and July 1868. Estimates of the number of Texas cattle feeding in central Illinois that year went as high as 35,000.[12]

John T. Alexander, on his great Broadlands Farm in Champaign County, received shipments of Texas cattle numbering from 109 to 600 head every week from May 31 to July 26, 1868. Most came from Cairo, but some were from Abilene, Kans., via Chicago. Alexander paid $35 a head for over 4,000 Texas cattle the next year and after six to nine months on grass and some corn expected to sell them for $70 a head.[13] That he expected a profit of at least $100,000 indicates why many men wanted to process Texas cattle. They felt that a limited amount of capital would buy more Texas than native cattle for feeding; and though natives sold for more than Texans, the profit per head was not much greater and the larger number of Texans produced more total profit. But by 1870 some men, including Alexander, had second thoughts about the profitability of Texas cattle. Morgan County feeders claimed they lost money because of poor gains in weight, while Alexander apparently concluded that it was now a "losing business" and he would go back to feeding only native cattle. Even so, by November 1870 Chicago receipts of Texans had passed 200,000 head for the year; the northern fattened Texas beef sold at a "good price," from $4.50 to $5.75 per hundred pounds.[14]

Coincident with the reappearance of Texas cattle, the southern or Texas cattle fever reappeared also and caused severe losses in native cattle as it had in the 1850s. Whereever natives mixed with Texans or occupied yards, roads, or fields soon after, the native cattle usually died. Armed Missouri farmers prevented Texas droves from passing through to Illinois or Iowa and caused the cattlemen to seek better ways to reach the northern feeding areas. In 1867 Joseph McCoy (an Illinois cattleman) and others established a western railroad shipping point at Abilene, Kans. Texas drovers could reach Abilene without crossing settled territory or encountering many native herds, and the cattle interests overcame local opposition by offering a bond for damages and packing a public meeting in their favor. Though this procedure satisfied Missouri farmers, it did not prevent the spread of Texas fever in Illinois from cattle brought by way of the Mississippi River to Cairo.[15]

Native cattle in the same field with Texans died about two months after contact. Cattlemen found it difficult to understand the nature of the disease because it did not affect the Texas cattle, yet there was an obvious connection between their presence and its incidence. Consequently, it was difficult to suggest remedies or propose procedures for handling Texas cattle short of excluding them. The Illinois legislature passed a quarantine law in 1867, but it proved ineffective and native cattle losses increased in 1868. In that year, Champaign County, Illinois, lost $150,000 worth; Ford County lost 500 head, Grundy County 100 head. Vigilance committees in some areas resorted to "calling on" Texas cattle importers to see that they made restitution for native cattle lost.[16]

In 1868 amid charges, countercharges, and calls for

more information about the disease, a convention of cattle commissioners from twelve midwestern and eastern states and the province of Ontario met in Springfield, Ill. Eastern states had already (or were preparing to do so) quarantined live cattle shipments from Illinois in order to prevent the spread of the disease. Joseph McCoy visited the convention as an honorary delegate and did his best to defend Texas cattle and prevent exclusion as the only recommended means of combating the fever.[17]

The convention debates produced reports on the latest observations of the fever and prevailing opinions as to the cause and the treatment to be followed. There were three views on the subject. Texas and Kansas shippers refused to acknowledge the existence of a fever spread by Texas cattle. They admitted that northern native cattle died, but it was argued this was the result of a constitutional weakness. The "superior" Texas cattle, they pointed out, never died of the fever. Practical northern cattlemen assumed that ticks found on Texas cattle were related to the problem. Professional medical doctors and veterinarians, however, dismissed the cattle tick and sought the cause in spores and their transfer by breath or waste matter from one animal to the next. In 1868 a report submitted to the Illinois State Agricultural Society contained all the needed information about the fever and ticks to connect the two conclusively but concluded that the tick theory was the "last and perhaps least" of various explanations.

All reports and discussions at the convention made it clear that fever appeared only from contact with Texas cattle or places where they had been, that they carried ticks if shipped from Cairo on the railroad, but that such cattle coming from Abilene hardly ever had ticks because adults fell off during the long drive (which may account for the relatively little excitement over Texas fever in Iowa).[18] In central Illinois, at least, no Abilene cattle had brought fever to natives in the area. Discussion revealed the facts that ticks grew to large size on cattle, dropped off, and hatched young that infected pastures and climbed on cattle to suck blood and repeat the life cycle. John T. Alexander's farm manager had removed Texas cattle to a second pasture after the adults had fallen off and before young ticks climbed back. Native cattle in the first pasture subsequently died, while those in the second pasture contracted no fever from the Texas cattle without ticks. Native cattle pastured with tick-infested Texans for two weeks but removed before the ticks fell off did not die of fever, but those remaining in the lot four weeks or more subsequently contracted the disease. Men observed that frost in the fall arrested the fever and that any Texas cattle wintered over in Kansas did not carry it.[19]

When the medical doctors (not veterinarians) and cattle commissioners from various states presented their observations and opinions, the solid conclusions that could have been drawn about ticks were lost in a morass of contemporary medical ignorance, misinformation, hearsay evidence, and elaborate theories founded on only one observation. Frequently, under questioning by members of the convention, that one bit of "evidence" for a theory turned out to be inaccurate. The whole procedure illustrated the power of the "expert" in the form of medical doctors to perpetrate theories that might have been intriguing or in vogue but ran counter to the logic of the observed evidence. The most famous of the medical experts, the English veterinarian John Gamgee, headed a scientific commission to investigate Texas fever and his ably written and deceptively exhaustive report of 1871 set back research on Texas fever by a decade. Gamgee said that the tick theory "was not only impossible but ridiculous. A little thought should have satisfied anyone of the absurdity of the idea."[20]

In fairness to Gamgee, it should be mentioned that to advocate the tick theory with the prevailing knowledge of veterinary medicine, biology, and pathology in 1870 would have made him a laughingstock in the medical world, which did not yet understand the function of an alternate host. It was all the United States Bureau of Animal Industry could do in the 1890s to gain acceptance of the tick theory as fact. The first bulletin of the bureau published the findings of Theobald Smith and F. L. Kilbourne that a protozoan parasite in the blood caused Texas fever but that the parasite could travel from one animal to another only through an intermediate host, the tick. A study of the life cycle of the tick showed that it could be eliminated by dipping cattle periodically. By 1898 all restrictions had been removed on transportation of Texas cattle if they were dipped.[21]

But the solution came long after the immediate problems of 1867 and 1868 when the only seemingly feasible response was a quarantine on Texas cattle from spring until the fall frosts. The convention of cattle commissioners so recommended, much to the disgust of McCoy, who saw it all as a plot to ruin him and the dealers in Texas cattle and to protect native cattle raisers. McCoy labeled the convention, perhaps rightly for the wrong reasons, "as a collection of quondom quacks and impractical theorists and imbecile ignoramuses, . . . without an equal." The medical men especially brought to McCoy's mind "the ancient royal commission of sage scientists who spent many days and weeks investigating and profoundly debating the all-absorbing question of natural history, to wit: 'which is the butt end of a billy goat?'"[22]

McCoy was not alone in his feeling that the law was being used to the advantage of native cattle feeders, but pressure favored the local interests; and in 1869 the Illinois legislature passed a bill excluding Texas and Cherokee cattle from March 1 to November 1, with the exception of cattle wintered in the states of Kansas, Nebraska, Missouri, Iowa, or Wisconsin, attested by a certificate from the clerk of the county where they were wintered. As a result, Kansas county clerks issued certificates liberally.[23] The debate over Texas fever continued to the 1890s, but the outbreaks became less severe after cattle coming from Texas to Cairo by boat were more effectively prohibited by the 1869 law or segregated, if allowed through, under later procedures. Texas cattle that came by way of Kansas lost most of their ticks even if fraudulently certified and so posed less of a danger to native cattle. Feeders also made efforts to segregate Texas cattle, though outbreaks of fever occurred in Illinois in McLean, Morgan, and Sangamon counties in 1872; Champaign County in 1876; and Perry County in 1885 and in the 1890s.[24] Railroads disregarded the law, which was eventually held unconstitutional by the Illinois Supreme Court in 1879 after a ruling by the U.S. Supreme Court that a similar Missouri quarantine law was an unconstitutional restraint of interstate commerce. Some cattlemen thought they could make a lot of money buying Texas cattle cheaply and selling them in Chicago after a summer on Illinois and Iowa bluegrass, and they did not allow quarantine laws and subsequent proclamations of state governors to effectively hinder their importation of Texas cattle to the Middle West.[25]

By 1886 Texas cattle could be brought into Illinois legally for immediate slaughter, if segregated from other cattle while being shipped, and for feeding, with permission of the state livestock board after a 90-day quarantine.[26] Also by the 1880s range cattle from Kansas, Nebraska, Wyoming,

Dakota, and Montana replaced those shipped directly from Texas and Cherokee country as feeder stock.

Texas fever killed many native cattle and disorganized the cattle feeding business in particular areas. Less dramatic but more significant, the competition of Texas and western range cattle brought changes in Illinois and Iowa beef feeding practice. This was a matter of choice to the feeder depending upon how much corn or grass, capital, and native cattle he had. Beef canners and packers bought western cattle as a cheaper more abundant source of meat than the better grades of natives used for live exports and dressed beef. There was a fairly good demand for western feeder stock, and even western grass-fattened cattle were sold directly to packers without an intermediate stop in the Middle West. The growing size of this latter group, which went to market from June through November, caused problems for Corn Belt feeders. In 1867, for example, 120,630 Texas and Colorado cattle arrived in Chicago during the six-month range season. This amounted to about 10.9% of total cattle receipts for that year.[27] From 1882 to 1895 Texas and western receipts in Chicago fluctuated between 36% and 25% of the total. The number dropped off sharply after 1895. In 1900 range cattle accounted for only 13% of total receipts. (See Table B.12.) Generally in the 1880s and 1890s, prices for western cattle ranged from about 16% to 25% less per hundred pounds than for native 1,200- to 1,500-pound steers. Prices of Texas cattle were the same as western cattle or a bit lower. Texans wintered in the north sold for more than direct shipments from the south. Northern-wintered Texans also created some problems in nomenclature in the market, Were they Texans or westerns?

As early as 1868 men began to discuss the consequences of western competition. The cheap cattle weakened prices, especially for the poor offerings of native cattle, and editors and market reporters urged feeders to hold their stock until properly fattened at the age of four years or more when they would bring better prices.[28] But the remedy lay not with fattening common stock at a later age, but in upgrading the quality of midwestern cattle and fattening them at a younger age for a faster turnover of capital. In 1870 the Department of Agriculture anticipated that within a decade there would be "an immense improvement" in midwestern stock as the "more intelligent cattle-raisers" supplanted their common native cattle with improved breeds. Individual complaints supported the department's view that native unimproved stock would become unprofitable for Illinois and Iowa feeders.[29]

As Illinois and Iowa land values increased, men who fattened cattle on grass were hard pressed. Many felt they must turn to grain to realize a sufficient return per acre. As early as 1872 some Morgan County, Illinois, grazers broke up their pastures and planted corn.[30] The Commissioner of Agriculture suggested in 1876 that the best way to organize the resources of the Middle West and the range country was to raise cattle to two years on the ranges and then send them to the Middle West for a year's fattening on corn.[31] This is what Illinois and Iowa feeders were doing in the 1870s and early 1880s. They fed improved local and range cattle and sent them to market at three years instead of four, while those who did not improve their stock turned to grain farming. By 1880 the Illinois Department of Agriculture reported that only the "choicest specimens of early matured steers" would successfully compete with the better grade range cattle.[32] As cattlemen improved their range herds with Shorthorn and Hereford bulls and shipped increasing numbers east, the competition, together with the developing centralization of the packing industry, lowered average prices in Chicago from an 1882 high of $6.25 per hundred pounds (1,200- to 1,500-pound steers) to $5.90 by 1884, $4.70 in 1888, and the low of $3.90 in 1889. Prices did not go above $5 per hundred pounds again until 1900.[33]

The agricultural press and farm leaders continued to exhort beef feeders to improve their animals and feeding procedures; they assured that "scientific and methodical application" of principles of good feeding "must necessarily serve to a large extent in overcoming the disadvantage of raising stock in high priced lands."[34] Writers stressed the ultimate advantage of combining Illinois and Iowa corn and improved cattle "as there will always be a demand for them at high prices"; and urged that only "good cattle" fed for the top market would do for the corn states.[35] As profit margins narrowed, it became more important to have improved cattle that would gain 100 pounds on fewer bushels of corn than the average available in the past. The profit came from cattle that converted corn into beef most economically.

Financial arrangements of feeders were an important aspect of the business. Increasing specialization in the production of prime beef cattle brought changes in the financing of feeding operations. Considerably more capital was involved in feeding improved grade cattle than in handling natives or Texans. Toward the end of the century the need for extra capital helped the development of cattle loan companies and the business of city banks discounting cattle loans made by rural banks.

In the years just before and after the Civil War some packers advanced money to feeders. Isaac Funk had occasionally fed cattle under contract to a packer who supplied the capital. But this could not have accounted for much of the total number fed because packers experienced chronic shortages of operating capital as they expanded their operations after 1875. Alexander, Strawn, and others before the 1860s borrowed extensively from local farmers or took cattle on credit and paid for them after they were marketed.[36] In the last half of the century much of the operation might have been similar to this. In some areas men who wanted to feed more cattle than they could raise or buy with their own funds could borrow from well-to-do farmers if their reputations were good.[37] Unfortunately, this is the least documented side of cattle feeder financing.

Country bankers backed some of the sizable operations. George Wilson, a Geneseo, Illinois, banker "with that rare quality called 'cattle sense,'" backed Alexander's purchase of 3,000 cattle during the panic of 1873. Local buyers supplied feeders on credit to trusted individuals, which allowed farmers without capital to feed cattle.[38] Rural sources of capital seemed adequate until the late 1880s when cattle loan companies appeared, many of which were backed by the packers or stockyard banks and commission firms. The financial panic of 1893 and its aftermath placed an added strain on rural credit. Men found it difficult to borrow from normal local sources. In 1896 a Prairie Farmer inquiry found that local banks were short of funds and more men would be feeding cattle if they could borrow money.[39] A more formal arrangement came with the new loan companies. Feeders had to secure loans with chattel mortgages on the cattle rather than a personal note. They also had to supply financial information about themselves and sometimes have their operations inspected before their personal reputations were established with the loan company. By the early twentieth century larger loans come from cattle loan companies while small feeders borrowed from local banks that sold the cattle notes to a city bank. Even with the more formal arrangements individual reputation played a great part in securing loans.[40]

The railroad was the major means of moving Illinois and Iowa cattle to market after 1865, and although they reached market faster than on foot, the problems of loss of weight and damage from overcrowding still existed. The first cattle cars were a makeshift lot that did not provide adequate protection. The air brake and an improved coupler helped conditions in the 1870s, but overcrowding and crippling continued. Concern over excessive loss of weight (as high as 15% of gross weight on the run from Chicago to New York) caused the enactment of the federal 28-hour law in 1873. This law limited interstate shipment of animals to a period no longer than 28 hours without unloading for food, water, and rest for a period of five consecutive hours. The law was defective and remained a dead letter until about 1905.[41]

The fluctuating rail rates between Chicago and New York affected prices paid to Illinois and Iowa feeders by livestock shippers or the dressed beef packers. Buyers effectively passed the higher rates on to the feeders by bidding less on cattle to be sent east. Periodic relief came in the form of rate wars between competing rail systems, but little relief was likely for the Illinois or Iowa farmer who had no choice but to ship on the railroad that served his area. The most frequent complaint against railroads, however, concerned the shortage of cars or the length of time for the trip to Chicago. Though the accusation was denied, feeders claimed that the railroads perpetuated the three-day market in Chicago by not furnishing cattle cars on an even distribution through the week. The railroads made efforts to come to agreement among themselves as to time of runs and rates from western points into Chicago. If they did not agree on rates, they often agreed on lengthening the time of the run, which in effect increased the rates.[42]

The three-day Chicago market was another problem for feeders. Most receipts came into Chicago on Monday, Wednesday, and Thursday with Monday being the dominant day. The uneven distribution of receipts strained market facilities, put uneconomical pressure on railroad equipment, and gave buyers a price advantage. With the bulk of weekly receipts coming on Monday and Wednesday, buyers could beat down prices by waiting until the yards filled up and men became worried about carrying cattle over to the next day. Feeders said that the railroads would not supply cars or put cattle on through trains except for the Monday or Wednesday markets. On the other hand, some feeders claimed that they received better prices on days of heavy receipts because of competition among buyers and suffered financially every time they tried to even market receipts. The railroads and the packers insisted they preferred a five-day market but that the farmers refused to even shipments if given the opportunity. There the matter stood until the government forced changes by a market zone system as part of World War I food controls.[43]

Farmers could lessen their costs by shipping carload lots and obtaining the most recent market information. Very little accurate information was available to the end of the century about how many cattle were in feedlots across the Middle West. Feeders and packers alike operated without sufficient information to plan very far ahead. Newspapers occasionally conducted surveys to estimate the number of cattle on feed and then advised their readers.[44] The Illinois Department of Agriculture in its crop report bulletins included assessors' returns of cattle on hand on May 1 as well as a list of the estimated number of cattle marketed and on hand in December compared to the previous year. This was of little value, however, because the estimate was always placed at 20% of the number on hand regardless of other factors. It was far better to do as William Stevenson did. This Tama County, Iowa, feeder subscribed to the weekly _Drovers' Journal_ and received the latest market reports by telegraph the morning he intended to ship cattle to Chicago. He did business through the commission firm operated by the brother of "Tama Jim" Wilson (well-known Iowa farmer and U.S. Secretary of Agriculture).[45]

A good commission firm was probably the feeders' best source of market information despite attacks by the Grange and newspapers. In the 1860s when some commission firms sold cattle for country buyers, these firms tried to prevent reporting of stockyard sales prices in order to keep farmers from having a source of information other than the local buyer. The _Prairie Farmer_ retaliated by printing the names of members of the "Ring" to keep sales secret so that farmers would know who not to patronize.[46] Commission firms came in for much abuse as a monopoly and were frequently blamed for problems beyond their control. In the mind of the militant agricultural press, the commission houses were part of the Union Stock Yards monopoly, which was on the opposite (and wrong) side of every issue from the standpoint of the meat producers. Through most of the last half of the century, the commission firms charged 50 cents a head for handling and selling cattle. This did not seem an exorbitant charge on a steer worth about $55 in the late 1890s.[47]

The better cattle feeders in the Corn Belt tried to produce prime, choice, or good steers, while the majority produced good or medium steers and indifferent feeders sold common grades. Cattle grades set at the stockyards were still not uniform and precise by the end of the century;[48] but there seemed to be general agreement among buyers for the packers and livestock reporters on the major characteristics of each grade. The grade assigned in the stockyards was important because it affected the price. Men had to be able to judge which grade their feeder cattle could be prepared for. Poor quality animals could not attain a prime grade no matter how they were fed, and the cattleman who could not see the limits and possibilities of those he bought would soon lose money. It cost more to feed cattle for the top two grades than for good or medium grades. There were usually too few of the top two grades and too many of the bottom three grades of beef on the market. The desirable feeder cattle for top grade beef had short legs, broad flat backs, and straight top and under lines that were nearly parallel. When the animal was properly fed, the outline remained the same while weight and total size increased. Lesser grades departed from the desired image in greater degree as one descended through the list, with lower grade cattle having longer legs, thinner hips and chest, and higher flanks. By the end of the century, few Illinois and Iowa cattle in the better grades exceeded 1,600 pounds, and many prime baby beeves went to market at weights from 800 to 1,100 pounds. This contrasted with the 1850s and 1870s when extra-fine cattle were four- to six-year-old steers and heavyweights over 1,500 pounds received better prices.[49] Western competition had forced Illinois and Iowa feeders to concentrate on quality beef cattle because range men could feed common to good steers on grass at less cost than midwestern farmers could.

Not all feeders sold fat stock of the quality necessary to compete successfully with range cattle. Prairie Farmer market reports frequently noted the lack of first-rate beef in the market and the oversupply of half-fattened stock; the editors urged feeders to ship only fully fattened cattle. In periods of drought or poor corn crops many feeders did not want to carry stock over the winter and therefore sent it to market early and in less than prime condition. Astute feeders profited from poor planning on the part of others and bought these partially fattened cattle and fed them a short time for the midwinter market.[50]

Illinois and Iowa men who fattened cattle on grass sold them in the fall in competition with range cattle and in the late spring before the latter came onto the market. In the fall they needed choice or prime quality animals to compete successfully.[51] Corn-fed cattle could go to market at any time they were ready, though this was generally during the winter and spring.

During the 1880s and into the 1890s corn feeding increased over grass, and the ages and weights of marketed fat cattle gradually declined. That is, the prime steers were more in the two- and three-year-old range and generally less than 1,600 pounds, while for the common and medium grades average weights probably increased from the previously low level.[52]

Although accounting for only a small percentage of the market, prime and choice cattle two years old or less, known as "baby beef," came to be a specialty of leading Illinois and Iowa cattlemen. The result of years of breeding and feeding for early maturity, these were prime examples of what corn and feeding skill could do to compete with range cattle in a market of falling prices. The Iowa Homestead said in 1891 that "the time will come, and it may not be very far in the future, when it may be desirable to feed everything off at from eighteen to twenty months." Leading cattle feeders produced baby beef by starting with top-quality calves and feeding them as much grain, grass, and supplements as the animals would take. In December 1895 ten 1,200-pound steers twenty months old brought the top price in the market. The next year heavy steers over 1,500 pounds sold at a discount regardless of grade.[53]

The majority of Illinois and Iowa cattle feeders did not produce baby beef nor always send prime or choice cattle to market. However, they were influenced by the general trends that more often than not made grass feeding and four-year-old cattle unprofitable in the Corn Belt. Joel W. Hopkins (of Putnam County, Illinois) was an example of prosperous Corn Belt farmers who fed cattle as a major part of their farm operations. By the end of the century, cattle feeding was not seasonal. Hopkins bought and sold cattle in practically every month of the years 1897 through 1899.

In 1897 Hopkins bought 797 cattle and sold 702. The next year, he bought 482 cattle and sold 515, and in 1899 he bought 607 and sold 641. A good many were bought and sold locally, but Hopkins periodically sold one and two carload lots in Chicago (carload = 16 cattle). In 1897, he sold one carload each in February, April, and May; two carloads in March, June, November, and December; and six carloads in July. He generally received from $4.45 to $4.65 per hundred pounds, but one March carload sold for $3.75 and the December lot sold at $4.80. At no time did prices of his steers exceed the reported high price for the month, but they were about average. The other two years were not notably different except that, with the general trend, prices received moved higher into the $5 and $6 range. Where shown, average weights of Hopkins's steers ran from 1,300 to 1,400 pounds. Hopkins bought about as many cattle in Chicago as he sold there, but he also bought and sold lots as large as 20 head locally.[54]

Hopkins had been feeding cattle for years and had not curtailed his operations as some did during the decline in cattle prices after 1882. Average beef cattle prices had risen in 1882 to the high of the last quarter of the century because of the growing demand for meat as well as the demand for cattle to stock ranges. But the trend soon reversed as the range filled and began sending large numbers to market. Overexpansion and periodic liquidations in the range cattle business and increasing consolidation of the packing industry kept the price trend moving downward after 1882, and the business depression pushed prices lower in the early 1890s. Not until 1896 did the trend of beef prices start up again; in 1902 they surpassed the 1882 high. (See Fig. B.2.) Within the long-term price trend, there were seasonal variations, with declines in the summer and fall at the time of heaviest receipts from the range and rises from December through May or June. The increase in corn fattening of range cattle in the 1890s diverted some from market in the fall and added to market receipts in the spring. This tended to reduce seasonal variations in prices.[55] The price rise in the 1890s came after a long period of low prices and a two-year liquidation of feeder stock in 1893 and 1894 when feed had been scarce. From year to year, prices could be affected by the size of the corn crop. Cheap corn encouraged men to feed it to cattle and swine; however, a high price encouraged less demand for feeder cattle or for the marketing of fat cattle at lighter weights.[56] The extra-large corn crop of 1895 increased the demand for available feeder cattle, which pushed prices almost as high as those for fat cattle. This further depletion of feeder stock supplies meant a shorter supply of fat cattle, and this brought market receipts of them more in line with demand. By 1896 cattle prices started a steady climb upward. Chicago receipts had reached a fifty-three-year high of 3,571,796 in 1892 and then declined and did not surpass that figure until 1918.[57]

In the 1880s and 1890s some Illinois and Iowa cattlemen claimed that they lost money. In 1886 W. Whitson (Pottawattamie County, Iowa, feeder) said he lost $600 feeding cattle in the previous year and thought he would lose money again. In 1898 E. E. Chester (at the Champaign County, Illinois, Farmers' Institute) claimed that cattle feeders in the county had lost money for the previous five years. Many of the men in east-central Illinois who had fattened cattle on bluegrass gave it up and turned their pastures into cornfields in the late 1880s and early 1890s. On the other hand, James N. Brown's sons said that they only fed cattle because they had bluegrass pastures, but that it would not pay them to fatten cattle on corn. The Bureau of Animal Industry concluded in 1885 that livestock feeders generally gave "little careful thought" to figuring operating costs.[58] It was also evident that individual ability made the difference between profit and loss during periods of low prices. Prime cattle still brought good prices for those who could produce them. J. W. Steele (Monona County, Iowa) made $16 a head on a carload of steers sent to Chicago in 1888. Fred Stowe (Grimes, Iowa) sold five carloads of cattle in 1891 at fully $2 per hundred pounds above average prices for the month, and others did the same. In times of good prices the rewards for prime cattle were even greater. In 1899 L. H. Kerrick (McLean County, Illinois, and son-in-law of Isaac Funk) sold two carloads of grade Angus steers for $8.25 per hundred pounds. These prime cattle brought 75 cents more than the highest price paid in 16 years and almost $2 above average prices for the month.[59] Good cattle feeders could get better than average prices in almost any market.

Many suggested that the Chicago packers caused the price decline. The _Iowa Homestead_ claimed that the "dressed beef combine," when possible, forced down the price of good beef to the level of range cattle prices.[60] A U.S. Senate committee investigated beef prices from 1888 to 1890 and charged that if men refused the first bid on their cattle in the Chicago market, the packer buyers would later force them to accept a lower one or offer none at all. The committee found that whether by design or not, packers acted together in buying cattle, setting the retail cost of beef, agreeing on retail market areas, bidding on government contracts, and underselling local butchers. The committee disputed Philip Armour's opinion that low prices came from too many cattle in the market by pointing out that in the fall of 1886 when receipts fell, prices fell, and that in 1890 low beef prices existed at the same time that there were fewer beef cattle in the country, a larger population, and higher exports than in the early 1880s.[61]

The Union Stock Yard's Annual Report and the market columns of the Prairie Farmer followed the more orthodox version of the price decline; that is, that it came from overexpansion on the range, large receipts of medium to poor grade cattle in the late 1880s, and the generally poor business conditions of the early 1890s. With the decline in the number of cattle and the increasing consumption of beef as economic conditions improved, prices began to rise after 1896.[62] In 1897 A. B. Groat (president of Illinois Cattle Breeders' Association) pointed out that the "good old days" of cattle feeding were gone and that farmers had to make the most of the fact that they had no effective control over prices. Feeders, therefore, must pay more attention to the costs of production, quality of cattle, and management of the feeding program.[63]

Cost-conscious feeders must assess the situation as to the price of feeder cattle, number on feed, cost of feed, and probable level of fat-cattle prices at the time they expected to market their animals. About the only figure they could ascertain with certainty was the current cost of feeder cattle. Feed costs over the months varied with size and quality of corn and fodder crops; other variables they could only estimate. A dry summer on the range might flood the fall market with under average quality cattle and bring down prices. A mild or severe winter could make a difference of about two months either way as to when fat range cattle first came to market in the late spring and early summer and would have an effect on prices. Feeders could have more than one carload of cattle on feed so that they could pick and choose from the feedlot to make up a load of fairly well-matched cattle. A carload of even quality beef brought a better average price than one of mixed quality. A few poor quality cattle in a lot would bring down the lot price proportionately more than a few above average cattle would raise it. Also the feeder could sell his carloads directly through a commission merchant in Chicago rather than selling a few animals at a time to a local buyer who paid less than the Chicago price in order to make his own profit.

By the end of the century Illinois and Iowa cattle feeding had changed from beginning the fattening of cattle for market after the age of three to completion of the process almost always before that age. A small but growing percentage of fat cattle were going to market at 18 to 24 months by 1900. Most of these young animals had a high percentage of purebred blood as did most of the feeder cattle raised in Illinois and Iowa. The range cattle brought into the feedlots and fattened on corn had a higher proportion of improved blood than those of two decades earlier and were far superior to the range cattle from Texas and Oklahoma that appeared in the 1850s and 1860s.

By 1900 several methods of cattle feeding operations were followed in Illinois and Iowa. Some men bred and raised their own high-grade cattle for fattening. Another large portion of feeders bought young range cattle in the Kansas City or Chicago markets and fattened them, and a third group fattened cattle raised on the farm as well as those brought from the range. But whether from the range or calved on farms, the majority of top-grade beef animals spent some time in feedlots eating corn.

Range competition and declining profit margins in the 1880s and early 1890s caused reactions among Illinois and Iowa feeders and caused many of the Illinois men who had fattened cattle on bluegrass to plow it up and turn their attention to grain farming. Nevertheless, efficient cattlemen in Illinois and Iowa combined their cheap corn and knowledge of fattening with the young western range cattle to continue producing beef at a profit. These men had survived by the end of the century because they produced cattle to match the changing market structure.

C H A P T E R 7. REGIONAL CONCENTRATION OF ILLINOIS-IOWA CATTLE FEEDERS

In the last half of the nineteenth century changes occurred in the areal concentration of cattle feeding operations, which shifted from central Illinois to northwestern Illinois and central and southwestern Iowa. Though only a partial measure, federal census statistics give some indication of the distribution of beef cattle on a county basis and reveal changes in this density in Illinois and Iowa.

From 1850 to 1890 the decennial census takers counted oxen, dairy cows, and "other" (beef) cattle on farms as of June 1. This generally excluded spring calves, but not always, depending upon how local enumerators interpreted their instructions. In 1900 the census schedules further subdivided cattle into calves under one year of age, steers from ages of one to two, two to three, and three and over, bulls one year old and over, heifers from one to two years old, dairy cows over two years old, and other cows. Because of the new divisions, figures for 1900 are not fully comparable with the earlier years. Also the June 1 date posed problems as a guide to feeder stock on farms because many fat cattle had already gone to spring market. In addition, Illinois tax assessors came in May or June, which induced pennywise farmers to have fewer cattle at that time of year. Within the limitations of ten-year intervals, a June 1 reporting date, and the county as the reporting unit, census figures for beef cattle per one hundred acres gave some indication of the magnitude of cattle feeding across Illinois and Iowa.[1]

In 1850 the highest density was in west-central Illinois in Morgan County, which had 4.7 beef cattle per hundred acres. (See Map A.7.) Scott County, immediately to the west, was second with 4.2, while surrounding counties to the south, east, and north had over 3. The eastern range counties of

Coles, Edgar, and Vermilion and the northeastern counties had from 1.5 to 3.1 beef cattle per hundred acres. McLean County in central Illinois, already noted for its corn and grass and settled by such cattlemen as the Funk family, had 2 beef cattle per hundred acres. The area from Morgan County south to Randolph showed a concentration from 1.5 to 2.7, while the area to the west and north of Morgan had more than 1.5 with the exceptions of Hancock, Mason, and Schuyler counties. Four counties in the southeast corner of Iowa had more than 1.5 beef cattle per hundred acres. These Iowa counties were the major area of older agricultural settlement.

In 1860 counties did not show major increases over the highest number of beef cattle in 1850, but many new counties had more than 1.5 per hundred acres. (See Map A.8.) The highest density was in Kane County in northeastern Illinois, which had 4.9 per hundred acres, while three southeastern Iowa counties had passed the 3.1 mark. Southern and northwestern Illinois counties showed increases in beef cattle, in many cases doubling the number since 1850. Increased settlement and the advance of the railroads accounted for much of this growth. In 1855 Benjamin F. Harris estimated that the construction of the railroad advanced the price of livestock 10% in one year.[2]

The Iowa growth was more uniform in progressing outward from the southeast toward the west as far as Appanoose County and north as far as Dubuque County to form an arc (turning on Washington County) with a density of from 1.5 to 3.1 beef cattle per hundred acres.

In 1870 Illinois had begun to consolidate into the general pattern it would follow for the rest of the century. (See Map A.9.) Morgan County dominated, but in the second magnitude the balance had shifted to north and west of the Illinois River. The number of beef animals in southern Illinois declined, and from then on that portion of the state never figured prominently in the fat-cattle business. McLean, Logan, and Sangamon counties in the center of the state still stood out from the surrounding area as centers of activity. In eastern Illinois, there was another center of concentration on the fringes of the upper Wabash valley. Iowa continued much as before as it shifted the orders of magnitude one step to the west. The old arc encompassing counties with from 1.56 to 3.10 beef cattle per hundred acres now generally contained from 3.12 to 4.66. Three counties (Clinton,

Louisa, and Muscatine) passed the 4.6 mark. Four western Iowa counties in the Missouri River valley formed an area of beef cattle concentration not contiguous with any other with the same number of "other" cattle. (See Table B.13.)

In 1880 northwestern Illinois had become the dominant part of that state, though McLean, Morgan, and Sangamon counties stood out in the center and the prominence of the upper Wabash valley area was still noticeable. (See Map A.10.) Carroll County in the upper northwest took the lead in Illinois with over 8 "other" cattle per hundred acres. Iowa no longer fit the arc description because the east-central counties of Cedar, Clinton, Delaware, Jones, Linn, and Muscatine formed the dominant area. Marshall County in the center of the state had the same number. The southwest counties again showed a higher concentration of cattle than the surrounding region.

From the Illinois map for 1890 a continuation of the trend toward the northwest can be seen. (See Map A.11.) The central and eastern counties increased in absolute numbers of beef cattle, but so did the northwestern counties and in eight cases by two steps in the scale. In the west-central part of the state Cass, Christian, Menard, Morgan, and Scott counties increased by more than one step, while in the east-central part, only De Witt County increased by two steps. By 1890 in some southwestern Iowa counties, the number of beef cattle doubled and even tripled over the number in 1880. While Illinois had only two counties with over 10.9 beef cattle per hundred acres, Iowa had 26 counties with over 10.9. The area of highest concentrations in Iowa stretched from Clinton County in the east to Jasper County in the center and then to the southwest through Warren, Clarke, and Ringgold counties to the Missouri River. The density patterns graphically show what census statistics reveal; Iowa had become the leader of the two states in numbers of beef cattle. This represented a continuation of the pattern noted earlier: the area of cattle feeding shifted west as eastern Illinois farmers concentrated on more cash-grain production.

By 1900 the number of beef cattle had declined in nine counties of central and eastern Illinois and in the west-central Illinois counties of Morgan, Sangamon, and Scott. (See Map A.12.) In several of the northwestern counties numbers had increased, but none had over 12.5 per hundred acres. In Iowa, the number continued to increase in many parts of the state. Twenty-seven counties had over 12.5 beef cattle per hundred acres and eleven of those had over 14. Though there were two areas of concentration in the eastern part of the state, southwestern Iowa had become the major center. The concentration also extended to the north as far as Cherokee, Buena Vista, and Pocahontas counties in the western half of the state and to Franklin County in the eastern half. The northern counties of Osceola, Lyon, Kossuth, Palo Alto, and Allamakee contained the fewest beef cattle per hundred acres as they usually had in each of the previous decades.[3]

Some of the reasons behind the areal distributions of beef cattle are relatively easy to find. In southern Illinois, the soil with a claypan underlayer did not drain well and was considerably less productive than the prairie soils to the north of the Shelbyville moraine. Farmers in the area grew less corn and more wheat, kept fewer animals of poorer quality, and made less money to invest in improvement. Few breeders of improved cattle lived in the area and local farmers thought such stock cost too much. As late as 1878 in Wabash County some people ridiculed purchasers of improved stock. P. E. Michaels (Jackson County farmer) wrote in 1880 that "I never was in a section of the country where the farmers, as a rule, took so little care of cattle as they do in this. It is no wonder that the cattle here are called 'scrubs,' for most of them are treated badly enough to make scrubs out of any thing."[4] In general, the area south of a line from Clark County in the east to Jersey County in the west was less prosperous and less advanced in agricultural technique than the rest of Illinois.

Mason County in west-central Illinois never had over 2.6 beef cattle per hundred acres because practically the whole county was covered with a sandy soil of low productivity, and offered little attraction to settlers when there was so much good land all around. Cook County in the northeast contained Chicago; its urban development hindered farming, though distilleries in the area fattened cattle on mash.

East-central Illinois grazers became less concerned with fattening cattle after 1880 as land values increased. Owners profited by draining the land and planting grain. In Champaign County, for example, from 1879 to 1899 organized drainage districts encompassed 30% of the land area. Further encouragement for the change to grain came when cattle prices started to fall after 1882. The several railroads running south from Chicago provided ready access to the city market, and grain elevators appeared at convenient intervals.[5] Owners of large tracts (especially in the second generation) who did not want to supervise large-scale livestock operations probably found it easier to rent their land to tenants. The tenants in turn found it easier to meet the rent requirements with a cash-grain crop than with livestock, and the landlords found it less risky to ask for a cash payment or shares of a grain crop.[6] Indeed, Thomas Scully (the great midwestern landowner) rented his Logan County lands for only one year at a time from January 1. This in addition to a cash rent requirement discouraged cattle feeding by his tenants because January 1 was not as logical a termination date for feeding operations as for grain production. Perhaps one reason Logan County did not have more beef cattle despite the example of John D. Gillett was that by 1890 52% (1900, 60%) of the farms in the county were tenant operated. As early as 1870 Scully alone owned 30,000 acres in Logan County, or about 7.56% of the land.[7] Successful cattle feeding required good judgment and skill in addition to steady work. While practically anyone could raise a corn crop in east-central Illinois, many could not fatten cattle at a profit in the 1880s and 1890s.

Not all landlords encouraged grain raising exclusively. Samuel Allerton (Chicago meat packer, businessman, and large-scale landowner in Piatt, Vermilion, and Henry counties, Illinois) encouraged his tenants to raise cattle in order to have a good rotation system between pastures and corn and to produce manure to help maintain soil fertility.[8]

The early Illinois concentrations of beef cattle occurred in Cass, McLean, Morgan, Sangamon, and Scott counties where from the 1830s men had been interested in fat and purebred cattle. James N. Brown and his sons in Sangamon County advocated bluegrass pasture and purebred cattle to all who would listen. The Browns, through sales of purebred cattle, provided a way to improve central Illinois stock as well as a visible example for cattlemen to follow.[9] Alexander came to Illinois from Ohio where he had learned the cattle business. He bought land in Morgan County in 1848 and almost immediately made a name for himself as a cattleman. In company with Strawn, Alexander made Morgan County well known in the New York and Chicago markets. During 1870 at the high point in his career, Alexander supposedly marketed over 50,000 head of cattle.[10] Strawn also learned the business in Ohio before he moved to Morgan County in 1831. Strawn was said to dominate the St. Louis cattle market for many years before 1850 and also sent cattle to Chicago, New York, and New Orleans. In the 1850s he was one of the first in the area to switch from grazing steers on prairie to feeding them on shocked corn. When he died in 1865, his estate was valued at about $1 million.[11]

Besides such well-known men as Alexander, Funk, Gillett, Harris, and Strawn, many smaller scale operators, to a certain extent influenced by these leaders, were the real backbone of the cattle feeding business. In Morgan County, for example, long-time resident John Henry credited Strawn's example with encouraging other farmers to feed cattle or to change their feeding practices. Strawn's "mode of farming was so simple that all could adopt it and see the great advantage derived from it."[12] Many in the Morgan-Sangamon county area came from Kentucky where there was an early cattle feeding center. The stall feeding method was often called the "Kentucky" method. For example, Samuel Wood came to Morgan County from Kentucky with his parents in the 1830s and became a cattle feeder and dealer, sometimes grazing 1,000 head of cattle.[13]

Other men fed cattle because their fathers had. John Burch was born in Morgan County in 1842 and reared on a farm where cattle were fattened. More representative perhaps than the better known owners of thousands of acres, Burch was a successful farmer and feeder who had accumulated 750 acres by the end of the century.[14] Immigrants unfamiliar with either Kentucky or Illinois cattle raising could follow the example of successful feeders. William Cleary, born in Ireland in 1818, came to Morgan County in 1838, where he worked as a farmhand and gradually bought land. Cleary sold his various bits of land in 1859 to buy one contiguous farm of a little over 500 acres and began raising stock and feeding cattle.[15] But even in counties with large numbers of beef cattle, not all farmers were feeders. Of the 557 men listed in an 1878 directory of Morgan County township (13 North, Range 8 West) 316 were identified as follows: farmer (187), farmer and stock raiser (57), renter (45), farmhand (26), and stock dealer (1).[16] The fact that stock raisers were mentioned separately indicates that they were considered different from farmers who raised just grain and perhaps a few livestock.

By 1880 the center of cattle feeding in Illinois had

shifted to the counties north and west of the Illinois River. Some credited this to the fact that English farmers settled parts of those counties and had "a fancy to improve their cattle," while the Scots turned to sheep and Yankees would do anything that paid. More recent studies question the influence of cultural and national origins on type of agricultural pursuits.[17] More influential in choice of type of farming in both states, it would seem, were the examples of leading farmers in the area who turned to the crop-livestock association that utilized their resources to best advantage and satisfied market demands. The more rolling terrain to the north and west of the Illinois River provided a good combination of fertile cropland for corn and more sloping areas for pasture. Illinois wheat met increasing competition from regions west of the Mississippi by 1870, which increased the trend toward corn as a grain that could be sold or fed.[18] It was possible too that those who grazed cattle at first because of terrain and distance to grain markets were so accustomed to handling them that they switched to feeding more grain and less grass over time even though their county was no longer primarily range. Large corn crops and low prices also could have had a temporary effect on the number of cattle on feed.[19] Mercer and Warren counties were the leaders in cattle feeding according to the census figures at the end of the century. In Iowa the pattern was not the same as in Illinois. No land area corresponded to southern Illinois in soil characteristics or farming types. The maps also do not reveal an area of Iowa similar to eastern Illinois that declined in importance after 1890. The number of beef cattle in Iowa surpassed the number in Illinois in 1880 and consequently the density in leading Iowa feeding counties tended to be greater. Clinton and Muscatine counties were leading cattle feeding centers by 1870 and retained that position to the end of the century. West Liberty in Muscatine County was the center of a region of Shorthorn and Hereford cattle breeders and feeders by 1870, while the surrounding counties of Cedar, Jackson, and Jones had such residents also.[20]

From this eastern cluster, cattle feeding concentration extended into east-central Iowa and into the southern and western counties of heavy corn production. Counties in the north-central area of the state extending south to Polk were of most recent glaciation and had drainage problems that hindered rapid farm settlement. They were grazing grounds for cattle before the late 1880s; but after they were drained, the farmers turned them to cash-grain combinations with oats predominating.[21]

In the east north-central area wheat was the important grain crop until the late 1870s, and farmers who raised cattle sold them at one or two years of age to feeders in central and southern Iowa. In 1877 J. H. Perry (Worth County) wrote the Prairie Farmer that "wheat has been the great staple here for years, and but little attention here is paid to raising corn." In northeastern Iowa dairying became increasingly important toward the end of the century.[22] The northeastern counties contained some of the most broken and hilly timberland of the state that served its owners better as dairy pasture than cropland.

In the corn-producing counties of Iowa, even more than in Illinois, it paid to feed corn rather than sell it because the transportation charges would take a good deal of the profit. The eastern and southwestern sections of Iowa came to be the epitome of Corn Belt farming where farmers fed their abundance of corn to cattle and swine. So ideal was Iowa for this crop-livestock economy that a publicity pamphlet of 1885 declared in effect, that stock feeding was divinely inspired and fostered since "the God of Nature made Iowa especially a corn-growing state . . . [and] it may safely be assumed" that the surplus grain was meant for feeding cattle.[23]

There was less publicity about Iowa feeders than about some of the widely known Illinois men who dealt in thousands of head of cattle. Men such as William G. Hann (Clinton County) and Oliver Mills (Cass County) introduced purebred stock before the Civil War to improve the beef cattle. By the end of the century others had followed their examples so that both counties were leading beef feeding areas.[24] In Adair County, which in 1870 had less than 1.5 beef cattle per hundred acres, farmers introduced improved stock in large numbers, and by 1880 the county had over 10.9 head per hundred acres.[25] Alexander John (a farmer in Taylor County, which had a concentration the same as Adair) started raising Shorthorns in the 1870s. By 1891 he had over 75 head of purebred cattle, held annual sales, and exhibited only at the local fairs. The 1890 census credited Taylor County with over 10.9 beef cattle per hundred acres.[26] In all of Iowa in 1879, only 441 farmers purchased or sold more than 300 cattle during the year, but 12,679 bought and sold more than 20 head. Most of these men resided in southwestern or east-central Iowa. Woodbury County had 17 large buyers (the largest number), but Muscatine County was the home of the largest dealer, Dennis Mahanah, who bought and sold 4,000 head during the year.[27]

George F. Green (Jackson County, Iowa, farmer) was perhaps typical of Illinois and Iowa feeders in the 1870s. He farmed 340 acres and handled from 100 to 200 head annually with the help of a son and two hired men. He bought small lots from local farmers in the spring and summer and occasionally bought droves as large as 50 to 75 head from a nearby auction dealer.[28]

The droving and grazing common to Illinois in the 1840s still existed in north-central and northwestern Iowa in the 1880s, where land in farms did not exceed half the acres in the counties in 1880. (See Map A.4.) Men such as A. D. Arthur of near Webster City purchased surplus cattle from farmers in the surrounding region to graze on prairie. In 1873 Arthur had 600 head and the next year 1,200 on grass in Hamilton County. In 1879 he pastured 1,200 in Wright County.[29]

General conclusions about location of feeders were difficult. Environment played a part, as seen in the limiting factors of Mason County and southern Illinois and the influence of wetlands on agricultural pursuits. But most of Illinois and Iowa produced corn and grass in abundance with no particular advantage of one area over another. The location of purebred raisers probably influenced cattle feeding by making improved stock available. Central Illinois and eastern Iowa had concentrations of breeders from an early period, which undoubtedly fostered beef cattle production. But why did the breeders settle in the areas they did? There is no adequate answer. Chance certainly was a factor as well as the desire to be in an area where relatives or friends had previously settled. Morgan and Sangamon counties probably impressed early Kentucky immigrants with the similarity to bluegrass country where cattle were well established. Other men came to Illinois and Iowa from the Ohio cattle feeding country looking for cheaper land on which to duplicate Ohio practices, perhaps on a larger scale. However, immigrants with no past experience in the business also took it up when they came to the cattle feeding areas in Illinois and Iowa. They imitated what seemed to be a good way to make money and probably succeeded as well as native-born Americans, depending upon their individual skills.

It is difficult to ascertain why certain cattle feeding areas developed as well as why certain areas sustained the industry over the years. New settlers imitated the Alexanders, Browns, and Strawns who were already in operation, and the readily available purebred cattle helped improve local beef quality in the face of range competition. But why then did few men imitate the Funks in McLean County, Benjamin Harris in Champaign County, or John D. Gillett in Logan County? Some reasons have been suggested that recognize the percentage of tenant farmers in the county or the value of the land in relation to cattle prices in the late 1880s and early 1890s. Many feeders felt they could not make money with low cattle prices and turned to production of cash grains, while others obviously made money and continued feeding cattle.

The amount of investment certainly played a role. In raising purebred cattle, a man needed considerable capital, but most farmers could get a start feeding a few head without too much trouble. Where they went from there depended more on the talents of the individual farmer than on the available capital. If a judgment is to be made, it must be in favor of personal factors accounting to a large degree for where cattle feeders located and the degree of success attained. Financial returns motivated many, but others participated because of the personal drive and satisfaction. In 1880 E. T. Brockway (Washington County, Iowa, feeder) wrote that he had "a passion for blooded stock," and though it might be easier to purchase feeder cattle, there was more personal "satisfaction" in raising cattle on his farm.[30] Personal skill in judging cattle by sight and touch and in the management of profitable operations were perhaps the main determinants in who and where feeders were. A reporter's description of Jacob Strawn: "The great passion of his life was and is—steers—cattle; to drive them, feed them, buy them, sell them; and then own other steers to go through the like process"; fits these cattlemen.[31] They had the drive to work, a good eye for cattle, a good head for figures, and pride in the finished product. Benjamin F. Harris had contests with his neighbors over who could produce the fattest cattle and then publicized his winning results.[32]

Cattle feeders were the elite of the farm world. Shiftless, lazy, or inefficient men could run a farm, but they could not succeed as cattle feeders. It took "a vigilant eye and a diligent hand" to fatten cattle.[33] Joseph McCoy said that "no right-thinking man can regard [cattle feeders] other than public benefactors."[34] If, claimed an Iowa speaker in 1876, he who made two blades of grass grow where one grew before was a benefactor of mankind, then the cattle breeders and feeders who added "an inch in thickness to the sirloin of a beast" were also.[35] In 1901 James ("Tama Jim") Wilson said that the "test of a farmer's ability comes when he handles live stock."[36] Indeed it did for many. The natural abundance of the Middle West led some settlers, economists, and historians to believe that anyone in the nineteenth century who had the capital could make a go of a farm. But the cattlemen knew better; they had a profound sense that the quality of the man influenced his success as an agriculturalist. The cattle feeder was a skilled craftsman, "a sculptor working not with such coarse tools as hammer, chisel, and brush, but with life forces, heredity, and the potencies of feeds."[37] In addition, there was a great pride and sense of personal satisfaction in feeding cattle. As J. H. Oakwood (Vermilion County, Illinois) put it: "Beside the gain, it is a great pleasure to a farmer of good taste, to care for and feed a herd of broad backed, round bodied, and beautifully formed cattle."[38]

The reverential deference paid to cattlemen was still apparent in the 1952 presidential address of Edward N. Wentworth to the Agricultural History Society. Wentworth (Armour's Livestock Bureau) recalled his first introduction to "higher agricultural motivations in 1900." As a cattle clerk at the Iowa State Fair he met the "master breeders" he had read about in the *Breeders' Gazette*, and it made him feel "as if I were seeing in a dream or vision the patriarchs of Genesis."[39] Though these men were probably not akin to Abraham, Isaac, and Jacob, it was nevertheless true that more was needed than merely capital and the idea of being a farmer to feed cattle successfully. In the cattlemen's view, skill, character, the intangible quality of the "eye," and the proper appreciation for one's duty to mankind were required to raise or fatten better cattle at a profit. As Wentworth stated, "the *unselfish* work of serving one's neighbor ministered to one's own selfish advancement."[40] A fortunate coincidence for many cattlemen.

CHAPTER 8. AGE OF IMPROVEMENT I: STATE AND PRIVATE INTERESTS

During the course of the century a large number of people, as individuals or as members of organizations, encouraged Illinois and Iowa cattle feeders with advice and practical information. In general, these people considered the changes suggested to be improvements. Some advisers and their ideas have been mentioned, but a number of groups or institutions that championed "improvement" should be discussed because they functioned over most of the period from 1840 to 1900 and were a constant if not consistent force for change. Chief among these were editors of agricultural newspapers; but state and local agricultural societies also sponsored informational meetings and fairs, and private cattle breeders' associations encouraged change. (Federal institutions and the agricultural colleges are discussed in Chapter 9.)

John S. Wright (Chicago businessman, promoter, and member of the Union Agricultural Society of northeastern Illinois) published the first of the notable midwestern farm periodicals. In 1841 he started the *Union Agriculturalist and Western Prairie Farmer* in Chicago. It became simply the *Prairie Farmer* in 1843 and except for a brief period in 1858 and 1859 has retained the name to the present. Though at first a monthly, from 1856 through 1900 (the years of development for the cattle industry) it was issued weekly. Although always a farm paper from the great livestock market of the nation, the emphasis of the *Prairie Farmer* changed as management changed, sometimes being very personal, showing the editor's interests on practically every page and at other times impersonal, with no identification of the writer. The preeminent position of Chicago to the agricultural economy of the Middle West gave the *Prairie Farmer* an audience far beyond northeast Illinois. Other Illinois farm newspapers came and went. Some such as Emery's Journal of Agriculture (Chicago, January 1858 to December 1858) and the Illinois Farmer (Springfield, 1856-1864) merged into the Prairie Farmer. The Western Agriculturist of Quincy (1858) (which evolved into the Live Stock Journal of Chicago), the Drovers' Journal (1873), the Breeders' Gazette (1881), and the Orange Judd Farmer (1887) sustained publication through the end of the nineteenth century. Newspapers connected with a cause, including the "Granger" papers in the 1870s and 1880s, usually did not last, but they carried useful information in addition to their polemics.[1]

Mark Miller (Wisconsin and Iowa publicist) began the major farm newspaper in Iowa as the Iowa Homestead in 1862 when he consolidated the Northwestern Farmer and Horticultural Journal with the Iowa Farmer and Horticulturalist. Henry Wallace entered Iowa agricultural journalism in a small way in 1879, then edited the Iowa Homestead from 1883 until 1895, when he began editing Wallaces' Farm and Dairy renamed Wallaces' Farmer the next year.[2]

Concerned with the overemphasis on grain crops, editors wrote their own articles or borrowed from other papers in an attempt to influence farmers to pay more attention to cattle raising. "This idea of growing grains from a farm continually, with no stock to keep it in heart ought by this time to be exploded," the editor of the Prairie Farmer stated in 1850. "It will ruin a farm or a country which practices it. It is the continual taking out, without ever putting in anything. . . . In short, if a farm is to be kept up, it must be stocked with animals of some sort."[3] Whenever possible, periodicals stressed the value of cattle manure for fertilizer as an additional reason for raising stock. The farm journals also argued that cattle feeding offered a more profitable use for corn. Cattle prices did not fluctuate as much as grain prices, and to some this made cattle fattening seem a less risky and more dependable business; however, it required more capital investment than grain farming. The stock grower was more "likely to be a rich man at the end of twenty years than a grain grower. He may not be as lucky for a single year; but his gains are more steady and certain."[4]

In the last half of the century the Prairie Farmer carried numerous articles urging farmers to raise corn instead of wheat and to use the corn to fatten livestock.[5] Wood engravings, depicting improved cattle and notable fat cattle raised in the Middle West, accompanied articles (original and reprinted) that urged improvement of breeds and discussed the points of good cattle and how to recognize them. Often newspapers listed names and addresses of cattle importers or owners of fine herds in the state as well as carrying pages of livestock breeder advertising.[6] In the 1870s and 1880s the quality of illustrations improved when lifelike engravings of good animals began to be used in the continuing campaign to convince farmers to work toward improvement of beef breeds. Writers agreed that it took just as much to feed a poor beef animal as a good one and stressed the point that profit was much greater in the good animal. Breeders inserted more advertisements, and editors generally mentioned notable cattle sales.[7]

Farm papers devoted space to discussions of care and feeding of cattle and sometimes reprinted articles from other papers or proceedings of English agricultural societies. Some of these articles conveyed little useful information to midwestern cattle feeders. One was a verbose treatise on "flesh forming" and "heat producing principles" taken from a paper read to the Royal Agricultural Improvement Society of Ireland. If a farmer read all the way through and gleaned anything, it was probably that one should keep cattle warm in winter because they would use less of their food ration to generate body heat and could use more of it for growth.[8] That this came from England might have made it seem more authoritative than a simple statement to the same effect in articles or letters from farmers, but probably its original purpose was to fill out that issue of the paper.

More often, advice on feeding was offered in straightforward language and directed to feeders in Illinois and Iowa. Articles and letters offered advice on how to make corn fodder for feed and how to haul it from the field, warned of the dangers of turning cattle into cornfields without sufficient water, and gave information on feed supplements. The letter writers also offered rules on feeding: "always be regular and systematic in feeding your stock, . . . neither too little nor too much, too often nor too seldom,"[9] and instructions to compare costs and value received for various feeds.[10]

As important as what and how to feed stock was the type of animal to be fed. The Prairie Farmer instructed readers that the desirable steer for economical fattening would have a sharp muzzle, broad chest, straight level back, prominent placid eye, and well-sprung ribs, all of which showed a "disposition to fatten."[11] Farmers were warned not to be fooled into thinking that offspring of fancy cattle shown at the fairs would look that way naturally. The biggest and best-combed animals got the most attention at fairs, but the Prairie Farmer warned, "there is a point beyond which size cannot be gained without the sacrifice of that which is more important. Quality and size go together for a certain distance, but they part company after a while, and size then takes to itself another ally, viz, coarseness."[12] Exhibiting overfed cattle, the editors thought, did a disservice to the breeders and to the farmers visiting fairs because men who bought improved breeds would expect every calf to grow up to look like the show animals they had seen.[13] The Prairie Farmer summed up its feelings on stock management in an article that ridiculed "The Mismanager" by stressing the following points: breeding was significant, young stock should be well cared for because they would never get over a poor start, it cost as much to feed a good type of steer as a bad type, and cattle needed winter shelter to be in good shape in the spring.[14]

Though the farm papers were well intentioned, they did not always succeed in providing pertinent practical information. Well into the 1870s the Prairie Farmer carried reports of English Cattle Association meetings discussing feeding, bean meal, mangels, trifolium, and tares, which had little useful application to Illinois and Iowa farmers with their fields of corn and bluegrass pasture.[15]

Similarly, newspaper columns were opened to controversies based on theoretical principles rather than any practical experience with cattle feeding. One such argument in the 1860s over "natural physiological principles" concerned the use of salt. Some men argued that cattle had a "natural" desire for salt, and others argued they did not. Some even worked through a system of reasoning that proved that salt was "never" beneficial to animals.[16] So much advice was offered on so many aspects of fattening cattle that one farmer complained, "valuable as your paper is, yet it brings confusion in one's head sometimes."[17]

Editors also had favorite cattle management practices they continued to urge upon feeders without much success. Such was the case with cattle barns. Editors and correspondents generalized from human experience and concluded that cattle liked to live in buildings and would gain more weight and be healthier if stabled, therefore it must be cheaper to feed cattle in barns. "We have had too much out-door feeding and the time is not distant when we hope to see a change in this respect."[18] This attitude seldom took into account the cost of the barn, the extra cost of handling feed for stabled cattle, the gains made by hogs that followed cattle in a feedlot, and the fact that one man could feed more cattle in less time if they were in a lot rather than a barn.

For a number of years, in fact into the late 1870s, many editors advocated cooking cattle feed. Articles offered figures to prove it was cheaper to achieve a pound of gain with cooked feed, but their authors did not consider the increased amount of time spent in using this method of feeding. Other letter writers insisted that cooking food for cattle in addition to stabling them did not pay. Some articles on the subject were probably inspired as advertising for manufacturers of steam cookers. Regardless of the fact that corn and grass remained the major cattle feeds in price and economy of use, papers seriously discussed root crops and steamed forage as cattle feeds for the Middle West until the 1880s.[19]

The farm journals carried much solid material for those who really wanted information and practical illustrations of stock feeding operations and results. Feeders wrote about their experiences and gave their feeding plans. Articles cited experiments at the state agricultural colleges, and professors at agricultural schools occasionally wrote articles embodying the results of their experiments. Also regular county correspondents reported on current cattle feeding practices.[20]

When reporting on the controversy over Texas fever and other diseases, the Prairie Farmer tried to be impartial. But when pleuropneumonia spread westward in the 1880s and the magnitude of the disease control problem became more widely known, the Prairie Farmer editors strongly supported attempts to establish federal disease control measures. When attacked as alarmists for discussing the spread and possible consequences of pleuropneumonia, the editors retorted that their critics should help fight the disease, not deny its existence.[21] Other papers closer to the plains cattle interests insisted pleuropneumonia was a "fraud" and "racket" for the profit of veterinarians.[22]

From 1849 on, farm journals generally carried market reports giving Chicago receipts, shipments, and prices for different grades of cattle and often included marketing advice or a discussion of probable market conditions in the weeks or months ahead. Sometimes, especially in the 1850s, the market reports included the names of the men who fed shipments of good quality cattle. The journals also carried articles on local and state fairs and the American Fat Stock Show in Chicago (begun in 1878) as well as reporting on some of the meetings of cattlemen's associations, breeders' groups, and local farmers' institutes.

In keeping with an injunction printed across the front page: "Farmers Write for Your Paper," the Prairie Farmer carried much material in the form of articles, addresses given at meetings, and letters from farmers offering or seeking information on a particular subject. These letters and speeches called for and gave the results of the "experience of intelligent, close observing practical farmers," as a speaker at the Peoria County, Illinois, fair said in 1885. He noted that book learning was becoming respectable and was helpful, but there was no substitute for experience. "Knowledge is power," he said, "but this must be practical knowledge—a knowledge of the proper mode of doing things—not a mere vague notion of how they should be done."[23] Farm papers offered themselves as a means of spreading this practical information about cattle feeding as well as other subjects. "We know that there is a good amount of enterprise, and no inconsiderable skill among [cattle feeders]. The question is cannot this skill be made available to others—cannot this spirit of enterprise be extended?"[24] The agricultural press attempted to disseminate information on the skills and spirit of enterprise of the cattle feeder.

In the Prairie Farmer and the Iowa Homestead, space was also given to the dairy, veterinary, household, and editorial departments in addition to articles on other types of livestock and grain crops. Later in the last half of the century, the Breeders' Gazette and the National Livestock Journal catered more to livestock breeding and feeding but were not, from the feeder's standpoint, noticeably different in content from the Prairie Farmer and Iowa Homestead.

The source and content of letters to the Prairie Farmer and Iowa Homestead indicate that many read the papers and followed them for advice, but circulation figures give some quantitative, though by no means exact, evidence of their popularity and use. Farmers could and did pass papers around, so that only one reader per subscription is a minimum number. In 1880, the Prairie Farmer claimed a circulation of 12,000; in 1891, 35,000; and in 1900, 25,000. For the same three years the *Iowa Homestead* claimed a circulation of 9,000, 10,000, and 40,075. In 1891 and 1900 Ayer's *American Newspaper Annual* credited the *Drovers' Journal* weekly edition with a circulation of 18,200 and 14,310 and *Breeders' Gazette* with 12,000 and 22,721. Bardolph concluded from his study, carried to 1870, that farm journals were more effective forces for improvement during the period than agricultural societies.[25]

In addition to periodicals, local and regional associations were important in promoting farm interests. They had exhibitions and awarded prizes to farmers with good cattle and held meetings to discuss matters of interest to cattle feeders. The fairs and meetings were reported in newspapers beyond the local area. In 1853, after several abortive attempts, leading Illinois and Iowa farmers organized state agricultural societies as quasi-public bodies autonomous from the state but governed by a legislative act and given periodic financial support. The first major activity of these societies was to organize an agricultural fair, which they started in Illinois in the fall of 1853 and in Iowa a year later. With the exceptions of 1862 and 1893 in Illinois, both states had annual fairs from then on. The agricultural press heartily supported these as a way to educate farmers. There they could see the best representatives of fat animals and excellent crops. The *Prairie Farmer* urged farmers: "Go to the Fairs then for *improvement*—to see to question and to reflect. It is the place to which you should take the best of the herd, the flock, the field, the garden, and the household. . . . Make them *pay* by what you may see and learn."[26] Improvement was the watchword of the day, but farmers also liked entertainment; and fair committees soon found that they could not draw good crowds with only fat stock standing around. Despite the disgust of the *Prairie Farmer* editors, ladies' equestrian exhibitions, horse races ("speed trials"), and carnival booths made an early appearance and stayed.[27]

Cattle shows relatively quickly turned into contests among rather small groups of men in the pure-breeding business, with Shorthorns dominating fairs until the 1880s when Hereford and Angus cattle became commoner. The Iowa State Fair offered prizes, not restricted to purebreds, for fat cattle each year to 1900, and Illinois did so until 1878 when this type was dropped for lack of interest. Purebreds were to provide examples against which farmers could compare their own cattle and toward which they could work. However, the beauty and bloodline interests of judges of purebred cattle ignored their merits for the beef market, and thus prizewinners could provide poor examples for cattle feeders. Also it was questionable as to how much fairs encouraged general cattle improvement throughout the state. Not many different breeders exhibited until the 1880s, and the same familiar names won the prizes year after year; but the state fairs at least offered competition among some breeders and through publicity kept the subject of cattle improvement before the public.[28] This helped feeders by encouraging better cattle breeding within the state. Purebred bulls produced better grade steers that converted corn into beef more economically and matured at an earlier age than common cattle.

Agricultural society reports and transactions distributed throughout the state to county clerks, local agricultural societies, farmers' clubs, and political friends of the legislators carried extensive information on the fairs including lists of exhibitors and their addresses, prizes offered and awarded, and reports of the awarding committees. In addition the reports printed prizewinning essays on topics such as breeding and feeding cattle and included statistics on crops. By the 1880s, state societies issued regular crop reports that gave feeders some idea of the coming volume of the corn harvest. The magnitude of circulation and usefulness of this information was hard to estimate. It was directly available to some and formed the basis for newspaper articles that carried it to a wider audience. The legislature authorized only 3,000 copies of the first issue of the Illinois society's Transactions, 8,000 of the second volume, 3,000 of the third, and 7,000 to 10,000 of the others to 1870. Bardolph concluded that farmers valued them, though the volumes were hard to obtain.[29] The legislature limited the Iowa society's Report to 3,000 copies from its beginning to at least 1873. By 1900 both state agricultural societies had been reorganized. The Illinois State Agricultural Society was superseded in 1870 by the State Board of Agriculture, which administered the new State Department of Agriculture. The same thing happened in Iowa in 1900.

Of perhaps more importance for cattle improvement were the local agricultural societies that formed a loose confederation supporting the Illinois state organization. In Iowa, presidents of local societies elected the state society officers. Both states encouraged county agricultural societies by voting moderate subsidies to such groups. In 1860 Illinois had at least 68 and Iowa 36 local agricultural societies. By the end of the century Illinois had at least 72 organizations that sponsored fairs, and Iowa had over 101 county or district societies that qualified for state aid.[30]

The county and district societies functioned on much the same lines as the state groups. They sponsored fairs and held meetings where local men or occasional outside speakers read papers on topics of local interest or to encourage something new. For example, the Buel Institute meeting of October 1850 in Granville, Illinois, held a fair at which the number of cattle, though short of expectations, "were generally of a superior stock and reflected great credit on those who raised them." The society also heard an address by Arthur Bryant (noted horticulturalist from neighboring Princeton and brother of poet William Cullen Bryant) who discussed prominent faults of agriculture in that part of the state. One of the faults was the "too small amount of stock kept on most of our farms." "The beef in Illinois," he said, was "equal to any in the world," and he urged the farmers to raise more of it.[31] This was not an isolated sentiment. Across Illinois and Iowa in the 1840s and 1850s speakers at local meetings called for improvement of cattle or offered papers dealing with cattle feeds, or for example, "Winter Care of Stock."[32] A Bremer County man reported to the _Iowa Homestead_ in 1869 that the "farmers are waking up to the subject of improving their herds by the introduction of thoroughbreds," and that they had organized a farmers' club that winter with 50 members.[33] In the 1870s such groups as the Madison County, Illinois, Farmers' Club and the West Liberty, Iowa, Stock-Breeders' and Farmers' Convention heard papers by representatives of their respective state agricultural schools, discussing in detail the rearing and feeding of beef cattle according to the latest theories and experiments.[34]

In the local societies as well, breeders of purebred cattle tended to dominate the cattle divisions of the fairs as they dominated state fairs with their beauty contests for pedigreed animals. Too often men forgot that purebreds were a means to an end and not the end itself, and breeders were unhappy if nonpedigreed stock won prizes. In Scott County, Iowa, the agricultural society had no requirement limiting exhibition to purebreds in 1868, and the prize for the best herd went to a group with only one purebred. In reaction to this, the writer of the Prairie Farmer article complained that a herd with almost all purebreds was passed over for the prize. The writer commented that fair committees would have to make rules eliminating nonpedigreed stock if breeders of purebreds were expected to exhibit.[35]

The Illinois State Board of Agriculture instituted the Fat Stock Show in Chicago in 1878 in order to have an exhibition in which cattle were judged on their merits as beef animals. The show was held each fall through 1891.[36] From the beginning, when judges were instructed to "award premiums to such animals as present the greatest weight in the smallest superfices—taking into consideration age, quality of flesh, and its distribution in the most valuable portions of the carcass," the Fat Stock Show was intended for the beef producers.[37] In time the show came to include dairy, poultry, and sheep and there were entries of purebred animals and established prizes for them, but only if they were shown as fat cattle intended for the butcher market. In that first year, five men (one each from Indiana and Kentucky, three from Illinois) entered 18 Shorthorns, two men (Will County, Illinois) entered 8 Herefords, three men (two from Illinois, one from Wisconsin) entered 5 Devons, and ten men (six from Illinois, one each from Iowa, Indiana, Kentucky, and Missouri) entered 53 grade or crossbreds. There were classes in each category for animals four years of age or over, three years and under four, two years and under three, and one year and under two. A group of yearling Shorthorn steers entered by James N. Brown's sons, which weighed from 1,275 to 1,480 pounds, had a claimed rate of gain from 1.9 to 2.2 pounds per day from birth.[38]

The Browns were leading the way in early maturity of beef and the Fat Stock Show helped make the idea popular. The Prairie Farmer observed of the Browns' winning yearling steers that the "quality of the flesh is very superior and much more ripe than in the average 3 and 4-year-old steers in our market, and proves conclusively that a better system of feeding for the more rapid growth and earlier maturity of young stock is imperative, if the greatest profit is obtained, and should be encouraged."[39]

The second Fat Stock Show offered prizes for carload lots and dressed carcasses. John D. Gillett won the prizes in both divisions. Gillett had been feeding fine fat cattle for many years, and the Fat Stock Show was an excellent place for him to show his skill. He had 54 of the total 95 grade or crossbred entries in the show. Gillett's winning carlots clearly showed the trend that came to be urged as the reason for early maturity. The yearlings averaged 1,313 pounds for a gain of 2.4 pounds per day, the two- to three-year-old carlot averaged 1,818 pounds for a gain of 1.8 pounds per day, and the lot over four years averaged 2,147 pounds for a gain of 1.34 pounds a day. As they grew older, the total gain became progressively less for the increased age, while progressively more food was required to keep the cattle gaining. Gillett won the carcass class with a three-year-old steer that dressed at 65.7% salable carcass to live-weight.[40]

For the 1880 show the state board decided that, in keeping with their aims of improving beef and encouraging early maturity, they would eliminate any classes for steers over four years of age. The Prairie Farmer endorsed the action with the time-worn phrase that it did not pay to feed old beef.[41] By the fourth show, though the majority of exhibitors came from Illinois, there were also entries from Canada, Kentucky, and Wisconsin.[42]

On the mundane level of financing, the Fat Stock Show continued to operate at a deficit, which caused some discussion of cancellation, but the president of the State Board of Agriculture felt that "the great benefit to accrue to farmers engaged in meat production by the continuance of this practical and important school of instruction is such as to make it necessary to continue the exhibitions. . . ."[43] The board continued to make up the deficit from state fair funds, and in 1882 they added a new classification—the cost-of-production competition. This was a needed addition, for some had complained that the wonderfully fattened animals that took the prizes were impractical for farmers to produce. The results from the cost-of-production competition would help settle the question. A comparison of costs in the first four years of the competition revealed that exhibitors had usually lost money on entries over two years old.[44] Thus the Fat Stock Show strove to be the "normal school for all interested in the production and consumption of meat."[45] "A Man of the Prairie" expressed a similar thought when he wrote, "I hope every stock raiser will do what I propose to do, get a copy of the State Board's report of this show, and study it for all it is worth. It should contain a great many important lessons for all of us."[46] The Prairie Farmer added its voice to the praise and said that the livestock interests of the country "owe a lasting debt of gratitude for the great work done through the Chicago Fat Stock Shows in elevating the standard of American meats, and for the resulting lessons in breeding and feeding for the market of the world."[47]

But not everyone was happy, especially the Hereford breeders and feeders who felt that the judges and management of the show were partial to Shorthorns. This was especially true after the state board put out a chart purporting to show "some value in determining the comparative feeding qualities as far as early maturity is concerned." The combined average weights and gains for all exhibits for four years appeared in a table showing that Shorthorns recorded the most rapid growth among purebreds; but because of the few Herefords shown during these years, there was no adequate representative base for a reliable comparison.[48] Hereford advocates specifically complained in 1881 and again in 1885 when the board received a long letter from C. M. Culbertson (Hereford breeder and feeder from Cook County). The letter cited "absurd" awards in past shows and complained that the show had originated as "a school, at which all might come to compete and learn." It was not serving this purpose because of poor judges, Culbertson complained.[49] T. L. Miller (pioneer Hereford breeder and long-time exhibitor at the state fair and Fat Stock Show) complained that the judges were butchers who only had experience with Shorthorns and their grades and favored them in judging. Better judges would recognize the merits of Herefords, according to Miller. He would not exhibit again until the board remedied the situation.[50] Miller and others were true to their words; the <u>Prairie Farmer</u> noted that the 1886 show was not as good in the beef department, as several noted feeders were "conspicuously absent."[51] However, a Hereford steer, which weighed 1,530 pounds with a carcass dressed at 67% liveweight, took first prize.[52]

Some breeders attempted to turn the Fat Stock Show into just another fair with prizes for purebreds only. In his report to the board, LaFayette Funk (of the well-known Illinois farm family and superintendent of the beef division of the show) said, "I believe the time is almost here when the exhibit of this show should be confined to the breeder. This idea is strongly urged by some of our exhibitors now." But in his address to the board, the next president discouraged any move along that line as "not in accordance with the principle upon which the show was founded."[53] Although strongly pressed for the next two years by various interests (Board of Trade, railroads, and Shorthorn breeders) to allow breeding cattle exhibits, the board repelled the purebred interests' attempt to subvert the purpose of the Fat Stock Show.[54]

The controversy over admission of purebreds, the establishment of a separate horse show, and the problems of finding adequate space for the exhibition combined to reduce public attendance and fat-cattle entries at the shows after 1889. There were no exhibitions in 1892, 1895, and 1896 and the Fat Stock Show came to a sad end in 1897. The manager (John A. Logan, Jr., son of the Illinois Civil War general who founded the Grand Army of the Republic) combined a society horse show, a fat-stock show, and a running battle with the board over management to produce a deficit of $30,705. At one time there were two janitorial staffs contending for custody of the door keys. From then on to the end of the century, the state board contented itself with sponsoring a ring for fat cattle at the state fair.[55] After much discussion as to what should replace the Fat Stock Show, the managers of the Chicago Union Stock Yards sponsored the International Live Stock Exhibition at the Stock Yards in 1900. With the first international exhibition, control of the show for fat stock passed to the livestock business interests in Chicago. This combination of packers, bankers, and railroads managed the show in terms of their interests in prime butcher beef regardless of breed. The prizes in the fat-cattle department of this first show went to Herefords and Aberdeen-Angus, with Shorthorns a poor third.[56]

The Fat Stock Show had also given exhibition space to beef by-products, which involved it in controversy and illustrated one of the problems of any attempt to achieve a unity of all agricultural interests. In 1885 the state board accepted the meat packers' offer to sponsor an exhibit of oleomargarine and brought down upon themselves the wrath of the dairy interests. That oleo made from processed beef fat was a legitimate product for a beef show seemed reasonable to the board, but the dairy producers, who exhibited at the Fat Stock Show also, looked upon oleo and butterine (leaf lard, milk, oleo oil, mixed with butter) as frauds and poisons in addition to being competition. The packers contended that oleo was a healthful food product so pure that "creameries" did "in fact use" it to manufacture what they sold for genuine butter, that packers sold a large part of their oleo to butter manufacturers, and that according to medical doctors oleo was a better product than low-grade butter. The dairy producers reacted within a few days. The Elgin, Ill., Board of Trade, representing the chief butter market of the nation, denounced with an unconscious pun, the exhibition of "bogus butter," as "not only in bad taste, but reprehensible, and merits the denunciation of every right minded man." Beef feeders responded with their own resolution backing the state board in encouraging every type of manufacture that aided the beef producers and stating that oleo was a legitimate food product and its manufacture of value to beef feeders, by giving more value to an otherwise waste product of their animals.[57]

In November 1885 the Ohio Dairymen's Protective Association resolved that the Illinois dairymen deserved sympathy for "the insult offered them, the surrendering of their rights and the betrayal of their trust by their State Board of Agriculture . . . in admitting bogus butter." The board, they charged, acted through either "*imbecility* or *treason*." The board stood its ground and refused to "enter into a conflict of preambles and resolutions with the Dairymen . . . nor . . . contest the palm for abusive billingsgate with the author of the [Ohio] resolutions; or [debate] the false assumptions and ill-natured criticisms and denunciations contained in the resolutions of the Elgin Board of Trade, and therefore content themselves with the following declaration of incontrovertable truths," asserting that it was not the duty of the board to suppress one form of food (oleo) for another competing variety (butter), that butterine and oleo when properly labeled and sold were not bogus but cheap wholesome substitutes for butter, and that the board was in favor of the farmer and laborer being able to get the best food for the least cost.[58] Illinois and Iowa beef feeders had a vital interest in the oleo trade because the more beef by-products packers could use, the more beef cattle were worth. The controversy went on in the country (and the world at large too) as dairymen the nation over attacked oleo and butterine retailed by the meat packers, but oleo continued to be exhibited at the Fat Stock Show and the dairy exhibitors tolerated it "instead of being there with the tomahawk and scalping knife" for which the board commended them. The dairy interests were not politically strong enough to effectively attack oleo until 1902.[59]

The Fat Stock Show offered an index to the changes made by the leading beef feeders of the Middle West. The winning cattle in the show forecast and documented the trend toward early maturity. Early shows offered a category for four years and older but soon dropped it, and in 1891 the show eliminated the rings for cattle three years old and under four, and led the way toward baby beef. By the early twentieth century two-year-olds dominated the markets and prime yearlings (baby beef) were encouraged, though by no means did they represent a large portion of the stock marketed. One commission firm estimated that baby beef represented only 1% of the fat cattle they marketed.[60] By the end of the century, however, leading Illinois and Iowa cattle feeders had completed the shift toward early maturity they had begun in response to western range competition. Though not accounting for a majority of cattle receipts in Chicago, Illinois and Iowa feeders did lead the industry in marketing at early maturity and in turning good range stock into prime beef in their feedlots.

The two basic requirements for early maturity were physical improvement of beef cattle by purebreeding and changes in feeding practices to accelerate weight gain. Though it came into prominence at the time of western range cattle competition, improvement of native cattle with purebred crosses had been going on for some time, though not as fast as many wanted. Fair associations and the agricultural press stressed the importance of "breeding up" stock by using quality bulls, but it was not always an easy matter to convince "our old fashioned farmers. . . , that any profits could be gained by costly and mixed blood." Some who experimented failed to do the necessary fencing to keep out stray bulls in order to maintain the benefits of having bred improved stock.[61] An Iowa farmer complained that it was akin to the "tulip mania" to pay high prices for purebred imported bulls, but the _Prairie Farmer_ replied that of course the bull as beef was not worth very much but the gain came in the quality of the offspring and the better prices for them as beef, and the paper continued to urge improvement.[62]

Some men such as Oliver Mills (Cass County, Iowa) knew the value of improved stock and did not need urging to utilize it. In the late 1850s Mills planned a cattle feeding operation on public range. Because he expected to purchase a large proportion of his cattle in the area, Mills distributed twelve purebred Shorthorn bulls free for local farmers to use in producing good grade stock, which he could buy and fatten on grass.[63]

Purebred Shorthorn bulls were a major element in beef cattle improvement. Though there were some Devons in Illinois and Iowa, they were a minor influence in improving beef cattle.[64] Men brought varieties of Shorthorn cattle to Ohio and Kentucky as early as 1800. An importation of English Shorthorns to Kentucky in 1814 established the bluegrass region as a major breeding center. Kentucky men brought Shorthorns into Illinois in the early 1830s and, in the next two decades Shorthorn importing companies or associations operated in Ohio, Kentucky, Illinois, and Iowa. From 1833 to 1855 Illinois men bought at least 118 registered Shorthorns calved in Kentucky or Ohio and 11 from New York and had produced 82 registered Shorthorns on their own farms. Iowa men had bought 33 Shorthorns from Kentucky, Ohio, or Illinois in the same period.[65] In 1855, 57 registered Shorthorns were calved in Illinois or imported. Most were found in the west-central counties of Morgan and Sangamon or around the fringes of Chicago. In 1855 only six Shorthorns were calved in Iowa or imported, and they were found in the extreme southeast in Lee County.[66]

Most of the breed improvements were individual projects in the early years. Some failed, as when a Colonel Oakley took a purebred Shorthorn bull and two cows into Tazewell County, Illinois. Farmers did not take care of them or make an effort to be selective in their crossbreeding. The superior blood lines were downgraded and lost.[67] On the other hand, by 1873 John Scott (prominent Iowa farmer) cited the fine grade of cattle in Cass County as evidence of the success of Oliver Mills's distribution of purebreds.[68]

Prominent central Illinois cattle producers formed the Illinois Stock Importing Association in 1857 to bring purebred cattle from England. In August 1857 the association sold its first shipment of 20 cows and 7 Shorthorn bulls at auction in Springfield. Nine of the cattle were sold to Sangamon County men; four and two were sold to men in the adjoining counties of Menard and Morgan respectively. Three each were sold to men in St. Clair and Will counties and one each to men in Cass, Champaign, Jersey, Madison, Pike, and Scott counties.[69]

By the middle 1850s "Long John" Wentworth (Chicago businessman and politician) made imported cattle available through his Illinois Breeding Association, and in 1857 the short-lived Ohio Stock Raising Company drove 60 purebred Shorthorns, which were soon dispersed, to Butler County, Iowa.[70]

By 1860 considerable concentrations of purebred cattle existed in Illinois and some in Iowa that exerted a local influence at least, but reports from county fairs continued to show few purebred entries and a very small number of different owners. Generally, less than a dozen different men won the Illinois State Fair prizes for Shorthorns, and they were the same men year after year.[71]

After the Civil War ownership of Shorthorns spread further. From 1855 to 1870, Illinois men bought at least 672 registered Shorthorns from Kentucky, Ohio, or New England and produced 2,445 on their own farms. Iowa men bought at least 226 Shorthorns from Kentucky, Ohio, New England, and Illinois and produced 210 on their own farms. However, in the last quarter of the century, Hereford cattle began to gain in favor as superior beef animals. Men found Hereford bulls especially suited for improving range cattle.[72] (See Maps A.13 and A.14.)

Breeders contended with each other at the fairs and in the press as to which breed was best for beef. In 1880 C. L. Hostetter (Carroll County, Illinois) wrote that he bred Shorthorns because they made the "best beef" while at the same time T. L. Miller (Will County, Illinois, breeder) reported that he kept Herefords because they were "pre-eminently the beef cattle of the world."[73] The farm periodicals urged cattle improvement through articles, editorials, and advertisements. Indeed the editor of the Iowa Homestead declared: "The advantage of improved blood in stock can be so satisfactorily proved that we shall not dwell on this point further."[74] From fairs, farmer associations, the Department of Agriculture, and the agricultural colleges also came urgings to improve beef cattle. By the last quarter of the century breeders had formed their own associations. In 1881 in Chicago prominent Illinois and Iowa breeders were the major force in creating both the American Hereford Cattle Breeders' Association and the American Short-Horn Breeders' Association, and in the next few years they set up state associations. The Iowa Shorthorn Association reported at least 142 breeders in 58 counties by 1883.[75] Muscatine County was a center of Shorthorns in Iowa. In the vicinity of West Liberty, there were 22 breeders with over 700 cattle.[76] Much of the breeder organization was beneficial, but there were also some disadvantages—financial speculation being one. Shorthorns suffered from this particularly, as men bid up the price of desired pedigree lines and priced Shorthorns out of the market. By the end of the century men bought Hereford and Aberdeen Angus bulls in preference to the Shorthorn breed which lost its commanding position in numbers of purebred cattle.[77]

There were also nonpartisan societies that included men interested in any kind of improved stock. The Iowa Improved Stock Breeders' Association, founded in 1875, held meetings that included presentation of papers on aspects of cattle breeding or discussions of cattle diseases, types of pasture, and the value of farmers' institutes.[78] Various Illinois breeder groups combined to form the Illinois Live Stock Breeders' Association in 1895. Its annual meetings had some programs devoted to cattle feeder problems and presented by Dean W. A. Henry of the College of Agriculture, University of Wisconsin, and Henry C. Wallace, editor, Iowa farm expert, and journalist. University of Illinois professors presented papers on such subjects as corn and cattle, feeding improved stock for the butcher, and the fundamentals of breeding.[79]

It is difficult to estimate how fast farmers improved their stock. Purebred cattle were available in Illinois and Iowa generally after the Civil War and many commented on the improved condition of cattle, but according to the census of 1890, 74% of Illinois beef cattle and 68.8% of those in Iowa still had less than half their bloodlines from improved cattle.[80] Even so, improvement of cattle was noticed in the beef market. In 1870 the _Prairie Farmer_ remarked favorably on the increasing number of quarter-, half-, and full-blooded Shorthorn steers coming to market from places that formerly had sent only native steers. An Iowa man commented that his steers were of better quality at a younger age after years of crossing them with Shorthorns.[81] County and state reports in the 1870s and 1880s noted improvements in cattle ascribed to the proximity of local breeders or the fairs, which gave men a chance to see good stock.[82] Farmer "S" reported from Poweshiek County, Iowa, in 1876 that "the number of fine cattle that have been brought into our county the last year makes some of us think that we will soon be up with any county in the state in raising No. 1 cattle for breeders and beef."[83] In Knox County, Illinois, that same year "L.S.J." said they had "a good feeling . . . in regard to the improved breeds of cattle."[84]

But large-scale improvement was a slow process, and the Prairie Farmer and others concerned called for greater effort for improving beef cattle "even if it may appear like useless repetition . . . to employ further argument to show the value of such improved over common stock." A ride through parts of Illinois disabused the editors of the Prairie Farmer of the idea that it was useless repetition because they saw evidence of a great need to "impress the facts so often presented in these columns on the minds of men who are not abreast with the advance of the age in this matter."[85] The Iowa Homestead urged farmers to upgrade their stock in 1869 and was still stressing the subject in 1890.[86] The editors urged men to visit fairs and attend cattle sales to see and buy good stock.[87] The markets of the world demanded better beef than common cattle produced and the day had passed when common cattle profited the feeder, wrote an Iowa farmer in 1880.[88]

The quality of cattle in Iowa improved more slowly than in Illinois, and there were not a large number of purebreds in the state until the 1880s. One reason for the slow pace of beef cattle improvement was the complaint by many farmers that the purebred bulls showed more fineness and style than they did satisfactory offspring. Breeders of Shorthorns, especially, became too interested in fine points of pedigree and show style in their cattle rather than the ability to sire beef animals.[89]

Advocates of improvement claimed that a good improved steer on the same feed would outweigh a common steer by 300 to 500 pounds at two years of age. Improved steers also brought a better price because of the quality of the meat. They simply were more efficient at turning grass and grain into beef.[90] Though the Prairie Farmer reported in 1879 that only one-fourth of the steers received in Chicago showed "good blood" and continued to complain of the abundance of poor cattle coming to market, increasing evidence of the improvement campaign showed up in the market and at the Fat Stock Show.[91]

In 1878, the best steer at the first Fat Stock Show was a grade Shorthorn, three years and seven months old, that weighed 2,185 pounds. The steer was nearly the "model of perfection for a choice butcher's bullock, that of an oblong square." The awarding committee described it thus: "The back of this steer was straight and broad from shoulder to loin, with flesh deep and even as a cushion; the ribs were well back, long and well covered, wide and deep chest; shoulders were well rounded, with a neat and short neck; head small, with fine expression. The hind quarters were loaded with flesh and free from bunches of inferior meat. This animal, combining so many good points in body, had short, fine and well tapered limbs, thereby giving the greater profit to the breeder and consumer, and the least offal to the butcher."[92]

Even then, Gillett was in the process of changing feeding practices. Previously, he had sold fat cattle at four years of age and at weights close to 2,500 pounds. A Scottish visitor who described Gillett's operations in the early 1880s ("the grandest bestial display I have ever seen in one man's possession") noted that Gillett was preparing to sell two-year-olds fattened for the export market.[93] In 1879, Gillett's prizewinning entry in the yearling class of the Fat Stock Show weighed 1,196 pounds with an average gain of 1.97 pounds a day since birth. This grade Shorthorn steer was described as "almost a prodigy in the nature of the species."[94] But it was hardly a prodigy; other entries in the Fat Stock Show revealed equal attention to the possibilities of early maturity. From 1878 to 1885 feeders entered yearling grade steers averaging from 1,000 to 1,500 pounds with average rates of gain from 2.17 to 2.97 pounds a day from birth. From 1882 to 1885 entries in the cost-of-production category for 24-month-old grade steers had average weights from 1,370 to 1,573 pounds, average costs per hundred pounds of gain from $4.42 to $4.78, and net profits from $18.61 to $20.59 per head. Similarly cattle at 36 months for the same period averaged 2,156 pounds at a cost of $6.89 per hundred pounds and a net loss of $21.70.[95]

By 1885, in the category for yearling dressed carcasses, liveweights averaged 1,232 pounds at a gain per day of 2.20 pounds and dressed carcasses averaged 65% of liveweight. From 1880 to 1885 all dressed carcasses entered averaged over 60% of liveweight.[96] Also by 1885 Shorthorns had begun to yield their monopoly of prizes to Herefords and Aberdeen Angus purebreds and grades, and there was lively competition and debate over prize awards to the various breeds.

As early as 1880 the *Prairie Farmer* favorably commented on "baby beef" and the examples at the Fat Stock Show, but not all were pleased. Critics of that kind of "progress" acknowledged that two-year-old cattle furnished a large portion of beef marketed but contended that some of the "so-called improvement" was a forced "unnatural degree of fat-ness." Attacks on early maturity drew the answer that it was silly to insist that only four- or five-year-old beef tasted good. Proponents of early maturity in beef cattle said it was an age of improvement and beef was only one of the elements that had benefited. They pointed out that improved varieties of fruits flowered sooner than they had 20 years before, and with no lack in quality. Young animals kept on full feed for their short life, it was asserted, would have better marbled fat rather than a layer of meat enclosed by fat put on at the age of three or four. The carcass class at the Fat Stock Show proved that good beef could be produced and marketed at a profit at two years. This was by no means considered the ultimate in early maturity because in 1885 the _National Livestock Journal_ predicted that men would soon be selling 1,400-pound steers at 20 months of age.[97] The _Breeders' Gazette_ concluded that the American public had agreed on one thing at least in 1886: they wanted tender beef in contrast to "something to bite at" which the British still seemed to prefer.[98]

Feeders found it paid to sell beef at a younger age. The quicker turnover reduced the risk of capital loss, the first two years of feeding put on proportionately more meat for the cost, and the same amount of capital could finance a greater volume of business by reducing the age spread of animals maintained on farms. By the 1890s many considered that a three-year-old steer was a "back number, and is not wanted in the feed lot," and feeders were producing first-class beef at 20 months.[99] Illinois and Iowa men who continued to feed three- and four-year-olds got "precious little encouragement" from the market because their cattle sold for the same price per pound as two-year-olds.[100] A number of sources commented on the fact that younger cattle dominated the market by 1900. The census reported that cattle went to slaughter at an earlier age because of improved breeding and feeding; and a commission firm summarized the December 1900 market as one showing a preference for 1,300- to 1,400-pound steers. "The heavy cuts are out of style and all classes of buyers would prefer lighter beeves." The drive for early maturity through cattle improvement had paid off.[101]

CHAPTER 9. AGE OF IMPROVEMENT II: FEDERAL INFLUENCE

The federal government aided farmers in general and beef producers in particular in a number of ways. This aid began inconspicuously in 1837 when the Patent Office started distributing seed of new or improved plants. That office issued its first report on agriculture in 1841, and by 1849 the report comprised a separate volume of the Patent Office Reports. In 1862 Congress established the Department of Agriculture and added divisions to it from time to time. Most important for the beef producers was the Bureau of Animal Industry (BAI), created in 1884 to supervise the veterinary service founded the year before. In 1890 the Federal Meat Inspection Service was added to the bureau.[1] Until the late 1870s and early 1880s, when the department began issuing a variety of publications, the annual report was the major vehicle for providing information to the public. The early printed reports were sent to state agricultural societies and distributed by congressmen to farmers and to newspapers, which frequently extracted items of interest so that information reached a wider audience than might be expected.

The Patent Office Reports frequently printed letters from selected correspondents scattered through the states and contained articles on specific subjects. The 1844 report contained reprinted articles on the amount of "fat-forming principle" in certain foods and the contemporary advice about cooked feed. Much influenced by British writings the editors for the Patent Office urged farmers to feed carrots, rutabagas, mangel-wurzels, and corn fodder to their cattle during the winter.[2] Though much of the information was impractical, the Patent Office had good intentions and urged farmers to raise more cattle to utilize grain and yet get manure to help maintain soil fertility. In the effort to foster beef cattle improvement by upgrading cattle through the use of purebred sires, the reports carried articles on types of purebred cattle, although there was emphasis on Shorthorns. One such article concluded that the great hope of the U.S. beef raisers lay in good grade cattle that "are thrifty, and lay on flesh rapidly and evenly, and are ready to turn off at two and a half or three years old."[3] The Department of Agriculture, when it superseded the agricultural branch of the Patent Office in 1862, issued similar reports. The new series of reports contained articles supporting cattle improvement, giving feeding instructions and the number of cattle in states, and analyzing various feeds. By the middle 1870s the reports began to rely less on European experiments for information and to print results from more pertinent sources. The 1873 report carried the results from an experiment at the Illinois Industrial University that concluded it was not advisable to cook feed for steers.[4]

The department reported on its own technical and scientific investigations. In 1880, for example, there was an article by department chemists on an analysis of the juices of cornstalks.[5] The department also began to print statistical reports on the number of cattle in the country, the volume of trade in live and dressed beef, market receipts, and price ranges; but these did not provide the kind of statistics that would help feeders in immediate planning, although they could appreciate the size and complexity of the market. In general, before 1880, reports of the Patent Office and Department of Agriculture were of more use to later historians than contemporary farmers. Limited circulation and often impractical information reduced their usefulness; but the printed correspondence indicated what farmers were doing, and agricultural editors gave wider dissemination to some information by borrowing from the reports to fill their columns.

Of more importance, in the 1880s and 1890s the department started to issue special series of publications. One bulletin discussed early maturity, the economy of feeding, the importance of good breeding, and how to select feeder cattle. There were illustrations of good and bad types and statistics from experiments to illustrate with specifics the points made about age and feed.[6]

The BAI played the most important part in government aid to beef feeders. Its publications dealt specifically with beef cattle problems such as correlations of estimates of numbers with estimates of population growth and price cycles or specific discussions of the "Livestock and Meat Traffic of Chicago" that would tell a feeder just how the market worked, what type of cattle the packers wanted for different purposes, and what influenced prices. In one report, Professor Morrow of the University of Illinois discussed cattle rearing and feeding in the Middle West; in another report a department inspector discussed the cattle business in Illinois with specifics on type and amount of feed, ages and weights of feeder cattle, and estimates of costs, to which, he concluded, not enough feeders gave careful thought. These were reports of practical value to men who had access to them and could see just what the trends were in cattle production in the Middle West. The publications became more specific and producer oriented soon after the turn of the century with such items as Bureau Circular No. 105 describing "Baby Beef" and telling how to raise it.[7]

Disease control was the most important of all services of the BAI and of major significance to beef producers as well as all other livestock interests in the nation. Originally begun in 1883 the veterinary division of the department was incorporated into the Bureau of Animal Industry in 1884. In addition to the long fight against Texas fever, the bureau played a significant part in eliminating pleuropneumonia, the other major threat to midwestern beef cattle in the last half of the nineteenth century. Much of the effort of the Department of Agriculture had been devoted to investigations of animal diseases, but it was not until the 1880s that public pressure and the alarming spread of pleuropneumonia persuaded Congress to give the department power to do something about the situation. In addition to the danger to American beef cattle, pleuropneumonia had been the justification for the British Order-in-Council of 1879 requiring slaughter at the port of entry of American live cattle sent to England.

Introduced into the United States in 1843, pleuropneumonia spread from New York and reached Massachusetts in 1859. The Massachusetts government by prompt and forceful action checked and eliminated the disease within the state by a quarantine and then forced slaughter of all exposed animals, but no other state took similar action. The Illinois State Agricultural Society expressed concern about the disease, expecially the Massachusetts quarantine, and in 1859 appointed an agent to investigate the situation. The agent's report to the society confirmed the existence of the disease in Massachusetts, but said it did not exist in Illinois. Pleuropneumonia continued to spread in the East outside Massachusetts and was very much in the news in the 1860s. A second report to the Illinois Society in 1866 concluded that the only remedy was quarantine and slaughter.[8] One of the most valuable results of the Department of Agriculture's statistical activities from 1866 to 1877 was proof of the magnitude of animal disease and its alarming spread. In 1877 Commissioner of Agriculture Le Duc delivered a massive report on diseases to the Senate and in effect started legislative machinery moving toward setting up the BAI. Despite strenuous objections of certain livestock interests in the South and Southwest; of packers; and of commission firms who argued that a federal disease control program violated states rights, that ignorant veterinarians would needlessly slaughter cattle at a great loss to the industry, and that pleuropneumonia was not contagious, Congress created the BAI. The combination of support generated by the reaction to the British Order-in-Council, by an outbreak of pleuropneumonia in Illinois, and by the success of Wyoming ranchers in convincing eastern investors that unchecked pleuropneumonia on the plains could literally kill off dividends and capital investment overwhelmed the last-minute personal lobbying efforts of Nelson Morris for the packers. The new bureau immediately found the disease in Ohio and Kentucky also. The first law gave the bureau only quarantine powers without authority to slaughter diseased cattle; but that was remedied in 1886, and from then on there was swift action against pleuropneumonia though not without much opposition. Because of the insistence that slaughter was the only effective method of control, many attacked the bureau in and out of Congress. The director, Doctor D. E. Salmon, one of the great men in American veterinary history, was labeled incompetent by the minority report of a House of Representatives committee, in spite of the fact that the second Illinois outbreak in 1887 and department tests proved that inoculated and apparently cured cattle could and did carry the disease to reinfect the area.[9]

The BAI was viciously abused by some of the agricultural press for its work that hurt short-term profits. *Field and Farm*, representing Colorado interests, charged that the pleuropneumonia scare was a "fraud" by "racket workers," "quack veterinarians," or "designing cranks" as a way of getting federal money for "riotous living." The editors continually maintained that pleuropneumonia did not exist and when federal appropriations ceased all the "sick" cattle would get well immediately, and in the meantime the scare hurt the cattle business. But the BAI pressed its work and within five years eradicated pleuropneumonia from the United States.[10]

At the same time that the BAI worked to control Texas fever and pleuropneumonia, it investigated foot-and-mouth disease, anthrax, and rinderpest, none of which was rampant but did pose threats, as well as blackleg, which was a problem for improved fat calves and yearlings. Blackleg came from a microorganism spread by spores that survived in the soil of a pasture for years, which accounted for its recurrence in areas for no apparent reason. The organism entered the animal through the mouth or some puncture wound in the skin or foot, grew only in the absence of oxygen, and was fatal within 36 hours. The French discovered a vaccine for blackleg, and the BAI distributed free vaccine in the United States from 1897 to 1922. Before the use of vaccine, the department estimated that 20% of calves in infected areas died, though often owners attributed the demise to anthrax or clover poisoning.[11] With such a high mortality rate, blackleg was a serious problem to cattle raisers in certain areas of Illinois and Iowa. A 20% loss could eliminate any profit to a cattle feeder for the year.

The BAI also tried to help cattle feeders by taking steps to eliminate foreign criticism of dressed beef. Through bureau efforts Congress passed a law requiring federal inspection of slaughtered dressed beef. The Meat Inspection Service started in 1890, and in 1893 inspectors passed as disease-free and tagged 1,036,809 quarters of beef for export and 10,534,102 for interstate trade; but because of lack of funds the inspectors could cover only the major packinghouses and stockyards. Congress did not pass a truly effective inspection law until 1906 after the controversy stirred up by Upton Sinclair's novel _The Jungle_, which attacked the packers in general and Armour in particular.[12]

The federal government had also aided the beef feeders indirectly with the Morrill Act in 1862, which established the land-grant agricultural colleges. In the latter quarter of the century, the government and the states combined with the agricultural colleges to create another institution of improvement, the agricultural experiment station. Here was an organization within each state at last able to conduct practical experiments in an attempt to find answers to questions that breeders and feeders had been debating for years concerning feeding rations, shelter, and costs. Authorized by the Hatch Act of 1887, the Illinois and Iowa agricultural experiment stations embarked upon significant roles in improving the agricultural practices of their states.[13]

For many years spokesmen for farm interests had demanded some practical aid to agriculture. In the 1850s, they had called for agricultural colleges or experimental farms to educate farm sons and daughters "in the science, beauty, and goodness of their occupation."[14] The Civil War interrupted the movement, but Illinois had an Industrial University and Iowa a State Agricultural College in the late 1860s bolstered by the "awakening . . . conviction in the minds of very many of our farmers, that our profession is not merely one of plodding drudgery, but that it was also conceived, born and had its being in a science having for its study and investigation the broadest expanse of domain and deepest research."[15] The universities pursued some experiments in the 1870s and 1880s that were discussed in reports to the legislatures, newspapers, or publications of the state agricultural societies. E. S. Lawrence (head farmer of the Illinois Industrial University) commented on rearing and feeding cattle in the Fifth Annual Circular of the university,[16] and the Prairie Farmer carried reports of university feeding experiments in 1874 and 1875. These two experiments concluded that beets did not make a good fattening food for steers and "that the best and cheapest mode of feeding is to feed shock-corn in a sheltered yard." However, the major interests of the scientific investigators at the Illinois and Iowa institutions were horticultural and grain problems. Again, farmer conventions called for federal support for experiment stations, and the National Livestock Journal, in demanding their establishment, asked why it was still an unsolved problem as to how much grain supplement should be fed when cattle were on grass.[17]

Part of the problem was in the conflict raging over the purpose of the Morrill Act, the expectations of farm spokesmen, the realities of the agricultural college budgets, and faculty aims. In both states, faculties had developed a general education program with most quality instruction in the sciences; they had not limited the program to vocational agriculture as many economy-minded legislators, agricultural editors, and farmer spokesmen wanted. However, after some competent work, the agricultural departments of both schools declined in the 1880s to only a handful of students and a meager staff. The agrarians attacked Iowa State College in the press and legislature and forced a change, although they did not get humanities and social sciences banned from the curriculum. The new administration in 1891 separated the experiment station management from the old professors, placed a leading critic, James ("Tama Jim") Wilson, in charge, and added new staff that revitalized the agricultural work. The University of Illinois suffered a similar crisis and reorganization of the agricultural work in 1894. After that the quality of the work, both agricultural and scientific, improved.[18]

The government had helped resolve the crisis in agricultural education by providing funds for expanded work through the Hatch Act of 1887 and a supplementary appropriation act the next year, which provided $15,000 annually to each state to operate an agricultural experiment station as part of the university. Illinois and Iowa established stations in 1888. Most of the Illinois station work concerned dairy, fruit, and field crops with important work done on corn. Of the 11 annual reports, 60 bulletins, and 17 circulars issued by the station from 1888 to 1900, only Bulletins 2, 9, 36, 43, and 46 directly concerned cattle feeding experiments. Other experiments on alfalfa, cowpeas, and soybeans contributed to the store of knowledge about cattle feeds in comparison to corn fodder and ensilage. Bulletin 26 in 1894 was perhaps the most important for interested feeders, as it tabulated 180 replies to 250 letters sent to Illinois livestock feeders soliciting information on types of livestock, kinds of feeds, length of feeding period, and evaluation of feeding as a business. Here was a cross section of opinion on the latest practices by the leading beef feeders for anyone who read the bulletin. Interest in beef experiments increased at the experiment stations after the turn of the century, and the legislature added significantly to federal funds to allow larger scale operations.[19]

The Iowa experiment station did more of interest to beef feeders than the Illinois station, including five beef cattle experiments and three related feedstuff investigations in its 43 bulletins and 11 annual reports to 1900.[20] Bulletin 2 offered chemical analysis of fodders but left the reader to draw his own conclusions about what it meant; Bulletin 6 concluded that feeding shelled corn to steers produced more gain cheaper than corn and cob meal did. One of the more useful, Bulletin 9 listed various feedstuffs with an analysis by percentage of digestible nutrients and offered instructions on choosing rations for different feeding purposes depending upon food cost and availability. Bulletins 24 and 33 dealt with comparisons of steers and heifers and concluded that it did not make any difference whether heifers raised for beef were spayed, that heifers made good beef as economically as steers, and that packers were not justified in the price discrimination against heifers. The experiment followed two different groups from beginning feeding to finished slaughter and shipment to retailers. This particular experiment received notice in the *Butchers' Advocate*, which printed a summary of the experiments as evidence of packer profiteering by use of price discrimination against heifers.[21]

In 1898 the Iowa station experience produced a general but short survey of cattle feeding in the form of U.S. Department of Agriculture Farmer's Bulletin 71, "Some Essentials of Beef Production," by the Iowa station director, Charles F. Curtiss. However, both beef states lost their laurels to Wisconsin for the first major book on feeding, by Wisconsin experiment station director and dean of agriculture at the university, W. A. Henry, who in 1898 published *Feeds and Feeding: A Hand-Book for the Student and Stockman*. Herbert W. Mumford, professor at the University of Illinois, did not write his *Beef Production* until 1907.

Experiment station bulletins and university circulars did not meet all the demands of the improvers. Men had to obtain copies of the publications and sometimes puzzle out the conclusions without being able to ask questions. It would be better yet, many thought, if the professors went to the farmers in person and gave talks and answered questions about practical farming problems. The regents of the university in Illinois had sponsored lecture meetings in various parts of the state in 1870 and 1871. Members of the faculty spoke on agriculture in general and then on specific topics, one of which usually had something to do with beef cattle. In 1873 there were eight weekly meetings around the state; but from then on to 1880, meetings (now called farmers' institutes) took place only at the university in Urbana.[22]

In 1882 the State Board of Agriculture started sponsoring institutes in each congressional district even though some failed for lack of attendance, and by 1889 it sponsored 9 district and 33 county institutes. That same year, the state appropriated $100 to each congressional district for institute purposes. From 1891 to 1897, the legislature provided each county institute with $50, and in 1895 created a statewide organization, the Illinois Farmers' Institute, to coordinate county activities and with authority to publish 10,000 annual reports a year. From 1897 the state institute had an annual appropriation of $7,000. It lasted well into the twentieth century.[23] The meetings stressed "good farming practices" such as urging men to keep livestock in order to feed their corn rather than sell it and discussed the need for early maturity and quality in order to compete with grade range cattle. The central institute attempted to coordinate activities, to supply speakers to the county meetings, and to encourage the submission of essays for its Annual Report. However, most of the county meetings consisted, for example, of the farmers gathering to hear a local minister invoke God's blessing, an address of welcome and response in "poetry," and a "song by Miss Callie Fisher" and to read papers to each other about rotation of crops, cattle feeding, and other topics.[24]

In Iowa, the State Agricultural College initiated institutes in 1870 and held them irregularly over the next decade. In 1880 the college gave institutes on a more regular basis, but there never was the degree of central organization developed in Illinois. The Iowa expenses rested on the local groups until the legislature granted $50 to each county in 1891. The Iowa State Agricultural Society reports of 1898, 1899, and 1900 included accounts of institute meetings and papers read at particular ones.[25]

Through the latter half of the century, men in periodical offices, government agencies, and the land-grant colleges tried to help Illinois and Iowa cattle feeders improve their beef cattle, their types of feed, and their feeding practices. The intention was to help men raise beef in the most economical way by providing information about the many variable factors involved in the profitable feeding of beef cattle. Men sought for answers to many questions: which, if any, breed of cattle gained weight faster than others on the same feed, what were the comparative rates of gain and costs for various feeds, was ground corn better than whole kernel corn? Although complete and correct answers to these and many other questions were not always available, and information was not always disseminated in a way to be practically useful, men in the institutions of improvement probably were a source of help to Illinois and Iowa cattle feeders.

CHAPTER 10. AGE OF IMPROVEMENT III: CULMINATION

The feedlot was the testing ground for cattle improvement because all the publicity, experiments, and capital invested in producing early maturation would come to nought if the feed and feeding routine did not complement the other parts of the grand design. This was particularly important in the late 1880s and early 1890s when cattle prices declined. The need for rapid and economical gains in weight forced men to be more conscious of relative costs of different feed combinations.

In the decade before 1870, feeding practices continued much the same as in the 1850s. In many parts of Illinois and Iowa, men still fed mostly grass, and all depended heavily upon summer pasture even when corn was the major feed.[1] Aside from grazing, feeding practices varied as some turned cattle into standing corn (though this seemed to be on the decline), while others cut the corn and fed it as fodder, ear and stalk together. Some feeders raised their own corn and frequently bought husked or standing corn from neighbors. The majority, however, seemed to adhere to "stalk feeding" in which, after grass failed in the fall, cattle were concentrated in small fields or lots and fed ear corn, stalks, and hay thrown on the ground. After a suitable time swine were run into the lots to scavenge after the cattle.[2] The basic plan had variations as men experimented. Morris Case, near Waterloo, Iowa, had 150 head of cattle in rows of stalls facing a track used to distribute feed ground by a steam engine. Others built troughs or feed bunks 3 feet off the ground and 10 to 14 feet long for snapped ear corn and hay, and a few advocated use of oil cake meal, bran, oats, and even turnips as supplements.[3]

There was still much argument over whether animals should be sheltered, a favorite journalistic subject in the 1840s and 1850s. Some writers still contended that feeding cattle in open lots was "radically wrong, inhuman, and unprofitable,"[4] but despite the preaching and the supposed economic advantages, very few men kept their cattle in barns. As one farmer put it, "where corn and cattle are both easily produced, we must adopt a plan by which to feed our corn with as little handling as possible."[5] The usual shelter was a grove of trees on the north and west sides of the feedlots, a board fence, or a rude pole shed with straw piled on it.

Summer grazing continued on undrained prairie and marsh areas. In 1871 a Lee County, Illinois, man reported the "farmers all turning their attention to corn, hogs, and cattle" and that during the summers over 8,000 cattle fed on the Winnebago "swamp." This free forage, he estimated, added about 300 pounds per head to three-year-old steers. In the 1870s, herd grounds still existed in eastern Iowa; in Kankakee and Iroquois counties, Illinois; and in adjoining Lake, Newton, and Benton counties, Indiana, but even these diminished in size each year[6] because of increasing settlement. Some range herding continued in Illinois into the 1880s, but the main range country was western Iowa, where men herded cattle for a season at $2 a head, though even that acreage shrank continually. By 1873 there was little open prairie east of the Des Moines River, and by 1880 only a very few herds grazed on prairie in the three tiers of counties in southwest Iowa. In the northwest men grazed cattle on open range until 1885, but railroads brought that to an end by opening the region for raising grain.[7]

With the loss of the range or prairie option, feeders turned to improving their fenced pastures or feeding more corn.[8] For those who kept grass as an important option in the feed regime, bluegrass and timothy replaced common prairie grass in the 1870s and later. Even so, a late spring or summer drought could have serious consequences for a man depending on pasture. Men often kept more cattle during the winter than they could full-feed on stored corn and hay beyond the usual beginning of spring, and a dry summer would reduce the carrying capacity of pasture quickly, particularly of common prairie. In a dry year, half-fattened stock had to be sold early in the fall, or cattle turned into cornfields in August.[9]

Bluegrass pasture in some cases also provided winter grazing. In the 1870s, one prominent central Illinois feeder who used bluegrass as the major feed for his fat cattle argued (but did not prove) that 100 acres of bluegrass could equal 66 acres of corn in fattening value. In the middle 1870s James N. Brown's sons sold 700 to 900 steers annually from their 3,000 acres of bluegrass. They bought feeder cattle in the fall and kept them 9 to 12 months for a gain of 350 to 400 pounds. They claimed that on only about 11 days during the year did they have to put out hay because snow had covered the grass.

The Browns fed Texas cattle from 1866 to 1873, but after that went back to grade Shorthorn steers bought at two and three years of age at 1,000 pounds. They sold their cattle at three and four years and cleared about $8 a head on the operation. But the Browns soon turned to preparing cattle for market at a younger age. In 1878 at the first Fat Stock Show, the Browns won first and second prizes with steers 21 and 22 months old that showed a gain of 2.28 and 1.9 pounds per day. The judges declared the first prize winner a "superior" beef animal with flesh "much riper than the average three and four-year old steers in our markets."[10]

The Browns were exceptions in depending on bluegrass for fattening cattle because most men fed corn to steers for at least several months in the late fall and winter. Many still turned the cattle loose in the cornfields to forage for themselves, especially if there was a light corn crop and it did not seem to pay to husk or shock it. Some Iowa men custom-fed cattle by renting out their cornfields with the fee based on the weight added to the cattle.[11]

The head farmer at the Illinois Industrial University insisted that the best way to feed beef cattle was to stable them and feed them cornmeal mixed with straw and hay, but few appeared to follow his advice in practice. All agreed, nevertheless, that corn was the grain to feed; and editors urged men to "never as a general rule, raise corn to sell, but feed it to good, thrifty stock of some kind."[12] When the price of corn fell, it paid farmers to feed it rather than sell it. Because corn more often than not was a surer crop than wheat, farmers produced more and more corn, which created an increasing demand for feeder cattle and hogs. During 1873 the Prairie Farmer market writer warned farmers not to ship half-fattened cattle to market. Because of cheap corn, he warned that there would be more cattle in the early spring market than usual, and as a result prices would decline for half-fattened stock faster than for fully fattened cattle.[13] The crop was so good one man even protested that it was "no use to tell farmers to feed the corn crop," there was just too much. The abundance even encouraged men to feed Texas cattle, in spite of the tick fever and their relative slowness to take on fat in the winter.[14]

While most Illinois feeders in the middle 1870s finished their beef on corn, some Iowans depended more on grass. Some men grazed cattle for two years on prairie or pasture and did not put them into the feedlot until the fall of the third or even fourth year. There they ate corn and hay for several months and were marketed when prices and finish were right for the individual feeder.[15]

The question of winter shelter for cattle continued to be ignored in the 1870s. Few farmers reported feeding cattle in a barn despite illustrated articles on barn building in the press or claims that it was cheaper in the long run. Northern Illinois and Iowa counties reported hard winters, which retarded the fattening process and caused as much as a two-month delay in spring marketing. The quality of winter feed, especially corn, determined how fast cattle gained weight and when they were ready for spring markets. Soft corn or a shortage affected the quality of the beef in the spring. In the case of a short crop many men fed what corn they had early in the fall and sent their beef and pork to market before winter to avoid purchasing grain at high prices.[16]

In the 1880s there were some changes in feeding practices as cattle raising declined, especially in Iowa, because farmers drained prairies and planted corn. Cattle fattening, however, increased as more men fed cattle not calved on their farms.[17] Illinois farmers reported feed practices ranging from use of grass alone to mixtures of ground corn and oats or additions of bran. Ear, shelled, or ground corn now prevailed over the old "stall" feeding method. A Henry County, Illinois, man reported a movable self-feeder holding a month's supply of shelled corn for a carload of cattle. Many men favored the third year for marketing, but some had begun to sell beef at 21 to 24 months of age at weights up to 1,600 pounds.[18] Although some Iowans still sold four-year-old beef, on the whole, practices did not differ from the changes taking place in Illinois, with the exception of more grazing in western Iowa. There in 1880, W. H. Fitch of Calhoun County marketed two carloads of grade Shorthorn steers that had been grazed on prairie during the summers.[19]

There was increasing evidence that feeders realized the

advantages of genetic improvement and feed changes for profitable marketing of cattle at a younger age. The greater capital investment of this system was balanced by the faster turnover in the capital. Tom C. Ponting (Illinois feeder and Hereford raiser) fed his prize 21-month-old steers for the sixth Fat Stock Show on pasture, hay, shelled corn, oats, and oil cake the last nine months.[20] John M. Stahl (a central Illinois feeder) argued that the increased cost of maintaining an improved steer made it impossible to first grow steers and then fatten them; they had to be fattened from birth. The best way to get quick maturity was to start feeding oats, bran, and pasture before weaning; good summer pasture the first year; then corn the spring of the second year. This routine would produce 1,500-pound beef steers by 24 months.[21] Of the six yearling entries in the cost-of-production ring at the sixth Fat Stock Show, in the first year all had been fed on milk, hay, and pasture; four on shelled corn also; three on oats; and one on oil meal. In the second year the standard formula was shelled corn, hay, and pasture; five had oats in addition; and one of those also had oil cake. Of the four steers over two and under three years, one had only shelled corn, hay, and pasture each year; two had oats, oil cake, and oil meal in addition; and the fourth had bran and shorts but no oats in addition to corn, hay, and pasture.[22] John D. Gillett said in an interview in 1883 that it did not pay to produce beef over three years old. He planned to wean calves on oats and add a corn ration before 12 months in order to obtain weights of 1,500 pounds between 20 to 28 months of age.[23]

University of Illinois experimenters concluded in 1885 that young animals made the best gains and that a combination of grazing and a finishing period of three months on grain made the best gains and profits. The <u>National Livestock Journal</u> answered a mailed question by saying that a ration consisting of a slightly dampened mixture of five pounds of cornmeal, five pounds of bran, and two pounds of linseed meal mixed with two bushels of cut straw or corn fodder would finish 1,000-pound steers for the fall market in about three months.[24]

By the 1890s "albuminoids, carbohydrates, [and] the nutritive ration" had replaced fancy bloodlines as the topic of discussion according to the <u>Iowa Homestead</u>. Cattle feeders were more aware of the properties of different foods as cattle prices remained low and feeders became more economy conscious. They learned more about balancing supplementary foods such as oil meal, crushed oil cake, bran, and molasses with corn and grass, although there was some danger of spending money needlessly on "condimental preparations for cattle" which were not basic foods according to *Wallaces' Farmer*. Feeders began to use clover and alfalfa and were urged to use chopped corn as economical fodder.[25]

By 1900 agricultural experiment station reports had produced some statistical results of feeding experiments which could help cattlemen even if only as confirmation of existing practices. In 1898 W. A. Henry (dean of the University of Wisconsin College of Agriculture) published a compilation of the results of feed experiments in the previous decade. (*Wallaces' Farmer* urged readers to order Henry's book.) Henry concluded that whole corn fed in the husk was the best feed when corn was cheap. Cornmeal made better gains on steers per day but was not good for fattening hogs that followed cattle, and in addition, the cost of grinding the meal added to the overhead. Oil meal was more expensive but gave better balance to the ration by adding protein to the carbohydrates in the corn. Grain fed in addition to good lush pasture did not appear to pay, but by using a grain supplement, the pasture needed for summer could be reduced by one-half.[26] By the end of the century, corn was still the major cattle feed and the feeder had to balance costs of feeder cattle, pasture, grain, and supplements with the rapidity of gain and the price of finished beef to decide which combination of feeds to use. A ration of cornmeal, oil meal, bran, and shorts with corn fodder required 28% less grain for 100 pounds of gain but cost 28% more than an ear-corn ration in the middle 1890s. The feeder had to decide which was the more economical way to fatten his beef in the given circumstances. Cattle made faster gains if fed cornmeal during the whole fattening period, but hogs in the feedlot could not utilize corn in the form of meal. Farmers had to balance the value of the better gains from cornmeal against the loss of gains in swine. Bran gave the feed ration bulk and protein, and when added to either whole or ground corn, it produced better gains than corn alone. Oil cake made from flaxseed was also an excellent feed supplement. Molasses was considered a good supplement, but it was not widely used. Locally grown clover and alfalfa also were coming into general use for fodder. These two legumes provided good protein and carbohydrate additions when fed as hay. In all cases of feed supplements, cattlemen had to balance the increased gains in weight against the increased cost of the feed. Depending upon prices for bran, oil cake, corn, and feeder cattle compared to beef prices, cattlemen could choose the type of ration to produce the most economical gains in weight.[27]

Indeed, as Henry concluded, "the ability to fatten cattle rapidly and profitably is a gift," and feeders who had it were able to balance alternatives, suit the food to the condition of the cattle, bring cattle onto full feed, and keep them there with a carefully established routine. By shortening the time available for producing prime beef, the margins for error shrank. Anything that disturbed cattle once they were started on the feed routine would reduce profits or even cause a loss. Dean Henry cautioned men to be regular in their feeding program. They should not change the time of feeding, the place of feeding, or the people doing the feeding, or it would put the cattle "off feed." The ability to take cattle from the range, place them in a lot, and then quickly bring them to full feed and keep them there for three to six months required management skill; the unskilled fell by the way.[28] A feeder with a "good eye" could tell by looking at his cattle if they were gaining as they should and what should be done to maintain this.

Good feeders understood the place of oats, oil meal, bran, and corn fodder in feeding rations; the advantages of different types of corn preparation; the fact that cattle gained less at a higher cost as they increased in age; and the importance of adjusting these combinations as the cost of feeds and price of beef changed.[29] They combined skill in feeding with cattle bred for early maturation to produce quality beef at an age their grandfathers would have thought impossible. By the end of 1900 the cattle feeding business had changed so much that the heavy cattle for which Gillett was famous would no longer bring awe and praise in the market. The "beautiful" heavy steers of the past had lost favor and indeed suffered the additional blow of price discrimination. Buyers did not want steers much over 1,600 pounds liveweight. As a leading commission firm at the stockyards reported, "the heavy cuts are out of style and all classes of buyers would prefer lighter beeves."[30] The improvers had succeeded in changing the industry.

C H A P T E R 1 1. ILLINOIS-IOWA CATTLE FEEDING IN THE NINETEENTH CENTURY

The development of the cattle feeding business detailed here occurred during the time when the United States went from an agricultural to an industrial economy and thus must be thought of in that larger context. Major changes in the transportation system and in population distribution and density created new business opportunities for both industry and agriculture that encouraged changes in production and organization. As industrial growth gave rise to production, distribution, and administrative change that encouraged corporations and "trusts," the Illinois and Iowa cattle feeding business also changed. In the immediate sense it grew up and changed with the Corn Belt, but it had also grown up with the national economy. The railroad-created national market and the growing urban complex demanded goods and services on an unprecedented scale, which encouraged and afforded the opportunity for change within agriculture.

The rich agricultural resources of Illinois and Iowa for producing corn and pasture made it an excellent area for beef cattle production. Settled by people familiar with cattle feeding operations and well positioned in relation to population movement and the expanding railroad network, Illinois and Iowa became the center of prime beef production in the nation by the end of the century.

As early as 1850 men had established a pattern of cattle production brought from Ohio and Kentucky. Cattlemen fed corn and grass to four- to six-year-old cattle, which were either driven to eastern markets or sent to some midwestern packer to be slaughtered, pickled, and shipped to retail markets in barrels.

In Ohio there had been a progression from range to feeding area and then a decline in the importance of corn fattening of cattle because of its rise in Illinois and Iowa. In the latter two states, however, there was little decline in fat cattle production as the industry spread to the Great Plains. The center of fat cattle production did not move westward again. Instead there was regional specialization as ranchers produced young cattle for the feedlots of Illinois and Iowa. The fact that these two states continued as the center of prime beef production was indicative of one of the significant changes in the development of the Corn Belt economy during the nineteenth century. Cattle feeders shifted their operations in response to technological and market changes during the course of the century so that the center of prime beef production remained in the same area from 1860 to well into the twentieth century.

Railroads (by centralizing markets and providing the nucleus for the great development of the beef processing industry in Chicago) and the development of the refrigerator car made possible the growth of the dressed beef trade and brought changes to the pattern of beef cattle feeding. Before the railroads and refrigeration, the market for Illinois and Iowa beef was limited to a slow expansion. Barreled beef was processed only during the cold months of the year and was not as appealing to consumers as fresh beef. The fresh beef trade was in the hands of the local butchers, who slaughtered cattle only as there was demand because of the problems of preserving fresh meat. Butchers obtained their beef from large stockyards or bought it directly from farmers. They took the whole animal and made up for the waste material such as hide, tallow, and bone, which they could not use, in the price of the meat offered for sale.

The expansion of the railroads, the rise of the urban market, and the development of techniques of refrigeration changed all this. It became possible to ship fresh meat raised in Illinois and Iowa and slaughtered at any time of year in Chicago to any place reached by a railroad. This made it possible and desirable to develop a central slaughter market in order to realize benefits in purchasing cattle on a large scale and utilizing the large volume of former waste material in the manufacture of by-products. At the same time that the Chicago packers developed a central cattle market and processing center, they also reorganized the wholesale distribution of fresh meat. The ability of the packer to quickly and cheaply supply almost any part of the country and the foreign market with fresh beef led to greater consumption of beef and increased demand for fat cattle.

In response to this change in market and increased demand, both historical and geographical changes took place in the cattle feeding industry. During the rest of the century Illinois and Iowa men improved the cattle they raised or bought improved range cattle to fatten. They also changed their feeding practices in an effort to produce better beef in a shorter time. With competition from more cheaply raised and grazed range cattle, Illinois and Iowa men concentrated on producing, in a short time, corn-fed cattle for the top beef grades in the market.

An important factor in this change in feeding practices was the close connection between cattle feeders and the institutions of improvement. Cattlemen participated in newspaper discussions of changes in feed routines, were in the forefront in the formation of associations for improved farming in general and livestock production in particular, and had close connections with state agricultural societies, fairs, agricultural schools, and ultimately with the agricultural experiment stations. Leaders in the movement toward breed improvement and early maturity in beef cattle contributed their knowledge to the farm community and benefited from the technical studies of feeds and feeding that became available toward the end of the century.

Within the two states there were also shifts in the areas of concentrations of beef cattle. On a county basis, the center moved from central Illinois to northwestern Illinois and east-central and southwestern Iowa by the end of the century. However, the geographical expansion of the cattle industry continued. Men saw the pattern of operations in Illinois and Iowa and modified it for use in eastern Kansas and Nebraska or introduced new factors. The major new element was the spread of cattle across western range country. The impact of this expansion in area and numbers of cattle influenced changes in cattle feeding in Illinois and Iowa toward the end of the century.

By the 1890s the midwestern cattle feeding industry had reached another period of stability in its organization and operation. The three-stage process of moving cattle from range through feedlots to market was well defined. Range areas specialized in producing and raising young cattle. Illinois and Iowa farmers specialized in fattening young range cattle on corn, alfalfa, oil meal, bran, and other feed supplements. Compared to the 1850s cattle were marketed at a younger age, and when slaughtered, they produced a higher percentage of beef to total weight. The animals themselves were different because of improvements in breeding. They made greater gains in weight for a given amount of feed and carried a better distribution of flesh on the carcass (which gave more of the better cuts of beef) than did cattle marketed in the 1850s.

This is not to say that the transition was a smoothly accomplished feat. That agriculture in general and cattle feeding in particular benefited from some of the changes in the national economy is not to be denied, but they were also victims. Farmers found themselves competing in a world market where supply and demand relationships changed quickly and the output of many agricultural units caused a price drop. But farmers could not respond as did some industrialists and organize "trusts" to control competition or to increase efficiency. Many farmers suffered through periods of low prices and incomes after 1870 when competition from Australia and Argentina and from lower ocean freight rates caused problems for American producers in a world market; and railroad development and the rise of a national market and relatively low capital requirements encouraged easy entry of domestic competition, especially in grain production.

Agriculture also suffered during the depression years of the mid-1870s and early 1890s. Within the broader context of American agricultural development, Illinois and Iowa cattle feeders were a small unit but shared many of the problems and benefits as well as having some special rewards. Cattle feeding was a high-risk enterprise with limits imposed by managerial talent and financial resources. But cattle feeders had some comfort in knowing that their numbers were limited by higher capital requirements than general farming and by the skill requisite to handling livestock successfully. Although unable to profitably combine into large-scale producing units, cattle feeders did not suffer as much as grain farmers in the plains from inability to create a producer "trust" in competition with the various processor "trusts" of the time. Industrial development, urban growth, and the railroad network all worked to raise demand for quality dressed beef. While generally at the mercy of the transport system, on occasion, cattlemen benefited from the competition of the industrial giants, for example, railroad rate wars in 1877, 1881, 1884, 1887, and 1888.

Within this context at the end of the century, the cattle feeders of Illinois and Iowa as a group found themselves in a more complex market that required management ability, capital, and some luck; but the market structures, cattle types, and feed routines were well defined. By the 1890s Chicago shared the market with Kansas City, Omaha, and smaller regional markets; but this did not change the system so much as to benefit the producer in western Iowa. The success of the grain-livestock combination in the Middle West may even have given the region a relative prosperity in the hard times often associated with farming in the 1880s and early 1890s; and when economic conditions began improving by 1896, the cattle feeders were in a good position to benefit. The efficient feeders survived the period of declining profit margins in the 1880s and early 1890s and established the dominant pattern of cattle feeding operations.

The fattening of prime cattle was an art as well as a science, and from 1840 to 1900 Illinois and Iowa cattle feeders had brought this to a state of performance and a pattern of operation that in many ways remained the norm for the industry until the 1950s. For indeed, cattle feeding did not decline in Illinois and Iowa after 1900. As late as 1959, cattle and calves accounted for 28.1% and 42.2% of the value of all farm products sold in Illinois and Iowa respectively, while hogs accounted for only 18.1% and 24.4% of the value of farm products sold in Illinois and Iowa.[1]

These figures and the foregoing discussion for the nineteenth century might well indicate a needed revision in the standard terminology of "corn-hog belt." There is no denying that hogs were important contributors to farm income in the nineteenth century, but in Illinois and Iowa hogs often were an adjunct to cattle feeding. Feeding cornmeal was never very popular because meal could not be utilized by hogs as whole corn could, and cattle feeders were quick to point this out. From the 1830s corn and cattle were a natural and rewarding combination for Illinois and Iowa farmers, and less experimenting was required in these two states than in others close by for farmers to find a commercial base for their agriculture. The ingredients for the corn-cattle belt were present from the early part of the nineteenth century, and farmers used them to continue cattle feeding as a major occupation to the present day by changing their operations to meet the changing market needs. Rather than being a passing phase of frontier agricultural development as earlier in areas to the east, cattle feeding in the Corn Belt was a commercial venture very early and remained a major element in the economy of Illinois and Iowa in the middle of the twentieth century.

A P P E N D I X A. MAPS

Map 1. Iowa and Illinois Counties

Map 2. Area of Wisconsin Glaciation

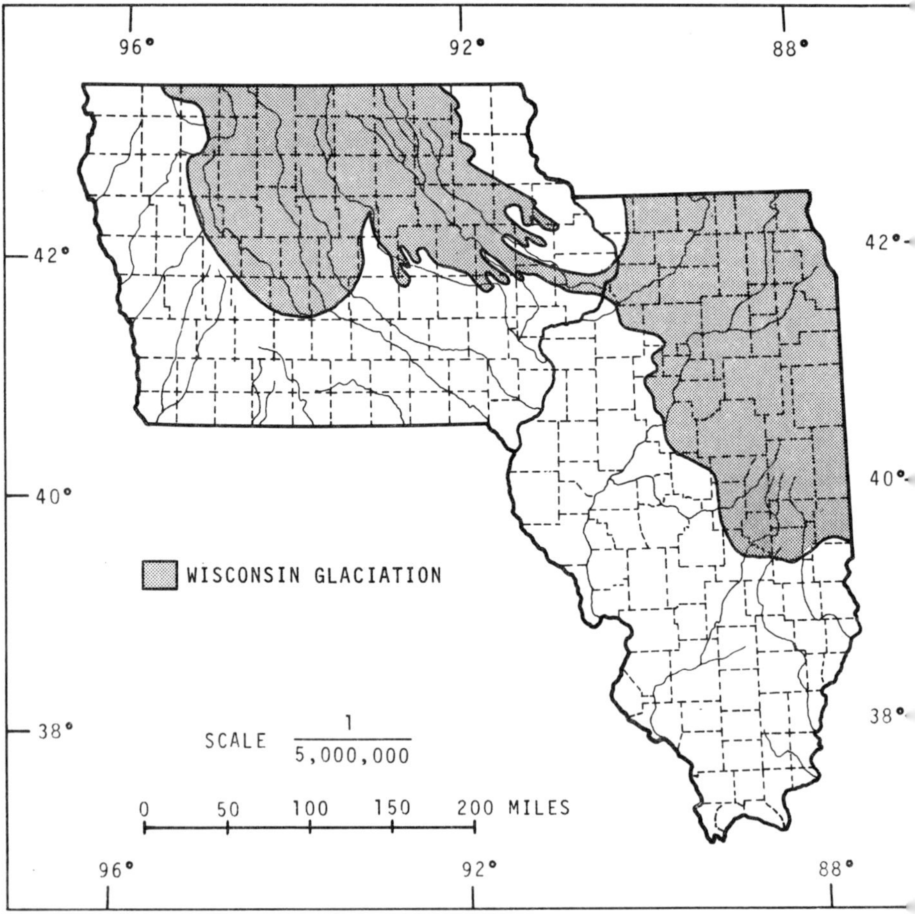

Map 3. Illinois-Iowa General Soil Types

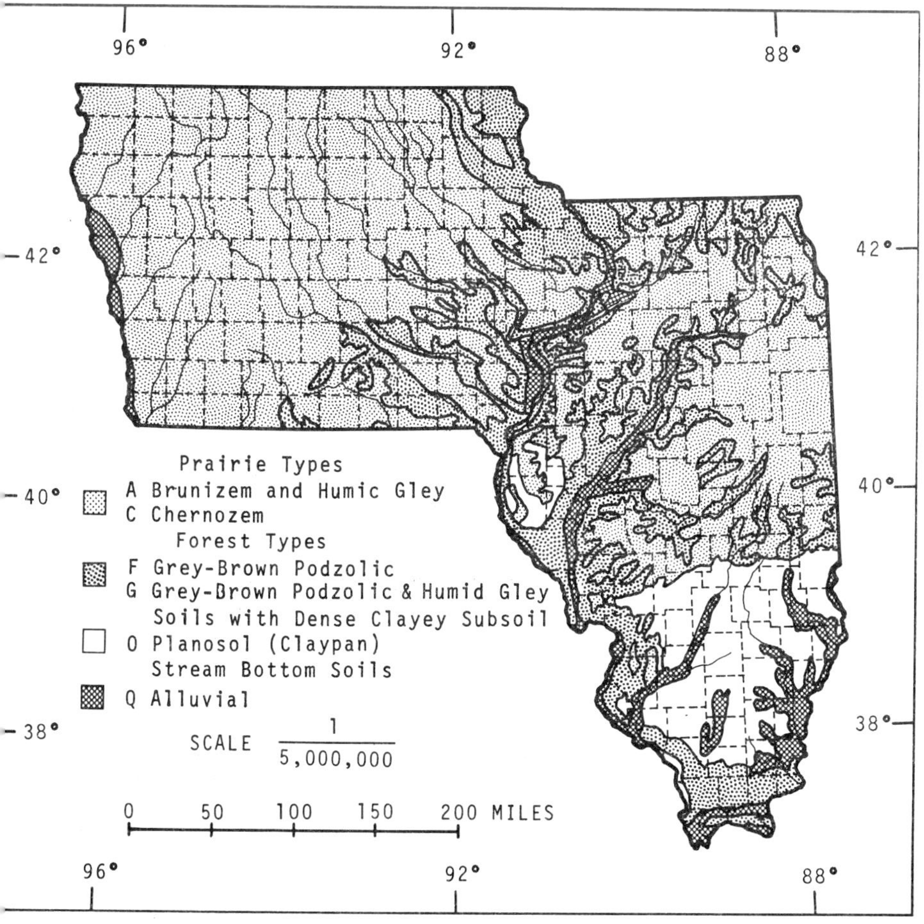

Map 4. Illinois-Iowa Land in Farms

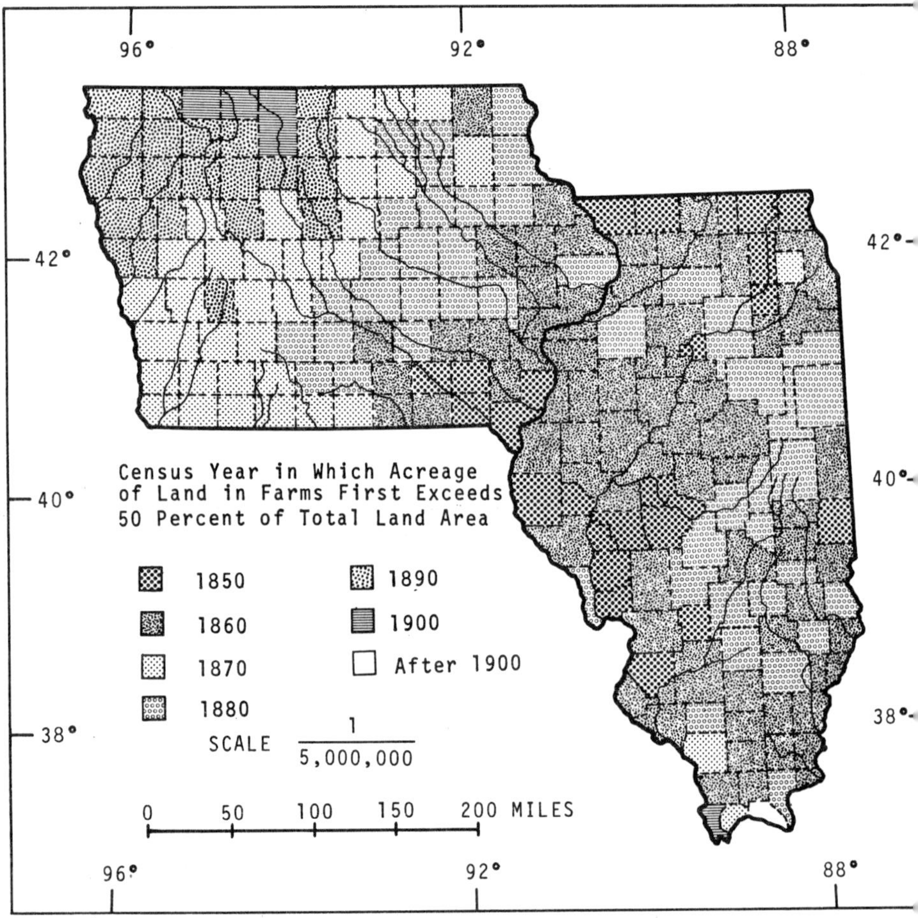

Map 5. Railroads, 1861

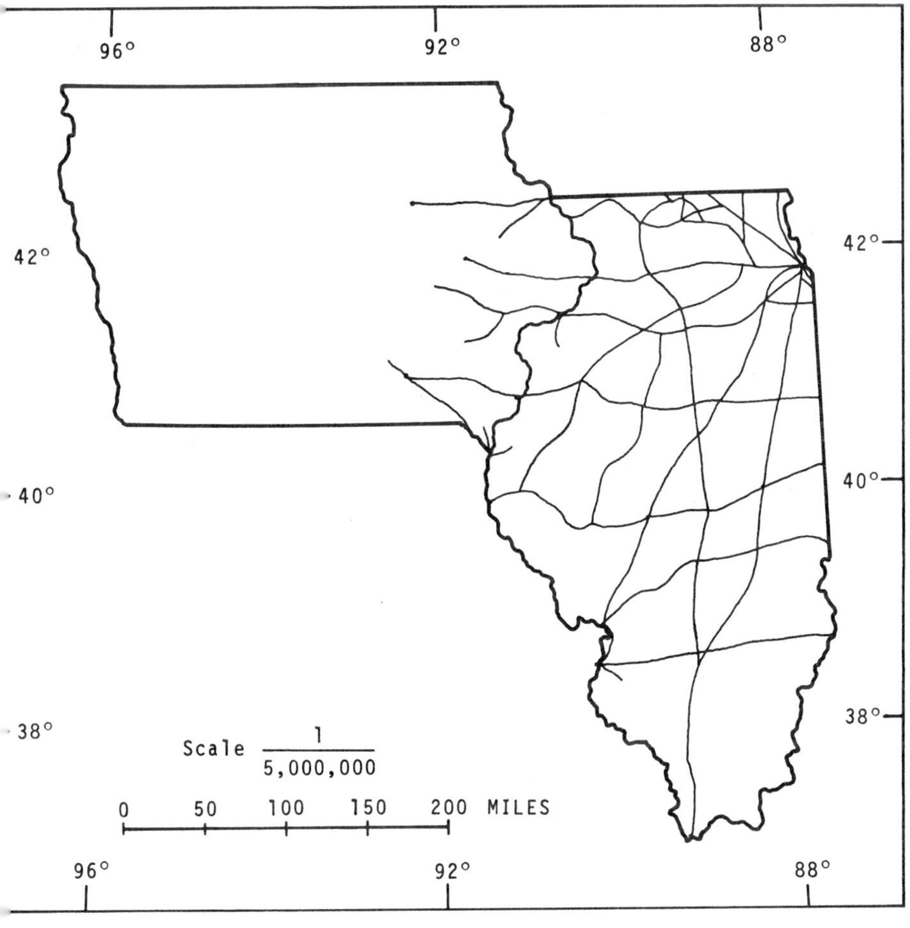

Map 6. Railroads, 1900

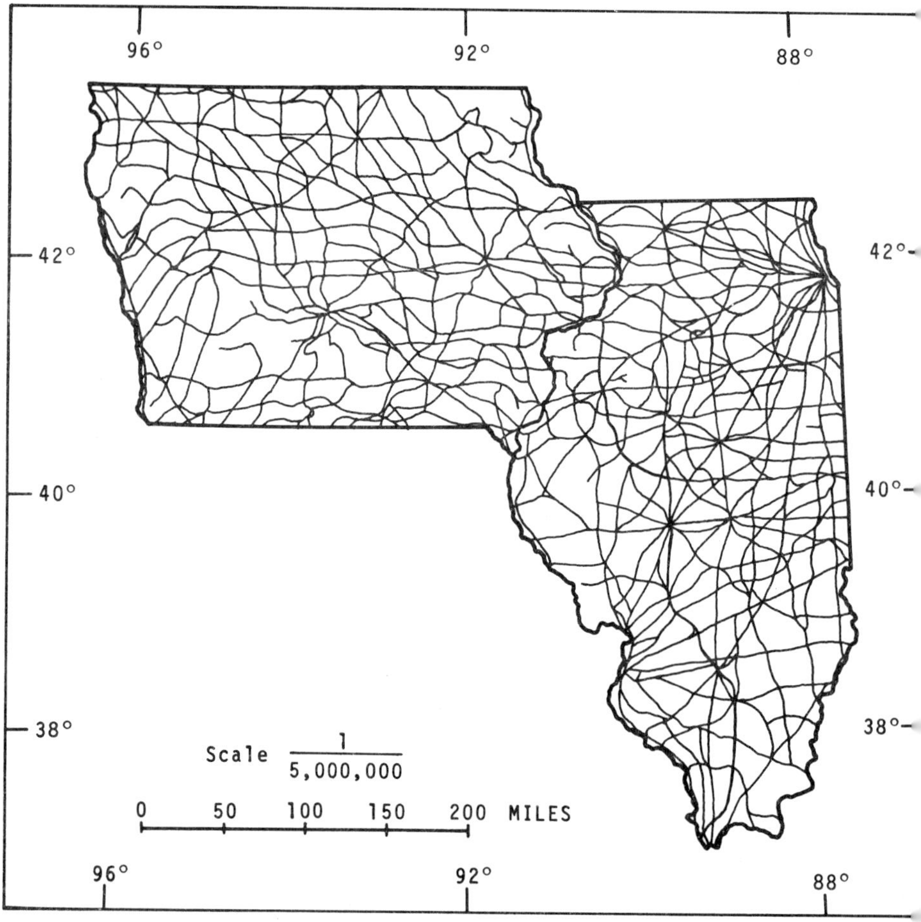

Map 7. Beef Cattle per Hundred Acres, 1850

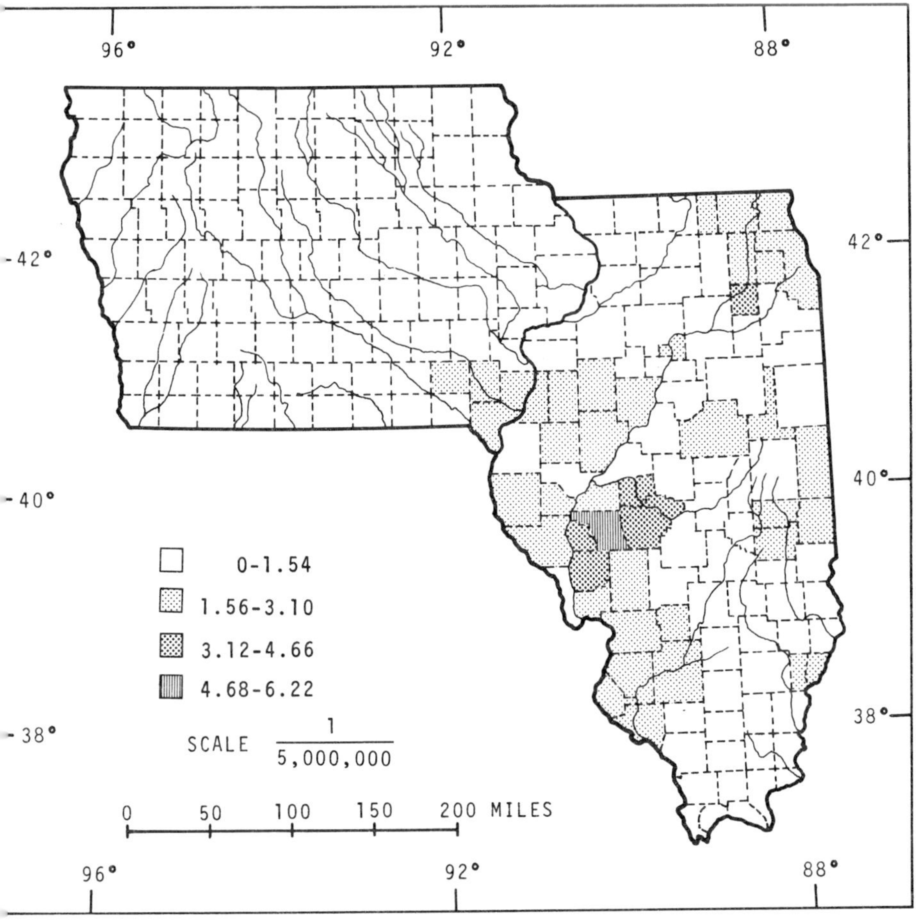

Map 8. Beef Cattle per Hundred Acres, 1860

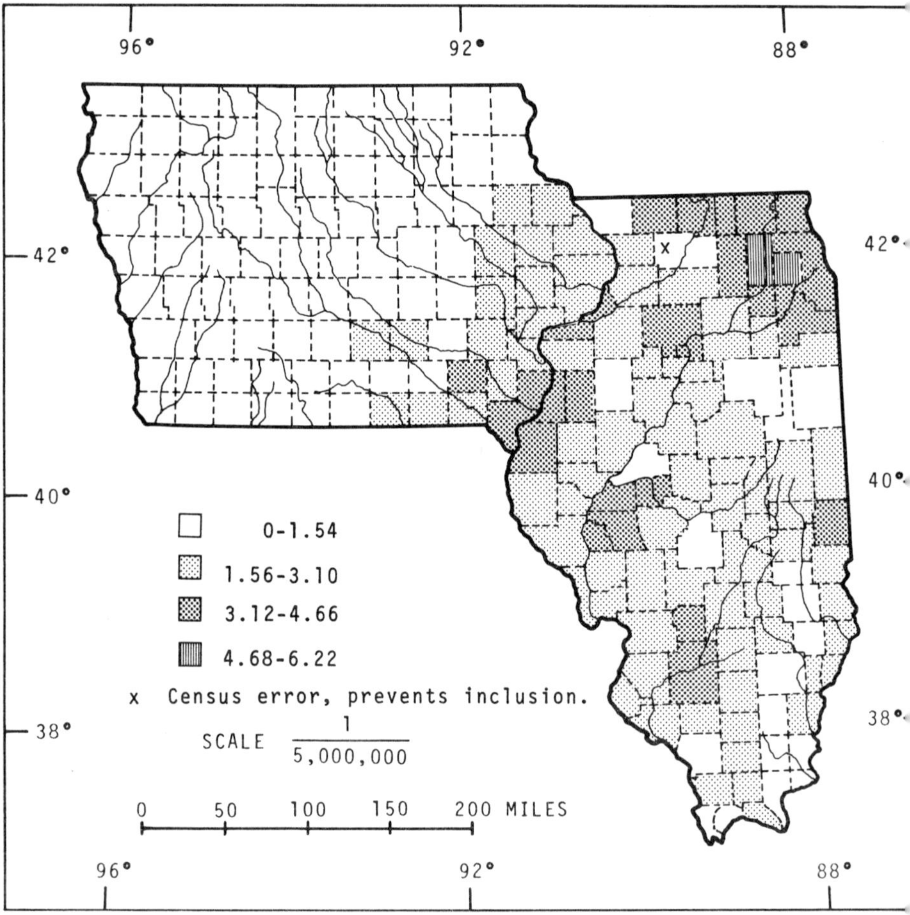

Map 9. Beef Cattle per Hundred Acres, 1870

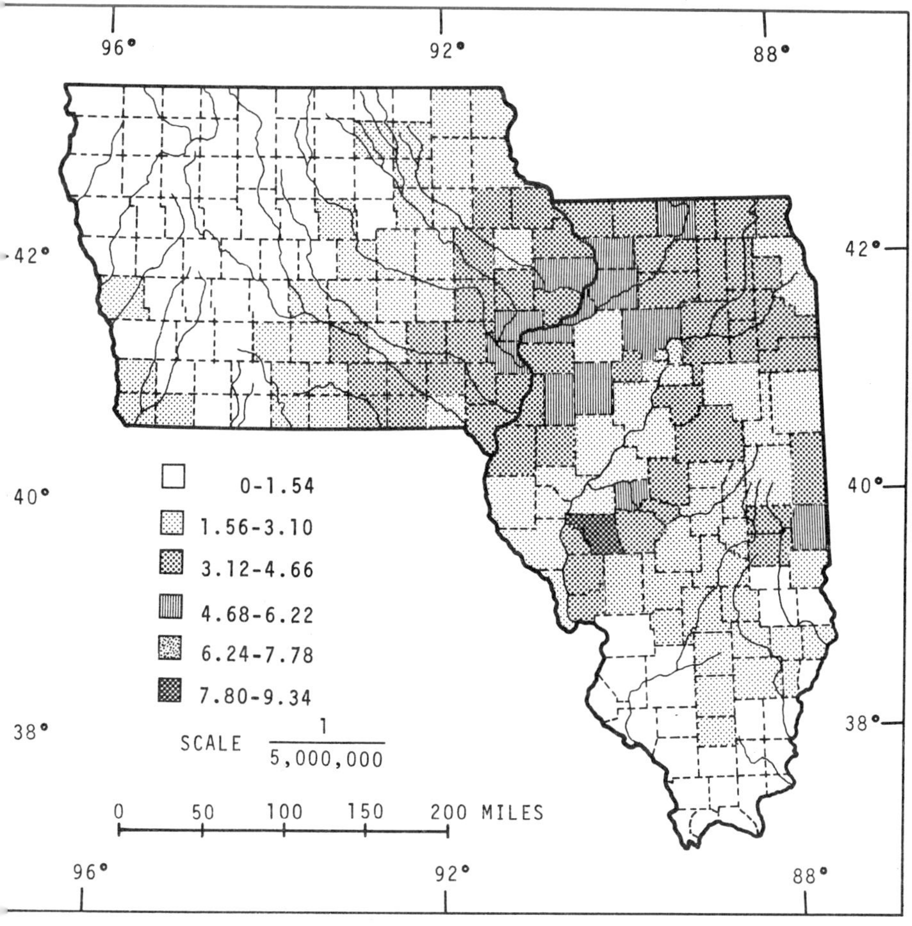

Map 10. Beef Cattle per Hundred Acres, 1880

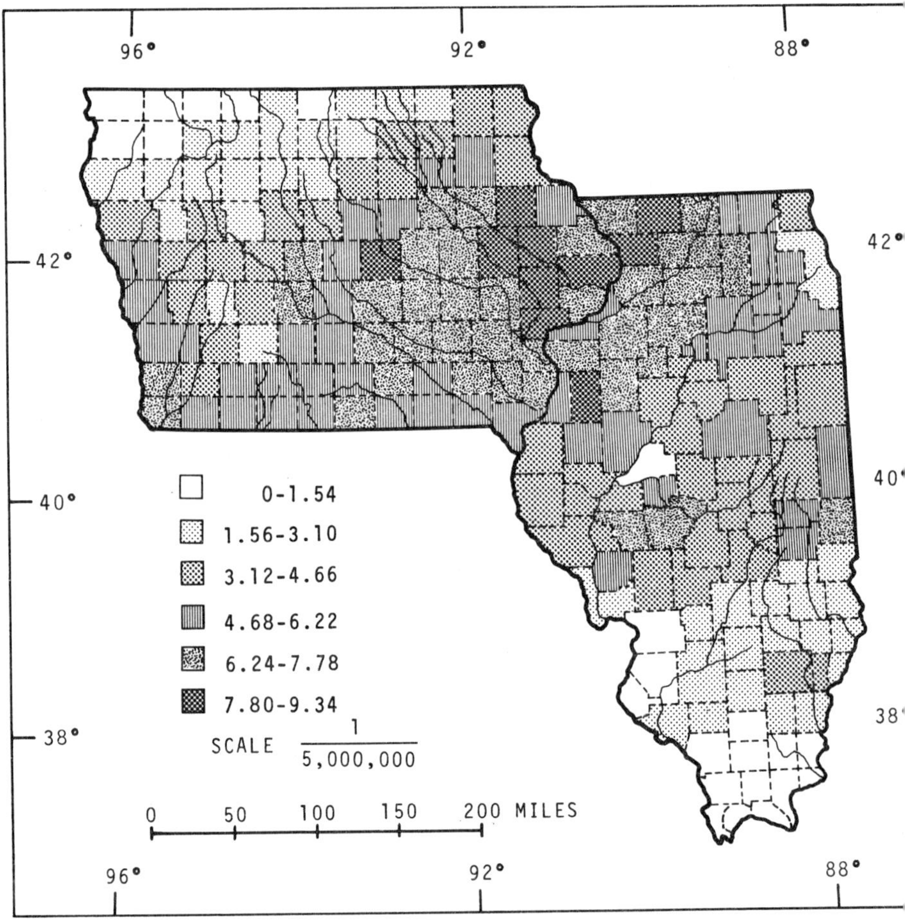

Map 11. Beef Cattle per Hundred Acres, 1890

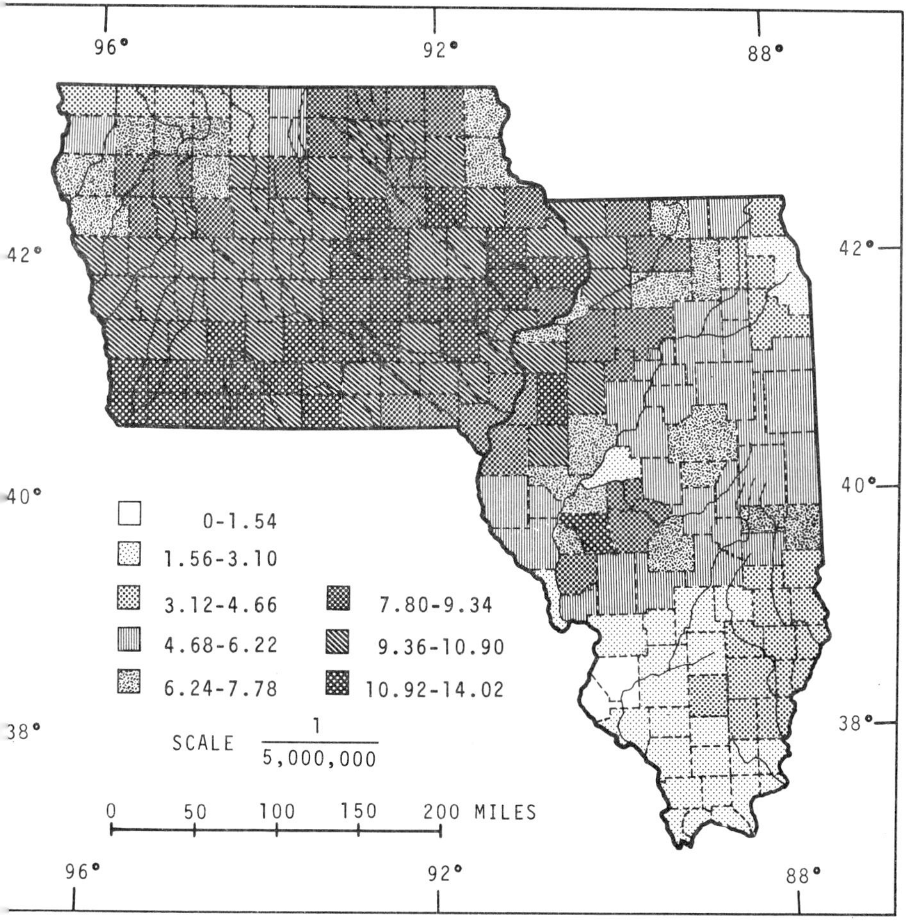

Map 12. Beef Cattle per Hundred Acres, 1900

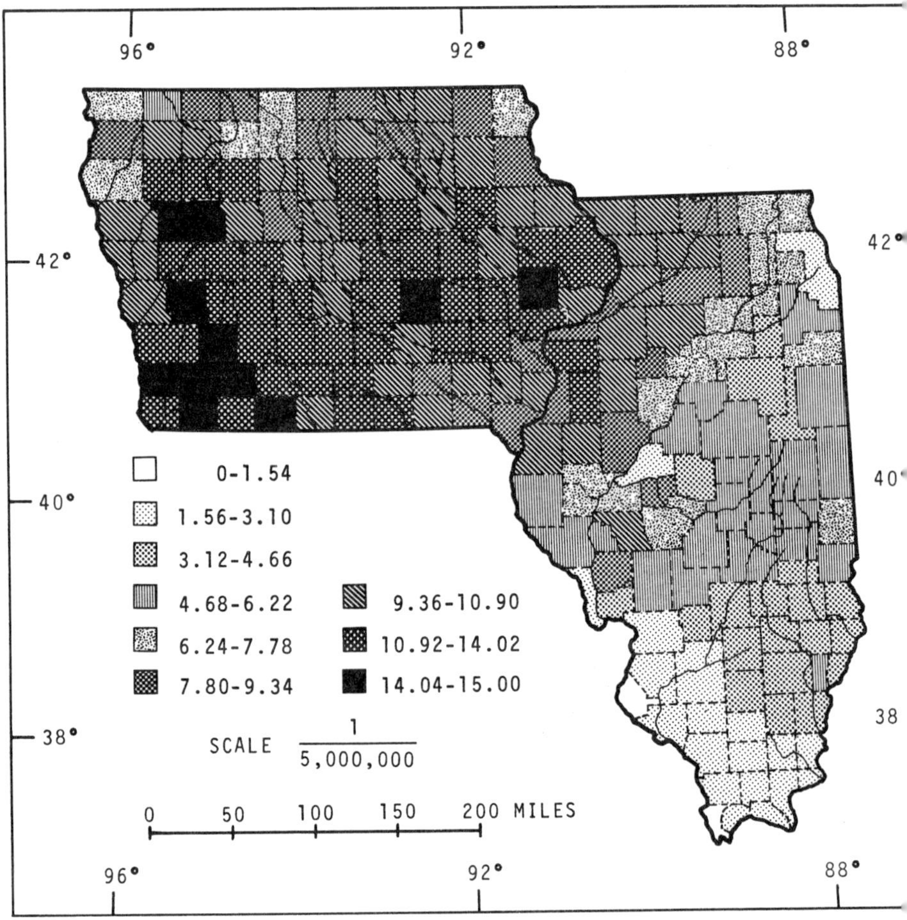

Map 13. Registered Cattle Calved or Imported, 1870 (Source: Brinkman, "Historical Geography," 1964; map 28, used by permission)

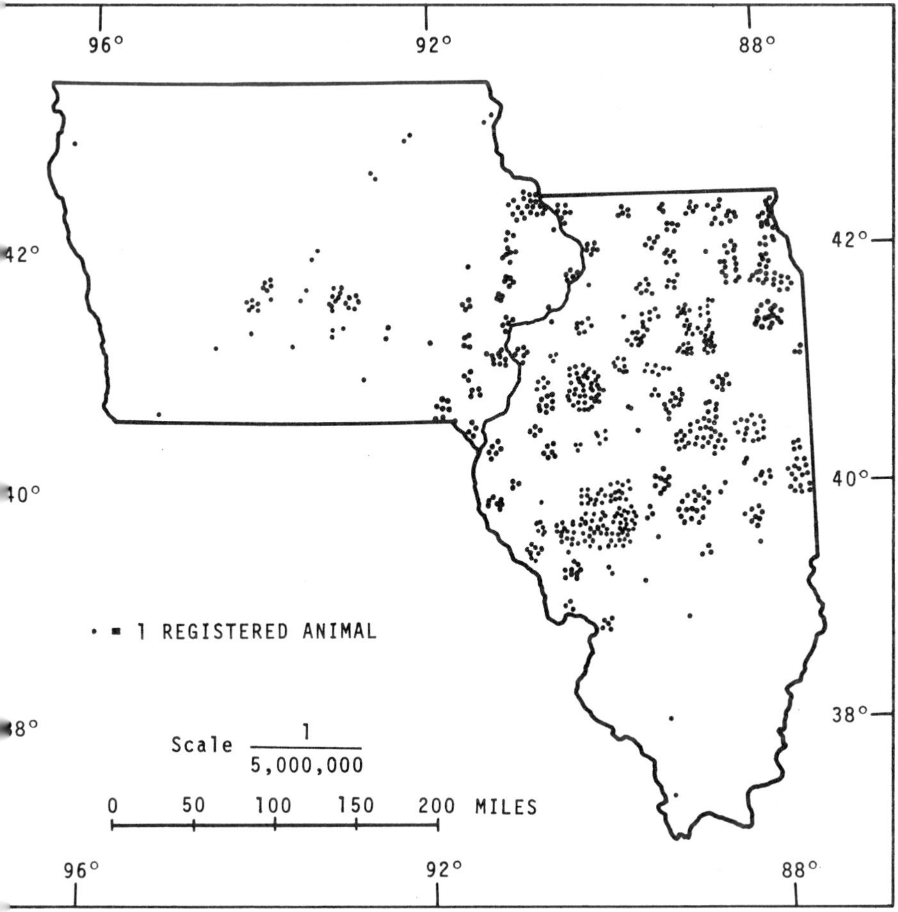

Map 14. Presumed Areas Influenced by Improved Breeds of Cattle, 1870 (Source: Brinkman, "Historical Geography," 1964; map 26, used by permission)

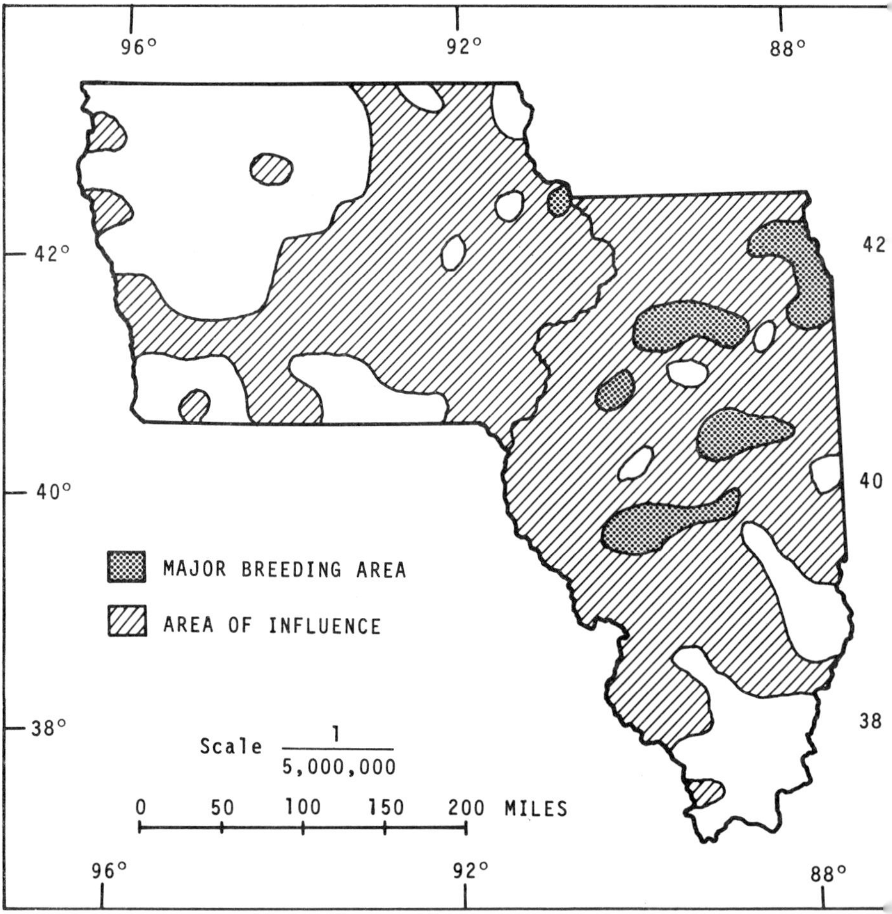

A P P E N D I X B. TABLES AND FIGURES

TABLE B.1. POPULATION GROWTH

State	1840	1860	1900
Illinois	476,183	1,711,951	4,821,550
Iowa	43,112	674,913	2,231,853

Source: Twelfth Census, Abstract, pp. 32-33.

E B.2. ILLINOIS-IOWA POSITION AS A PERCENTAGE OF TOTAL U.S. AMOUN
1900

l a	Popu- lation	Improved farm- land	Wheat harvested	Neat* cattle on farms	Swine	Corn harvested	Oats harveste
5	9.15	13.8	6.5	11.3	24.7	29.3	36.9

Source: Twelfth Census, Abstract, pp. 32-33, 234, 238, 245, 258-

*Neat = common domestic bovine.

BLE B.3. ILLINOIS-IOWA AGRICULTURE: NATIONAL POSITION, 1850

	Illinois		Iowa
em	Amount	Per-cent, U.S. total	Amount
umber of farms	76,208	5.26	14,805
nproved land in farms	5,039,545 acres	4.46	824,682 acres
orn	57,646,984 bu	9.7	8,656,799 bu
eat	9,414,575 bu	9.4	1,530,581 bu
ats	10,687,241 bu	6.9	1,524,345 bu
vine	1,915,907	6.3	323,247
Other" cattle	617,365	5.4	90,917

Source: Twelfth Census, V, pp. 704-5, 707 and VI, pp. 80-81, 8
2-93; Thirteenth Census, V, pp. 69, 73-74 (crop year 1849; farms,
nd livestock of 1850).

TABLE B.4. ILLINOIS-IOWA AGRICULTURE: NATIONAL POSITION, 1870

Item	Illinois (% of U.S. total)	Iowa (% of U.S. total)
Number of farms	7.6	4.3
Improved land in farms	10.2	4.9
Corn produced	17.1	9.1
Wheat produced	10.5	10.2
Oats produced	15.2	7.4

Source: Thirteenth Census, V, pp. 69, 73-74; Twelfth Census, VI, pp. 80-81, 84-85, 92-93.

TABLE B.5. ILLINOIS-IOWA AGRICULTURE: NATIONAL POSITION, 1900

Item	Illinois (% of U.S. total)	Iowa (% of U.S. total)
Number of farms	4.6	3.9
Improved land in farms	6.6	7.2
Corn produced	14.9	14.4
Wheat produced	3.0	3.5
Oats produced	19.1	17.8
"Other" cattle on farms	3.8	7.5
Swine on farms	9.4	16.8

Source: Twelfth Census, V, pp. 704-5, 707 and VI, pp. 80-81, 84-85, 92-93; Thirteenth Census, V, pp. 69, 73-74.

TABLE B.6. SELECTED CHICAGO BEEF PRICES, 1840-60

Date	Price per cwt	Date	Price per cwt
1840 April	$4.00-4.50	1852 January	$4.00-4.50
July	4.00-4.50	April	4.00-4.50
1841 January	4.00-4.50	July	4.00-4.50
April	3.00-4.00	October	4.00-4.25
July	3.00-3.50	1853 January	2.50-4.00
1844 April	3.00	April	3.00-3.50
July	3.00	July	2.75
1845 January	2.50-2.62	October	2.50-2.75
April	3.50-4.00	1854 January	3.00-3.12
July	2.50-3.00	April	3.75-4.50
October	2.00-2.50	July	3.50-3.62
1846 January	2.00-2.50	October	2.75-3.12
April	2.50-3.50	1855 January	3.00-3.25
July	2.50-3.50	April	3.50-4.00
October	2.37-2.50	July	3.50-4.00
1847 January	2.25-3.00	October	2.50-3.12
April	3.50-4.00	1856 January	3.00-4.50
July	3.50-4.50	April	4.25-4.50
October	3.00-3.50	July	3.00-3.75
1848 January	2.25-3.25	October	2.00-2.50
April	2.50-3.50	1857 January	3.00-3.75
July	3.50-4.00	April	3.87-4.25
October	2.50-3.50	July	4.25-4.65
1849 January	2.50-3.50	October	2.50-3.00
April	3.00-4.00	1858 January	2.25-2.75
July	3.00-4.00	April	2.75-3.25
October	3.00-3.50	July	2.00-3.50
1850 January	2.50-3.25	October	1.50-2.50
April	2.50-3.25	1859 January	2.25-2.75
July	4.00-4.50	April	3.00-3.50
October	3.25-3.50	July	2.00-2.50
1851 January	3.00-3.50	October	1.50-1.75
April	3.00-3.50	1860 January	2.00-2.75
July	4.00-5.00	April	2.50-3.00
October	4.00-5.00	July	2.50-3.00
		October	2.25-2.50

Source: Compiled from Aldrich for 1840-44, 1854-57; _Prairie Farmer_, for 1845-53; Chicago Board of Trade, Annual Report, for 1858-60. It is assumed these are prices for liveweights.

Note: Prices are not fully comparable because there are no grade definitions in the sources. Some figures represent actual sales and others are averages. The ranges can generally be assumed to represent figures for average quality beef.

TABLE B.7. ILLINOIS-IOWA RAILROAD DEVELOPMENT: SQUARE MILES FOR EACH MILE OF TRACK

State	1860	1880	1900
Illinois	19.8	7	5
Iowa	84	10	5.8

Source: Calculated from Chicago Board of Trade, Report, 1900, pp. 41, 169 and 1901, p. 169.

TABLE B.8. RECEIPTS AND SHIPMENTS OF CATTLE AT CHICAGO UNION STOCK YARDS, 1865-1900

Year	Receipts (head)	Shipments (head)
1865	613 (5 days only)	0
1866	393,007	263,693
1867	329,188	203,580
1868	324,524	215,987
1869	403,102	294,717
1870	532,964	391,709
1871	543,050	401,927
1872	684,075	510,025
1873	761,428	574,181
1874	843,966	622,929
1875	920,843	696,534
1876	1,096,745	797,724
1877	1,033,151	703,402
1878	1,083,068	699,108
1879	1,215,732	726,903
1880	1,382,477	886,614
1881	1,498,550	938,713
1882	1,582,530	921,009
1883	1,878,944	966,758
1884	1,817,697	791,884
1885	1,905,518	744,093
1886	1,963,900	704,675
1887	2,382,008	791,483
1888	2,611,543	968,385
1889	3,023,281	1,258,971
1890	3,484,280	1,260,309
1891	3,250,359	1,066,264
1892	3,571,796	1,121,675
1893	3,133,406	900,163
1894	2,974,363	950,738
1895	2,588,558	785,092
1896	2,600,476	818,326
1897	2,554,924	843,392
1898	2,480,897	865,642
1899	2,514,446	811,874
1900	2,729,046	934,649

Source: Drovers' Journal, Year Book, 1932, pp. 10-11.

TABLE B.9. DRESSED BEEF SHIPMENTS FROM CHICAGO

Year	Dressed beef
	(tons)
1876	8,865
1880	15,680
1884	182,067
1888	384,741
1890	482,067
1895	455,169
1900	589,475

Source: USDA, Report, 1876, p. 320; Nimmo, Report on the Internal Commerce of the United States, Exec. Doc. 7, pp. 157-58; Chicago Board of Trade, Report, 1886-1900.

TABLE B.10. LIVE CATTLE AND DRESSED BEEF SHIPMENTS

	Live cattle		Dressed	Total dressed	
Year	Chicago	St. Louis	beef (Chicago)	beef (Chicago)	Destination
		(tons)			
1880				15,680	
	222,262	---	---	---	New York
	127,060	35,840	15,680	---	New England
1884				172,824	
	191,736	57,227	34,916	---	New York
	55,996	7,532	120,922	---	New England

Source: U.S. Congress, H. Exec. Doc. 267, pp. 155-58.

TABLE B.11. U.S. EXPORTS: FISCAL YEAR ENDING JUNE 30, 1900

Commodity	Value	Number
Total U.S. merchandise	$1,371,000,000	...
Raw cotton	242,000,000	...
Wheat and wheat flour	141,000,000	...
Total meats	114,000,000	...
Live cattle	30,623,768	396,977 head
Dressed beef	29,643,830	329,078,609 lb
Canned beef	5,233,982	55,553,745 lb
Pickled beef	2,893,902	49,622,328 lb
Total	68,395,482	

Source: Chicago Board of Trade, Report, 1900, pp. xxxvii-xxxviii, lvi; Bureau of the Census, _Historical Statistics_, pp. 537, 546.

Fig. B.1. Value of U.S. Exports of Pork, Lard, Live Cattle, and Beef, 1866-1900 by Five-Year Averages (Source: USDA, <u>Yearbook</u>, 1906, p. 248)

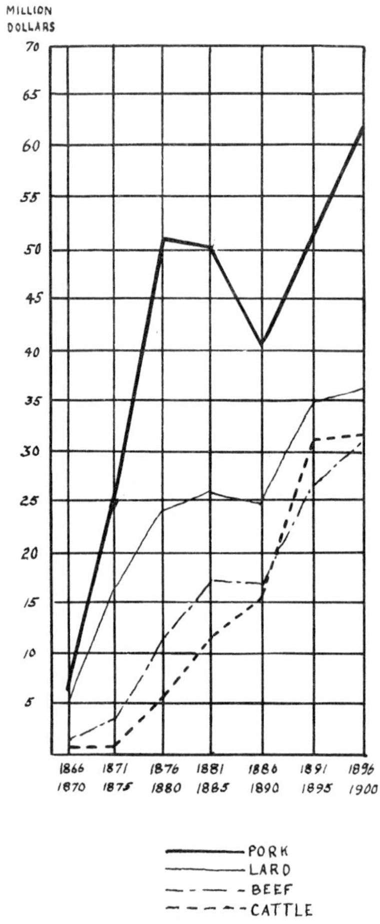

TABLE B.12. CHICAGO CATTLE RECEIPTS, SELECTED YEARS

Year	Texas and Cherokee	Western and northern range	Natives	Total	Range and Texas, % of total
1880	88,600	109,500	1,184,377	1,382,477	14
1882	346,300	229,700	1,006,530	1,582,530	36
1884	358,374	231,700	1,227,623	1,817,697	32
1886	320,830	238,520	1,404,550	1,963,900	28
1888	547,185	267,494	1,796,864	2,611,543	31
1890	657,053	229,494	2,597,733	3,484,280	25
1895	359,643	430,525	1,798,389	2,588,558	31
1900	194,726	147,000	2,387,320	2,729,046	13

Source: Drovers' Journal, *Year Book*, 1901, p. 22; ibid., 1932, pp. 10, 45, 46; *Prairie Farmer*, Jan. 4, 1890, p. 4.

Fig. B.2. Average Monthly Prices of Fat Cattle and Feeder Cattle, 1874-1900 (Source: Hopkins, Statistical Study, pp. 359-61)

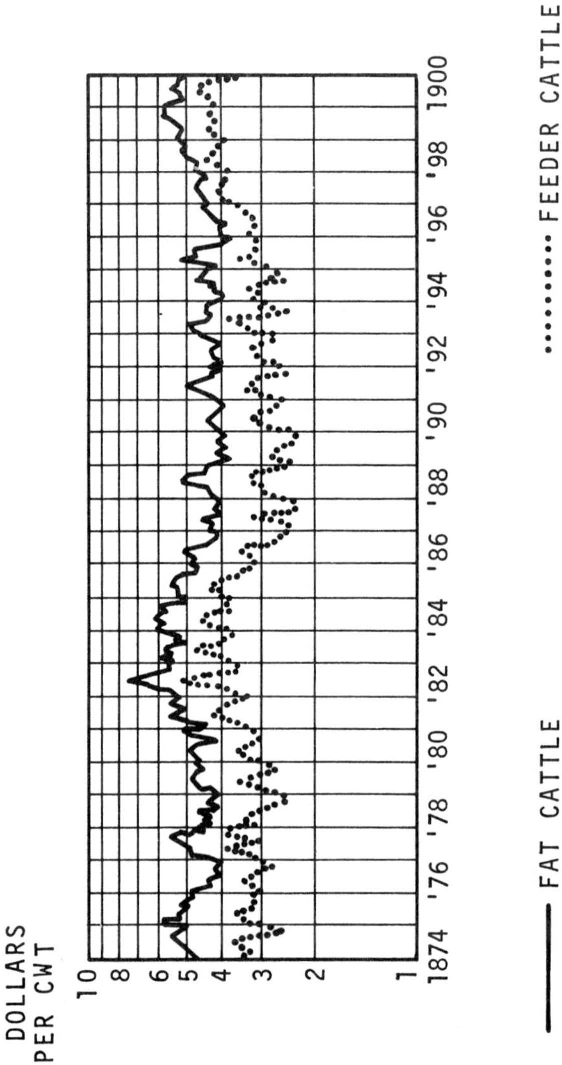

: B.13. BEEF CATTLE ON FARMS

e	1850	1860	1870	1880	1890	1900
ois	617,365	1,061,179	1,075,265	1,518,409	1,975,233	1,373,024
	90,917	350,286	636,244	1,757,849	3,397,132	2,653,703
		Percent of U.S. total "other" cattle				
ois	5.41	6.2	7.2	5.0	4.8	3.8
	0.79	2.0	4.2	6.4	8.2	7.5

Source: Twelfth Census, V, pp. 704-5.

NOTES

Preface

1. Thompson, History of Livestock Raising, USDA, Agr. Hist. Ser. 5, pp. 67-85, 99; Dunbar, Carolina Cowpens, Agr. Hist. 35(July 1961):125-31; Henlein, Cattle Kingdom in the Ohio Valley, pp. 1-10, 103-5, 110-13; Shannon, Farmer's Last Frontier, pp. 198-99.

Chapter 1

1. Wallace and Bressman, Corn and Corn Growing, pp. 89-92.
2. Garland, North American Midwest, pp. 6-11.
3. Odell et al., Soils of the North Central Region, Univ. Wis. Bull. 544, p. 7; Ross and Case, Types of Farming in Illinois, Univ. Ill. Bull. 601, pp. 7-9; for a more detailed discussion see Fenneman, Physiography, pp. 452-56, 472-82, 499-518, 559-60, 576-79, 594-605.
4. Odell, Soils of the North Central Region, pp. 6-9, 15-17, 25-26, 29-31, 36-37.
5. Kay et al., Pleistocene Geology of Iowa, Iowa Geol. Survey, pp. 11-15, 33-69; Odell, Soils of the North Central Region, pp. 6-7, 15-17, 25-26, 29-31, 36-37, 26.
6. Garland, North American Midwest, pp. 22, 25.
7. Nelson, "A History of Agriculture in Illinois"; Hawk, "Iowa Farming Types."
8. George Rogers Clark to George Mason, Nov. 19, 1779, in English, Conquest of the Country Northwest of the River Ohio, vol. 1, p. 453.
9. Shirreff, Tour through North America, pp. 451-52; Lea, Notes on Wisconsin Territory (Repr. State Hist. Soc. Iowa as The Book That Gave to Iowa Its Name, Iowa City, n.d.),

p. 28; Angle, _Prairie State_, passim; Harding, "Economic and Social Conditions in Iowa to 1880"; Bushnell, _Work and Play_, p. 236.

10. Twelfth Census, Statistical Atlas, 1903, pls. 4-8; Royce, Indian Land Cessions, H. Doc. 736, map 24; Pooley, Settlement of Illinois, Univ. Wis. Bull. 220, p. 274.

11. Pooley, Settlement of Illinois, p. 316; Royce, Indian Land Cessions, pp. 656-57, 664-67, 672-73, 678-81, 692-93, 696-97, 706-7, 722-27, 736-37, 750-51, 766-69, 778-79, 784-85, maps 17, 18, 24, 25; U.S. Commissioner of the General Land Office, Report, 1838, 25th Congress, 3rd sess., S. Doc. 17, map 10; idem, Report, 1843, 28th Congress, 1st sess., S. Doc. 15, p. 65; Lokken, _Iowa Public Land Disposal_, pp. 13-16, 26, 34, 45-46, 50-53.

12. Twelfth Census, Statistical Atlas, pp. 22-37, pls. 2-13; ibid., V, pp. 142, 322-37, 703-5, 707; ibid., VI, pp. 62, 68-69, 80-81, 84-85, 213.

13. Buckingham, Early Settlers, _J. Ill. State Hist. Soc._ 35:243; Goodwin, American Occupation, _Iowa J. Hist. Polit._ 17:89-92, 96-99; Census of Iowa for 1880, pp. 4-7; U.S., Statutes at Large, 2:58-59, 514-16, 743-47, 3:428-31, 536, 4:701, 5:10-16, 23-41, 9:117; Pooley, Settlement of Illinois, pp. 315-22; Thirteenth Census, 1910, V, pp. 73-74.

Chapter 2

1. Bogue, _Prairie to Corn Belt_, pp. 230-34; Danhof, _Change in Agriculture_, pp. 108, 121-25; Clark, _Grain Trade_, pp. 85-91, 100-101. Although Danhof made a major point that farmers had to change attitudes to become commercial while social pressure reinforced the attempt to be self-sufficient and thus inefficient (pp. 16-17, 19, 144), his comments about resisting specialization and slowness to be market oriented concern New England in the 1830s more than the midwestern prairies of 1850. When discussing the prairies, he acknowledged a quick shift to commercial agriculture (pp. 146-47, 150-52).

2. Lampard, _Rise of the Dairy Industry in Wisconsin_, pp. 47-57.

3. Springfield _State Register_, May 10, Aug. 2, 11, Nov. 15, 1849, Cole Notes; Anderson, "Agriculture in Illinois," pp. 148-49; Bogue, _Prairie to Corn Belt_, p. 234; _Iowa Homestead_, June 14, 21, Aug. 23, 1865; _Prairie Farmer_, May 15, p. 156, May 29, p. 176, Oct. 30, p. 360, Nov. 20, 1869, p. 384.

4. Cole, _Era of the Civil War_, pp. 82-83; Ill. State Agr. Soc. Trans., 1861-1864, pp. 317-82 passim in County Society Reports, hereafter this series will be cited as ISAST; _Northwestern Farmer_ (Dubuque), Jan. to Aug. 1858, passim; Twelfth Census, VI, p. 477.

5. Cole, _Era of the Civil War_, p. 381; _Prairie Farmer_, Oct. 10, 1868, p. 116, Apr. 23, 1870, p. 124, Apr. 15, 1871, p. 116.

6. Schoolcraft, _Travels in the Central Portions of the Mississippi Valley_, pp. 303, 308, 310; Flint, _Condensed Geography_, vol. 2, pp. 120-21, 128; Shirreff, _Tour through North America_, pp. 445-47, 239, 242; _Prairie Farmer_, Feb. 1850, p. 45; Carman, English Views, _Agr. Hist_. 8(Jan. 1934): 3-19; _Prairie Farmer_, passim; Oliver, _Eight Months in Illinois_, pp. 102-4; the average size of farms as determined by the federal census became progressively smaller from 1840 to 1890.

7. _Prairie Farmer_, July 21, 1859, p. 41.

8. Anderson, "Agriculture in Illinois," pp. 310-14; Twelfth Census, V, pp. 704-8, VI, pp. 80-81, 84-85, 92-93; Thirteenth Census, V, pp. 73-74.

9. Bogue, _Prairie to Corn Belt_, pp. 218, 221, 226; Newlove, "Economic History of Illinois Agriculture," pp. 28-29; Clark, _Grain Trade_, pp. 85-86, 153-55.

10. _Prairie Farmer_, Sept., pp. 298-99, Nov. 1850, p. 334, Sept. 1, p. 132, Oct. 13, 1859, p. 233; _Northwestern Farmer_, Oct. 1858, p. 322; Anderson, "Agriculture in Illinois," pp. 77-83, 185-88; Clark, _Grain Trade_, pp. 202-5; ISAST, 1869-1870, pp. 142-43; Ill. Dept. Agr. Trans., 1875, pp. 297, 308-9, hereafter this series, which superseded ISAST, will be cited as IDAT.

11. Salesman quoted in Anderson, "Agriculture in Illinois," p. 123; Buck, Pioneer Letters of Flagg, _Trans. Ill. State Hist. Soc_., 1910, p. 182; Cole, _Era of the Civil War_, pp. 374-75; _Mound City Emporium_, May 13, 1858, p. 2, Cole Notes; _Prairie Farmer_, Aug. 18, 1859, p. 101, Dec. 5, 1868, p. 184, Oct. 30, 1869, p. 360; see Power, _Planting Corn Belt Culture_, for a discussion of the relationship between cultural attitudes and types of farming.

12. Cole, _Era of the Civil War_, p. 375; _Prairie Farmer_, May 7, 1870, p. 138.

13. Anderson, "Agriculture in Illinois," pp. 86-87; Cole, _Era of the Civil War_, pp. 39-47, 373-75; Taylor and Neu, _American Railroad Network_, map.

14. Taylor and Neu, _American Railroad Network_, map;

Mildred Throne, Population Study, Iowa J. Hist. 57:317-19, 321.

15. Bogue, Prairie to Corn Belt, pp. 219-22, 227; Hawk, "Iowa Farming Types," p. 20; for a more detailed discussion of agriculture in both states see also Anderson, "Agriculture in Illinois," Nelson, "A History of Agriculture in Illinois," Newlove, "Economic History of Illinois Agriculture," and Ross, Iowa Agriculture.

16. USDA, Yearbook, 1940, pp. 228-32; see any Prairie Farmer in the 1850s for machinery manufacturers' advertisements and claims; Prairie Farmer, Aug. 25, 1859, p. 116; Rogin, Introduction of Farm Machinery, p. 35, and pp. 31-52, 85-119, 133-45 for changes taking place in tillage and wheat harvesting implements; Shannon, Farmer's Last Frontier, pp. 125-72.

17. Putnam, Illinois and Michigan Canal, pp. 99-125, discusses the economic influence of the canal and the quickly developed railroad competition; Iowa Census for 1880, pp. 12-27; U.S. Interstate Commerce Commission, Thirteenth Annual Report, p. 12.

18. Ross, Iowa Agriculture, p. 49; Twelfth Census, V, p. 702; Thirteenth Census, V, p. 69.

19. Hawk, "Iowa Farming Types," pp. 63-77, 99-100, 106.

20. Nelson, "History of Agriculture in Illinois," pp. 96-102, 127-57.

21. Twelfth Census, Abstract, pp. 250-51, 258-59; ibid. V, pp. 322, 326-27; John King of northeastern Iowa in the Chillicothe Advertiser quoted in the Prairie Farmer, Sept. 1845, p. 215.

22. Twelfth Census, I, pp. lxxxiii, lxxxiv, lxxxviii.

Chapter 3

1. Thompson, History of Livestock Raising, USDA Agr. Hist. Ser. 5, pp. 67-85; Dunbar, Carolina Cowpens, Agr. Hist. 35:125-31; Shannon, Farmer's Last Frontier, pp. 198-99; Henlein, Cattle Kingdom in the Ohio Valley, pp. 1-10, 103-5, 110-13.

2. Schoolcraft, Travels, p. 310; Flint, Condensed Geography, vol. 1, pp. 82, 228-29, vol. 2, p. 128; Solomon Koepfli to J. A. Prukett, Apr. 13, 1861, Flagg Papers; Bateman and Selby, Historical Encyclopedia, p. 899; Shirreff,

Tour through North America, p. 446; see also Oliver, Eight Months in Illinois, and Massey, "British Travelers' Impressions of Agriculture," pp. 89-91.

3. Shirreff, Tour through North America, p. 249; U.S. Commissioner of the Patent Office, Report, 1849, pt. 2, p. 293, hereafter referred to as Patent Office, Report.

4. Henlein, Cattle Kingdom in the Ohio Valley, pp. 58-65, 116-20, 170; Harris, Autobiography, Trans. Ill. State Hist. Soc., 1923, pp. 72-101.

5. Patent Office, Report, 1852, pp. 340-41; A Pioneer, Northern Iowa, pp. 7-8; Henlein, Cattle Kingdom in the Ohio Valley, pp. 58-65, 116-21, 170; Bogue, Swamp Land Act, Agr. Hist. 25:169-80; Kerrick, Life and Character of Isaac Funk, Trans. Ill. State Hist. Soc., 1902, pp. 161-65; Hopkins, Economic History, pp. 68-69; Eighth Census, II, pp. cxxx-cxxxi.

6. Prairie Farmer, Sept. 1850, pp. 278-79.

7. Patent Office, Report, 1856, pp. x-xii; see also Prairie Farmer, July 1847, p. 201, July 1849, p. 216, Sept. 1850, p. 265, Dec. 1851, p. 576, Feb. 1852, p. 93, May 1854, p. 166; Northwestern Farmer, May 1858, pp. 166-67, July 1858, p. 230.

8. Prairie Farmer, market reports, May 22 to Aug. 7, 1856. This is based upon reports from New York City stockyards on the number and origin of cattle received each week.

9. Burlend and Burlend, True Picture of Emigration, pp. 75-78; [Ponting] Life, pp. 5-19; Kerrick, Life and Character of Isaac Funk, pp. 161-63; Eli Strawn to Joel Hopkins, Sept. 20, Oct. 17, 25, 1860, Hopkins Papers; Prairie Farmer, Nov. 1845, pp. 427-30; Bateman and Selby, Historical Encyclopedia, p. 965; Shaw Smith to Amos Williams, Feb. 28, 1834, Williams-Woodbury Papers; McCoy, Historic Sketches, p. 238; Chicago Daily Democratic Press, Oct. 25, 1854, quoted in Anderson, "Agriculture in Illinois," p. 146.

10. Northwestern Farmer, July 1858, p. 231.

11. Prairie Farmer, Aug. 1855, p. 248, June 25, Oct. 15, 1847; Cattleman 35:34-35; [Ponting] Life, pp. 28-34; New York Tribune, July 4, 1854, p. 8; Springfield Journal, July 17, 1854, p. 3, Cole Notes; Chicago Daily Democrat Press, Sept. 10, 1855, p. 2, Cole Notes; Henlein, Cattle Kingdom in the Ohio Valley, pp. 177-79; McCoy, Historic Sketches, pp. 38-41.

12. Prairie Farmer, Nov. 1849, p. 356.

13. Sanders, James N. Brown, Trans. Ill. State Hist.

Soc., 1912, pp. 172-73; Prairie Farmer, Jan. 3, 1856, p. 2; McClelland, Jacob Strawn, J. Ill. State Hist. Soc. 34:195-97; Ross, Iowa Agriculture, pp. 45-46; Northwestern Farmer, May 1858, pp. 166-67; Brinkman, "Historical Geography of Improved Cattle," maps 1-4, 8, 10, 12.

14. Prairie Farmer, Jan. 1843, p. 4, Oct. 1848, pp. 332-33, Sept. 1850, pp. 298-99, Nov. 1850, pp. 330, 334, July 1851, p. 317, Dec. 1851, pp. 561-64, Nov. 1854, p. 404, Feb. 12, 1857, p. 1, Jan. 27, 1859, p. 51, Dec. 15, 1859, pp. 370-71; Patent Office, Report, 1854, p. 6; ibid., 1855, pp. 20-21; ibid., 1858, p. 107; Northwestern Farmer, Mar. 1858, pp. 94-95, Nov. 1858, pp. 369-70; Illinois State Register, Apr. 14, 1853, p. 5, Cole Notes; ISAST, II:256-57, 276-77, 294-95, 332, III:133-34; Brinkman, "Historical Geography of Improved Cattle," map 8.

15. Thomas Le Duc, Public Policy, Private Investment, Agr. Hist. 37:8; Prairie Farmer, Apr. 19, 1857.

16. Kerrick, Life and Character of Isaac Funk, pp. 164-65; Cavanagh, Funk of Funk's Grove, pp. 49, 66-67.

17. Newcombe, Alson J. Streeter, J. Ill. State Hist. Soc. 38:422; Shaw Smith to Amos Williams, Feb. 28, 1834, Abram Stansbury to Amos Williams, Mar, 30, 1844, Williams-Woodbury Papers; Rammelkamp, Memoirs of John Henry, J. Ill. State Hist. Soc. 18:70.

18. Cavanagh, Funk of Funk's Grove, p. 66; McClelland, Jacob Strawn, p. 188; Harris, Autobiography, pp. 81-82; Eli Strawn to Joel W. Hopkins, Sept. 20, Oct. 17, 25, 1860, Hopkins Papers.

19. Prairie Farmer, Feb. 1843, pp. 45-46, Mar. 1843, pp. 50-51; Patent Office, Report, 1845, p. 1012; some of the feed recipes given sound like pudding for cattle, e.g., to 156 pounds boiling water add 2 pounds linseed meal, boil five minutes, add 63 pounds barley or bean meal, then cool and feed 5 to 28 pounds per day per animal depending on other feeds. From Patent Office, Report, 1844, pp. 387-88.

20. Prairie Farmer, June 1847, p. 181; Anderson, "Agriculture in Illinois," p. 94; Patent Office, Report, 1853, p. 6. "Stall" could possibly be a corruption of "stalk feeding or derived from the sense of standing up to a feed bunk in an open lot. "Stall" feeding was also called the Kentucky or Ohio system.

21. McCoy, Historic Sketches, pp. 235-36, 238.

22. *Prairie Farmer*, Jan. 3, 1856, p. 2; ISAST, I:429-35.

23. ISAST, II:378, 376, 372-79.

24. Harris, Autobiography, pp. 74-79, 88-90; *Prairie Farmer*, June 1854, p. 239, July 1855, pp. 203, 207, Apr. 3, 1856.

25. Rammelkamp, Memoirs of John Henry, p. 70; *Prairie Farmer*, Nov. 1854, pp. 427-30; Bateman and Selby, *Historical Encyclopedia*, pp. 954-56.

26. McCoy, *Historic Sketches*, p. 236; Bateman and Selby, *Historical Encyclopedia*, pp. 806-7; [Ponting], *Life*, p. 28.

27. *Prairie Farmer*, Sept. 1846, p. 289, Feb. 1847, p. 57, Feb. 1849, pp. 48-49, June 1849, pp. 188-89, July 1852, p. 319, May 1853, p. 200, Dec. 11, 1856, Oct. 25, 1860, p. 258; ISAST, II:380-84; Patent Office, Report, 1851, pp. 439, 456; ibid., 1852, pp. 336, 340-41; ibid., 1854, p. 12; Anderson, "Agriculture in Illinois," p. 94; Pioneer, *Northern Iowa*, pp. 7-8; Porter, History of Harmon Township, *J. Ill. State Hist. Soc.* 10:609; Johnston, Sketches of the History of Stephenson County, *Trans. Ill. State Hist. Soc.*, 1923, p. 317; Wistar, *Autobiography*, pp. 348-50; McCoy, *Historic Sketches*, pp. 235-36.

28. *Prairie Farmer*, Feb. 1844, p. 35, Jan. 1845, p. 18, Apr. 1848, p. 127, Mar. 1854, p. 92, Sept. 18, 1856, Jan. 28, 1858, Feb. 10, 1859, p. 82; Patent Office, Report, 1850, pp. 198-99; ibid., 1853, p. 336.

29. *Prairie Farmer*, Jan. 1845, p. 31, Mar. 1845, p. 65, Apr. 1845, p. 94, Aug. 1845, p. 187, Nov. 1846, p. 337, Jan. 1847, p. 29, Apr. 1853, p. 154; *Rockford Register*, Sept. 19, 1857, p. 2, Cole Notes.

30. Oliver, *Eight Months in Illinois*, pp. 103-5; Patent Office, Report, 1843, p. 121; *Prairie Farmer*, Feb. 1843, pp. 45-46, July 1843, p. 152, Nov. 1843, p. 261, Feb. 1844, p. 35, Dec. 1844, p. 287, Jan. 1845, p. 18, Oct. 1845, pp. 238-39, Sept. 1846, p. 289, Jan. 1847, p. 29, Feb. 1849, p. 64, Feb. 1851, p. 71, Jan. 1852, pp. 9, 45, Feb. 1852, pp. 66-67, June 1852, p. 263, Mar. 5, 1857, Oct. 15, 1857, Feb. 10, 1859, p. 83, May 19, 1859, p. 313, July 7, 1859, p. 8, Aug. 25, 1859, p. 115, Sept. 15, 1859, p. 161; *Northwestern Farmer*, May 5, 1858, pp. 166-67; Harris, Autobiography, pp. 88-90.

31. Buckingham, Early Settlers, *J. Ill. State Hist.*

Soc. 35:238; ISAST, III:148; Beardstown Gazette, Nov. 26, 1851, Cole Notes; Springfield Daily Register, Aug. 15, 1851, Cole Notes.

32. Prairie Farmer, Oct. 1847, p. 305, Jan. 1848, p. 27, June 1845, p. 159.

33. Gem of the Prairie, Nov. 15, 1850, quoted in Anderson, "Agriculture in Illinois"; Kerrick, Life and Character of Isaac Funk, pp. 161-65; Prairie Farmer, May 1850, p. 166.

34. Patent Office, Report, 1854, pp. 13-14; Newcombe, Alson J. Streeter, pp. 421-22; Johnston, "Sketches of the History of Stephenson County, pp. 288-90, 314-15, 317; Eli Strawn to Joel Hopkins, Sept. 20, Oct. 17, 25, 1860, Hopkins Papers.

35. New York Tribune in Prairie Farmer, Apr. 17, 1856; Prairie Farmer, Apr. 3, May 22, 1856, May 7 to June 25, 1857; Bateman and Selby, Historical Encyclopedia, p. 957.

36. Prairie Farmer, Nov. 6, 1856.

37. Cavanagh, Funk of Funk's Grove, pp. 52-53, 68, 78, 84; Prairie Farmer, Nov. 1848, pp. 336-37, Dec. 1849, p. 367, Dec. 1850, pp. 381-82, Jan. 1853, p. 20; New York Tribune, Apr. 25, 1854, p. 8.

38. Cattleman 34:110; Prairie Farmer, Oct. 1847, p. 305, Nov. 1848, p. 336, Dec. 1853, p. 467, Jan. 1855, p. 36, July 1855, p. 203, Aug. 1855, p. 263, May 1856, Feb. 17, 1859, p. 100; Ottawa Free Trader, Feb. 22, 1851, p. 2, Cole Notes; Democratic Press, The Railroads, p. 66.

39. Prairie Farmer, Feb. 1843, p. 44, Oct. 1847, p. 304, Nov. 1848, p. 336, Jan. 1852, p. 36, July 1855, p. 207, July 25, 1857; Patent Office, Report, 1850, p. 356; ibid., 1851, p. 448; ibid., 1852, pp. 336, 340-41, 343; ibid., 1853, p. 6; ibid., 1854, pp. 12-14; Anderson, "Agriculture in Illinois," p. 94; Bateman and Selby, Historical Encyclopedia, p. 806; Quincy Whig, Oct. 31, 1855, p. 3, Cole Notes.

40. ISAST, I:434-35; Prairie Farmer, July 1851, p. 317.

41. Prairie Farmer, Apr. 7, 1859, p. 216.

42. Ibid., Sept. 1851, p. 381, Aug. 1853, p. 322, July 9, 1857, Nov.-Dec. 1857; for the packing season, see the Prairie Farmer market columns, 1845-60.

43. Eighth Census, II, p. cxxxiii.

44. Patent Office, Report, 1861, p. 258.

Chapter 4

1. Pierce, History of Chicago, vol. 1, pp. 44-51, 127-31.

2. Seventeenth Census, I, p. 46.

3. Lee, Transportation, Ill. State Hist. Soc. 10:31-33; Pierce, History of Chicago, vol. 1, pp. 76-79.

4. Chicago Board of Trade, Report, 1880, p. 34; Clemen, American Livestock and Meat Industry, pp. 92-105; Prairie Farmer, June 1848, p. 199, Jan. 1853, p. 38.

5. Pierce, History of Chicago, vol. 1, pp. 137-41; Andreas, History of Chicago, vol. 1, p. 556.

6. Prairie Farmer, Sept. 1846, p. 296, Oct. 1846, p. 328; Illinois State Register, Nov. 12, 1849, p. 2, Cole Notes.

7. Prairie Farmer, Nov. 1848, pp. 336-37. Packed beef was cured meat salted in barrels, the normal way of preserving beef until the 1870s when packers put beef in sealed tin cans, hence tinned beef, and shipped refrigerated fresh quartered beef as dressed beef.

8. Ibid., Dec. 1849, p. 367; Cavanagh, Funk of Funk's Grove, pp. 49, 62, 66-67.

9. Prairie Farmer, Dec. 1849, p. 367.

10. Ibid., Dec. 1850, pp. 381-82; Clemen, American Livestock and Meat Industry, pp. 105-9.

11. Pierce, History of Chicago, vol. 1, p. 118; Lee, Transportation, pp. 33-35, 37.

12. Yungmeyer, Excursion into Early History, J. Ill. State Hist. Soc. 38:10-19; Throne, Burlington and Missouri Railroad, Palimpsest 33:1-25; Pierce, History of Chicago, vol. 2, pp. 41-43, 50; Agnew, Rock Island Railroad, Iowa J. Hist. 52:203-22; Lee, Transportation, p. 38; Hopkins, Economic History, p. 33; Gates, Illinois Central Railroad, pp. 86-98.

13. Chicago Board of Trade, Report, 1901, pp. 41, 169.

14. Prairie Farmer, Jan. 31, 1856, p. 18.

15. Patent Office, Report, 1854, p. 12; New York Tribune, Apr. 28, 1854, Ann. Iowa 27:296; Prairie Farmer, Apr. 3, May 22, 1856.

16. Prairie Farmer, July 1855, p. 207; Harper, Railroad and the Prairie, Ill. State Hist. Soc. Trans. 1923, pp. 106-8.

17. Prairie Farmer, Aug. 15, 1863, p. 99, Feb. 20, 1869, p. 64.

18. Ibid., Feb. 1851, p. 56, Jan. 1853, p. 20; U.S. Commissioner of Agriculture, Report, 1863, pp. 617-18, 1865, p. 86, hereafter this series will be cited as USDA Report; Chicago Board of Trade, Report, 1858, p. 26, 1860, p. 35, 1880, p. 35. There is disagreement among statistical sources

as to cattle receipts, shipments, and numbers packed or butchered for city use before 1866. Figures differ because some sources report calendar years and others report packing season. Some sources even disagree internally; therefore, all figures before 1866 are only approximate guides to the magnitude of the cattle business. In most cases, Chicago Board of Trade figures are preferred here.

19. Chicago Board of Trade, Report, 1900, p. 45; USDA, Report, 1862, p. 326.

20. Pierce, _History of Chicago_, vol. 2, pp. 98-99; _Griffiths' Review_, p. 94. Plankington and Armour of Chicago and Milwaukee opened a slaughter plant in Kansas City in 1869.

21. _Griffiths' Review_, pp. 19-22; _Drovers' Journal_, June 28, 1939; Clemen, _American Livestock and Meat Industry_, pp. 83-85.

22. _Drovers' Journal_, June 28, 1939; Pierce, _History of Chicago_, vol. 2, pp. 92-94; ISAST, VI:314-24; _Prairie Farmer_, Jan. 6, 1866, pp. 1-2.

23. U.S. Federal Trade Commission, _Report on the Meat-Packing Industry_, pt. 3, pp. 194-95, hereafter cited as FTC, Report.

24. _Prairie Farmer_, Apr. 3, 1873, p. 109, Mar. 10, Mar. 17, June 2, 1877; Union Stock Yards and Transit Company, Annual Report, 1889, pp. 5-7, hereafter this series will be cited as Stock Yards, Report; _Western Rural_, Sept. 27, Nov. 22, 1879, Nov. 15, 1884, Dec. 20, 1890; _Prairie Farmer_, May 21, 1887, p. 329.

25. Stock Yards, Report, 1887, 1900.

26. Ibid., 1900, p. 31.

27. Pierce, _History of Chicago_, vol. 3, p. 204; Stock Yards, Report, 1900, pp. 67-69.

Chapter 5

1. Chandler, Beginning of "Big Business" in American Industry, _Bus. Hist. Rev_. 33:1-8.

2. Anderson, _Refrigeration in America_, pp. 24-26, 32-35, 45-61, 91-92, 123, 142.

3. Leech and Carroll, _Armour and His Times_, pp. 125-29; USDA Bureau of Animal Industry, Annual Report, 1884, pp. 265-66, hereafter this will be cited as BAI; Clemen, _American Livestock and Meat Industry_, p. 232; _Prairie Farmer_, Mar. 11, 1876, p. 83.

4. Swift and Van Vlissingen, *Yankee of the Yards*, pp. 6, 126-30, 178, 185-90; FTC, Report on Private Car Lines, pp. 27-31; Kujovich, Refrigerator Car and the Growth of the American Dressed Beef Industry, *Bus. Hist. Rev.* 44:460-82; Armour, in *Packers, Private Car Lines, and People*, pp. 19-26, claims his father, Philip Armour, saw the need for refrigerator cars first rather than Swift, but other sources do not support that assertion.

5. Swift and Van Vlissingen, *Yankee of the Yards*, pp. 27-30, 67-69, 130-32, 137, 205; FTC, Report on Meat Packing, pt. 1, p. 214.

6. *Prairie Farmer*, Nov. 4, 1882, p. 4; USDA, Report, 1876, p. 320; Nimmo, Report on the Internal Commerce of the United States, H. Exec. Doc. 7, pt. 3, pp. 157-58; Chicago Board of Trade, Report, 1886-1900; Pierce, *History of Chicago*, vol. 3, pp. 110-11; BAI, Report, 1884, p. 267; *Prairie Farmer*, Oct. 6, 1883, p. 640.

7. Clemen, *American Livestock and Meat Industry*, pp. 228-31, 237.

8. *Prairie Farmer*, Dec. 9, 1882, p. 9.

9. Ibid., Apr. 14, 1883, p. 228; Aldrich, Wholesale Prices, Wages, and Transportation, S. Rept. 1394, pt. 1, p. 525.

10. *Prairie Farmer*, Aug. 14, 1883, p. 228; Aldrich, Wholesale Prices, Wages, and Transportation, pt. 1, pp. 525-26; Kujovich, Refrigerator Car and the Growth of the American Dressed Beef Industry, pp. 460-82.

11. *Drovers' Journal*, Oct. 7, 1882, p. 4; Swift, *Yankee of the Yards*, pp. 67-69; *Prairie Farmer*, Dec. 9, 1882, p. 9; *Western Rural*, Sept. 18, 1886, p. 604; *Butchers' Advocate*, Dec. 21, 1890, p. 11, Jan. 28, 1891, p. 3; Clemen, *American Livestock and Meat Industry*, pp. 243-44; U.S. Congress, H. Exec. Doc. 267, p. 176; Kujovich, Refrigerator Car and the Growth of the American Dressed Beef Industry, pp. 466-68.

12. *Swift* v. *Sutphin*; *In Re Barber*; *Minnesota* v. *Barber*; Clemen, *American Livestock and Meat Industry*, pp. 245-51; *Butchers' Advocate*, Nov. 21, 1894, p. 9.

13. Leech and Carroll, *Armour and His Times*, pp. 51-52; Tenth Census, III, p. 1109; Pierce, *History of Chicago*, vol. 3, pp. 115-18; BAI, Report, 1884, pp. 262-64; U.S. Congress, H. Exec. Doc. 267, pp. 174-76.

14. Nimmo, Report on the Internal Commerce of the United States, pp. 286-88.

15. Zimmerman, Live Cattle Export Trade, *Agr. Hist.*

36:46-47; Griffiths' Review, 1876, p. 5.

16. Morris, Heritage from My Father, p. 209; Nimmo, Report on the Internal Commerce of the United States, pp. 266-67; Zimmerman, Live Cattle Export Trade, p. 47; North British Agriculturalist quoted in Prairie Farmer, Nov. 2, 1878, p. 349.

17. Prairie Farmer, Feb. 9, 1878, p. 45, June 15, 1878, p. 189, Aug. 3, 1878, p. 246, Oct. 26, 1878, p. 341, Nov. 2, 1878, p. 349, Apr. 26, 1879, p. 133.

18. Ibid., Oct. 16, 1880, p. 333; Chicago Board of Trade, Report, 1900, p. xxxvii; contrary to the impression in Zimmerman, exports of live beef cattle to England did not decline greatly after 1885. See Perrin, North American Beef and Cattle Trade, Econ. Hist. Rev. 24:431-32.

19. Western Farm Journal, Apr. 5, 1879, pp. 1-2; Iowa Homestead, Jan. 10, 1890, p. 3, Apr. 4, 1890, p. 3, Apr. 11, 1890, p. 5, Apr. 18, 1890, p. 1, May 8, 1891, p. 436; Prairie Farmer, Nov. 25, 1876, p. 380, citing Dundee Advertiser.

20. Prairie Farmer, Nov. 2, 1878, p. 349, Feb. 9, 1878, p. 48, July 27, 1878, p. 237; Griffiths' Review, 1877, pp. 5, 120.

21. Prairie Farmer, Oct. 15, 1880, p. 333; Chicago Board of Trade, Report, 1900, pp. xxxvii, 161; USDA, Yearbook, 1894, p. 10, 1906, p. 256; for an early British view of the dressed beef trade, see MacDonald, Food from the Far West, pp. 253-61, 277, 284-89; Perrin, North American Beef and Cattle Trade, pp. 434-44; Chicago Tribune, June 17, 1964, sec. 3, p. 7 under the heading "Beef from U.S. Bids for British Market," carried an article about an attempt of the American Meat Institute to sell U.S. packaged beef in England by opening a display of 7,000 pounds of beef from Cedar Rapids, Iowa, at the Smithfield market in London. The promoters hoped to develop a "selective trade" in American beef over the long term if favorable Atlantic freight rates could be arranged and if the British bought.

22. FTC, Report, pt. 2, pp. 13-17.

23. U.S. Congress, Report of the Select Committee on the Transportation and Sale of Meat Products, S. Rept. 829, pt. 1, pp. 1-20; Iowa Homestead, Oct. 30, 1891, p. 1007.

24. East, "Distillers' and Cattle Feeders' Trust," J. Ill. State Hist. Soc. 45:104-5; Peoria Board of Trade, Annual Report, 1880-1884; Gressley, Bankers and Cattlemen, p. 66; "Nelson Morris," Dictionary of American Biography, 13:217-18.

25. FTC, Report, Summary, pp. 33-35.

26. Ibid., pt. 2, pp. 43; U.S. Commissioner of Corporations, Report on the Beef Industry, H. Doc. 382, p. 7; IDAT, 1898, p. 300.

Chapter 6

1. Dale, Range Cattle Industry, tells the story of the growth, overexpansion, collapse, and reorganization of the range industry; Gressley, Bankers and Cattlemen, concerns the 1880s and 1890s; see also Prairie Farmer, July 14, 1883, p. 436, warning of overexpansion and speculation in range cattle.

2. Iowa City Republican, quoted in Ross, Iowa Agriculture, p. 72; Ross, Iowa Agriculture, pp. 71-72; Anderson, "Agriculture in Illinois," p. 267; Prairie Farmer, Nov. 28, 1861, p. 353, Jan. 4, 1863, p. 1, Apr. 16, 1864, p. 268, May 6, 1865, p. 356, June 24, 1865, p. 505, Oct. 7, 1865, p. 268, Dec. 16, 1865, p. 431, Apr. 27, 1872, p. 133.

3. Prairie Farmer, Dec. 12, 1863, p. 386, May 6, 1865, p. 356, Dec. 15, 1865, p. 431.

4. Prime, Model Farms, pp. 430-34.

5. MacDonald, Food from the Far West, pp. 143-44; Gates, Cattle Kings, Miss. Valley Hist. Rev. 35:395.

6. Prairie Farmer, Feb. 27, 1870, p. 64, Feb. 17, 1872, p. 56, Oct. 23, 1875, p. 344; MacDonald, Food from the Far West, pp. 120-22; Prime, Model Farms, pp. 282-84, 501, 503-4.

7. Prairie Farmer, Dec. 6, 1862, p. 353; MacDonald, Food from the Far West, pp. 140-42.

8. Bogue, Prairie to Corn Belt, p. 99; MacDonald, Food from the Far West, pp. 120-22; Prairie Farmer, Feb. 3, 1877, p. 33, Dec. 22, 1877, p. 408, Nov. 2, 1878, p. 349, Jan. 16, 1886, p. 366, Dec. 30, 1882, pp. 1-2; Johnson, E. S. Ellsworth, Ann. Iowa 35:24; Hopkins, Economic History, p. 224 n175.

9. Hopkins, Economic History, p. 181.

10. Iowa Homestead, Sept. 18, 1891, p. 862; Prairie Farmer, Feb. 15, 1878, p. 53.

11. Anderson, "Agriculture in Illinois," pp. 273-75; Baldwin, Driving Cattle from Texas to Iowa, Ann. Iowa 14:243-62.

12. ISAST, VII:411, 426; Anderson, "Agriculture in Illinois," pp. 275-76.

13. ISAST, VII:135-36; Prairie Farmer, Aug. 7, 1869, p. 249.

14. Illinois State Journal, Feb. 23, 1870, p. 3, Cole Notes; Prairie Farmer, Nov. 26, 1870, p. 369.

15. See McCoy, Historic Sketches, for the story of Abilene, McCoy's comments about Texas fever, and his battle against the state restrictions on Texas cattle. Dykstra, Cattle Towns, pp. 17-30, is the proper antidote to McCoy's view of his actions.

16. Prairie Farmer, Feb. 23, 1867, p. 114, Apr. 13, 1867, p. 234; USDA, Report, 1868, pp. 38-40; Prairie Farmer, Aug. 1, 1868, p. 36, Aug. 8, 1868, p. 44, Aug. 15, 1868, p. 50.

17. McCoy, Historic Sketches, pp. 221-29.

18. ISAST, VII:134-57; Iowa State Agricultural Society Report, 1872, passim, hereafter this series will be cited as Iowa SASR.

19. Report of the convention in ISAST, VII:173-273.

20. Bierer, Short History of Veterinary Medicine, p. 54.

21. Ibid., pp. 49-55, 60; USDA, Yearbook, 1956, pp. 2-4, 269; by 1940 the cattle tick had been eliminated from all of the United States through a federally enforced dipping program, except for a narrow strip of land in Texas bordering Mexico.

22. McCoy, Historic Sketches, pp. 222, 225.

23. Prairie Farmer, Mar. 2, 1867, p. 129; Illinois Revised Statutes, 1874, pp. 141-44; McCoy, Historic Sketches, p. 254.

24. Prairie Farmer, Aug. 17, 1872, p. 270, Aug. 24, 1872, p. 268, Jan. 11, 1873, p. 12, Sept. 30, 1876, p. 316, Sept. 26, 1885, p. 626; Illinois Board of Live Stock Commissioners, Annual Report, 1890-1900, passim.

25. Prairie Farmer, Aug. 15, 1879, p. 261; ISAST, VIII:252-53; Breeders' Gazette, May 13, 1886, p. 697; Prairie Farmer, July 27, 1872, p. 236, Feb. 25, 1888, Dec. 31, 1887, p. 851; Gates, Cattle Kings, p. 398.

26. Prairie Farmer, May 15, 1886, p. 310.

27. Ibid., Jan. 6, 1877, p. 8.

28. Ibid., July 25, 1868, p. 26.

29. USDA, Report, 1870, pp. 350-51; IDAT, XIV:105.

30. Prairie Farmer, Feb. 10, 1872, p. 48.

31. USDA, Report, 1876, p. 319.

32. Prairie Farmer, Dec. 1, 1877, p. 381; IDAT, XVII: 463, XVIII:470-71.

33. Prairie Farmer, Oct. 4, 1884, p. 628; Tenth Census, III, pp. 978, 991, 1000, 1005, 1008, 1011; Drovers' Journal, Year Book, 1922, p. 36; Dale, Range Cattle Industry, passim.

34. Address by Governor John Hamilton of Illinois at the Fat Stock Show, Nov. 1883, in IDAT, XXI:106-7.

35. Iowa Register in Prairie Farmer, Oct. 4, 1884, p. 628; Iowa Homestead, Mar. 27, 1891, p. 289, Sept. 18, 1891, p. 860.

36. Cavanagh, Funk of Funk's Grove, pp. 49, 66-67; see Chapter 3, pp. 23-24.

37. Loan ledgers in the Hopkins Papers show that a considerable amount of money was being loaned to neighboring farmers in the 1880s and 1890s.

38. McCoy, Historic Sketches, p. 242; Bogue, Pioneer Farmers and Innovation, Iowa J. Hist. 56:26.

39. Prairie Farmer, Sept. 19, 1896, p. 3.

40. Hopkins, Economic History, pp. 33, 40, 143-52, 225, n182, n183; Prairie Farmer, Feb. 1, 1890, p. 65; Stock Yards, Report, 1898, p. 10, 1900, pp. 68-71; USDA, Yearbook, 1918, pp. 101-8.

41. U.S., Statutes at Large, 17:584-85; Hopkins, Economic History, p. 171.

42. Pierce, History of Chicago, vol. 3, pp. 137-39; Hopkins, Economic History, pp. 171, 230-32; Prairie Farmer, Oct. 26, 1872, p. 337; Aldrich, Wholesale Prices, Wages, and Transportation, pt. 1, pp. 525-27; U.S. Congress, H. Doc. 855, vol. 144, pp. 128-33; ibid., S. Rept. 829, pp. 1-4, 18-20; Grodinsky, Iowa Pool, passim.

43. Prairie Farmer, market columns, 1880s, 1890s; Clemen, American Livestock and Meat Industry, pp. 635-39.

44. Prairie Farmer, Jan. 27, 1900; Iowa Homestead, Dec. 18, 1891, p. 1173.

45. Illinois State Board of Agriculture, Statistical Report, 1883-1900, passim; Murray, Shipping the Fat Cattle, Palimpsest 28:85-93.

46. Prairie Farmer, Oct. 27, 1866, p. 273, Nov. 17, 1866, p. 321.

47. Western Rural, Feb. 23, 1884, p. 124, Mar. 1, 1884, p. 136, Nov. 15, 1884, p. 732; Prairie Farmer, Nov. 3, 1866, p. 289; Sales receipts of R. Waugh, Commission Dealers, 1897-1900, Hopkins Papers.

48. Clemen, American Livestock and Meat Industry, pp. 485-89.

49. Prairie Farmer, Nov. 18, 1865, p. 373, Mar, 9, 1867, p. 157, Apr. 23, 1870, p. 128; MacDonald, Food from

the *Far West*, pp. 273, 277, 279; *Griffiths' Review*, 1876, pp. 21-23.

50. *Prairie Farmer*, Jan. 17, 1874, p. 24.

51. Ibid., Sept. 20, 1884, p. 605.

52. USDA, Report, 1883, pp. 281-82; *Prairie Farmer*, July 31, 1886, p. 495; BAI, Report, 1887-88, pp. 363, 410-13; USDA, *Yearbook*, 1894, p. 12.

53. *Iowa Homestead*, Jan. 9, 1891, p. 25; *Prairie Farmer*, Jan. 25, 1896, p. 2, Feb. 15, 1896, p. 1, Apr. 4, 1896, p. 7, July 4, 1896, p. 6.

54. Ledgers and commission house receipts, 1897-1900, Hopkins Papers.

55. *Field and Farm*, Oct. 6, p. 4, Oct. 13, p. 9, Oct. 27, p. 4, Nov. 24, p. 9, Dec. 15, 1888, p. 9; *Butchers' Advocate*, Apr. 8, 1891, p. 3; Drovers' Journal, *Year Book*, p. 36; Hopkins, Statistical Study, Iowa State Coll. Agr. Exp. Sta. Bull. 101, pp. 344-53, 359-61.

56. *Prairie Farmer*, Aug. 10, p. 256, Sept. 14, 1872, p. 296, Jan. 11, p. 13, Feb. 8, 1873, p. 48; *Field and Farm*, Jan. 7, p. 4, Jan. 28, 1888, p. 4; *Wallaces' Farmer*, Feb. 3, p. 85, Feb. 17, 1899, p. 125.

57. *Prairie Farmer*, Aug. 29, 1896, p. 6; Drovers' Journal, *Year Book*, 1922, p. 36; ibid., 1936, p. 10.

58. *Iowa Homestead*, Mar. 20, 1891, p. 265; *Prairie Farmer*, May 22, 1886, p. 326; Illinois Farmer's Institute, Annual Report, pp. 261-64; University of Illinois, Stock Feeding, Bull. 36, pp. 421, 426; BAI, Report, 1885, p. 376.

59. *Prairie Farmer*, May 18, 1888, p. 319; *Iowa Homestead*, Nov. 6, 1891, p. 1030; Illinois Farmer's Institute, Report, 1899, pp. 331-33, 337.

60. *Iowa Homestead*, Nov. 6, 1891, p. 1037.

61. U.S. Congress, S. Report 829, pp. 1-4, 6-7, 10-11.

62. Stock Yards, Report, 1903, pp. 7-9; *Prairie Farmer*, 1890-1900, passim.

63. Illinois Farmer's Institute, Report, 1897, pp. 223-24.

Chapter 7

1. This discussion is based on statistics taken from the U.S. Census 1850-1900 for the number of cattle on farms, and from Hopkins, *Economic History*, pp. 71-80. In 1900, "other" cattle was taken to be all but dairy cows two years

old or over. The author is aware of faults in this system and of other methods of indicating relative concentrations of cattle, as well as some problems in the accuracy of the printed census. Although attempts have been made to have more accurate figures for cattle in the census years for the nation and the states, any attempt to look at counties is dependent on the "uncorrected" census.

2. Prairie Farmer, July, 1855, p. 207.

3. Hawk in "Iowa Farming Types" delineated Iowa types of farming by "other" cattle per farm and came to some different conclusions. His 1880 map included a high-density area of beef cattle in the northwest which centered on Palo Alto and Pocahontas counties, which does not appear on my maps. He had an area of concentration in eastern Iowa that corresponded with my beef cattle concentrations and Bogue's corn and swine concentrations. Hawk's study ended with 1880. Bogue kindly supplied me with maps of some of his data on leading counties for corn, swine, and milk cows in Illinois and Iowa from 1850 to 1890. Bogue measured corn per improved acre and swine and milk cows per farm on a county basis so that comparisons are not exact. He found a closer relationship between leading corn and swine producing counties than there appears to be between my leading cattle counties and his leading corn and swine counties. For 1880 and 1890 there was a closer relationship between leading cattle and hog counties than between leading cattle and corn counties. There was a close correspondence between heavy numbers of milk cows and beef cattle densities in 1860 in northeastern urban Illinois, and there was some correspondence between the two in 1860 in northwestern Illinois, but not in northeastern Illinois (heavy number of milk cows, light number of beef cattle). Otherwise, in regard to milk cow-beef cattle densities, there were no remarkable correspondences.

There were a number of ways to map these statistics. Bogue argued that beef cattle per farm unit was better than per 100 acres. The former would indicate immediate intensity of endeavor perhaps, but is open to certain distortions too. One could also map beef cattle per improved acre, but that would produce maps showing densities of cattle in grazing counties where total cattle numbers might still be less than in counties where corn was the major feed. The decision to use beef cattle per hundred acres was arbitrary and its limits are recognized. It is offered as an approximate indication of cattle feeding areas. The delineation of these areas, in general, seems to be supported by information from other sources, though the delineation does not necessarily indicate the real relative magnitudes of cattle feeding in the counties. All these calculations are dependent upon census figures which are undergoing "correction" by scholars, although so far only for state totals and not the county totals necessary here.

4. Prairie Farmer, Apr. 19, 1847, May 29, 1869, p. 169, June 12, 1875, p. 192, Feb. 23, 1878, p. 61, May 11, 1878, p. 152, Apr. 7, 1883, p. 217, May 17, 1884, p. 307; ISAST, VIII:209; IDAT, XI:109, XIV:125, 174; Anderson, "Agriculture in Illinois," pp. 290-91; Prime, Model Farms, p. 277.

5. University of Illinois, Stock Feeding, Bull. 36, p. 1; Newlove, "Economic History of Illinois Agriculture," pp. 81-82; Bogue, Swamp Land Act, Agr. Hist. 25:169-80; Prairie Farmer, Jan. 15, 1869, p. 17, June 4, 1870, p. 172; Smith, "Historical Geography of Champaign County, Illinois," pp. 112-13, 129-30.

6. This would seem true for east-central Illinois where, by 1900, most of the ten counties with the highest percentage of tenants (54% to 62%) were located. Only one of those counties ranked as high as fifth in the nine levels of cattle density on the 1900 map. The evidence for Iowa is less conclusive because of the ten counties with the highest percentage of tenants (45% to 52%); three each were ranked at the sixth, seventh, and eighth levels of cattle density. One county ranked at the fifth level. However, even the county with the highest percentage of tenancy in Iowa ranked below the eleventh highest Illinois county so that the level of tenancy in Iowa was lower than in Illinois in 1900. Twelfth Census, V, pp. 1, 73-75, 79-81.

7. Gates, Cattle Kings, Miss. Valley Hist. Rev. 35: 399-406; idem, Frontier Landlords and Pioneer Tenants, J. Ill. State Hist. Soc. 38:178-89, 204; Scully Papers.

8. Allerton, Systematic Farming, passim.

9. Sanders, James N. Brown, Trans. Ill. State Hist. Soc., 1912, pp. 172-75; Prairie Farmer, Sept. 3, 1864, p. 146, June 30, 1877, p. 205, Sept. 4, 1880, p. 285; ISAST, I:429-36, II:372-78.

10. McClelland, Jacob Strawn, J. Ill. State Hist. Soc. 34:201-5.

11. Ibid., pp. 182-95.

12. Rammelkamp, Memoirs of John Henry, J. Ill. State Hist. Soc. 18:70.

13. Bateman and Selby, Historical Encyclopedia, p. 976.
14. Ibid., p. 777.
15. Ibid., pp. 793-94.
16. History of Morgan County, pp. 537-66.
17. Little, "Early Days of Henry County," in Little Papers; Bogue, Prairie to Corn Belt, pp. 193-215, 237-38; Hawk, "Iowa Farming Types," pp. 92-98.
18. Wallaces' Farmer, Sept. 1, 1899, p. 701; Prairie Farmer, Feb. 28, 1880, p. 69; Anderson, "Agriculture in Illinois," pp. 287-89; Prime, Model Farms, p. 262; Prairie Farmer, Dec. 6, 1873, p. 392, Nov. 15, 1872, p. 368; IDAT, XIII:193.
19. Prairie Farmer, Aug. 10, 1872, p. 256, Sept. 14, 1872, p. 296, Jan. 11, 1873, p. 13; Wallaces' Farmer, Sept. 1, 1899, p. 701.
20. Brinkman, "Historical Geography of Improved Cattle," Appendix, maps 14, 24, 25.
21. See Maps 7-12; Bogue, Prairie to Corn Belt, pp. 12, 220-23.
22. Iowa SASR, 1875, p. 438; Hopkins, Economic History, p. 107; Prairie Farmer, Feb. 3, 1877, p. 33; Bogue, Prairie to Corn Belt, pp. 229-30.
23. Bushnell, Iowa Resources and Industries, p. 59.
24. Prairie Farmer, May 1848, p. 164, Apr. 12, 1878, p. 120; USDA, Report, 1873, p. 392.
25. Prairie Farmer, Mar. 28, 1874, p. 104.
26. Iowa Homestead, Jan. 30, 1891, p. 101.
27. Bogue, Prairie to Corn Belt, p. 96.
28. Ibid., pp. 97-98.
29. Bogue, Pioneer Farmers, Iowa J. Hist. 55:24-25.
30. Prime, Model Farms, pp. 430-34.
31. Prairie Farmer, Nov. 1854, pp. 427-30.
32. Harris, Autobiography, Trans. Ill. State Hist. Soc., 1923, pp. 90-91; Illinois State Register, Mar. 24, 1853, p. 1; Urbana Union, Nov. 10, 1853, p. 2, Cole Notes.
33. USDA, Report, 1866, p. 319.
34. McCoy, Historic Sketches, p. 243.
35. Prairie Farmer, Dec. 23, 1876, p. 413.
36. National Live Stock Association, Proceedings, p. 177.
37. Conway, Cattle Handbook; Wallaces' Farmer, Mar. 3, 1899, p. 175.
38. Prime, Model Farms, p. 47.
39. Wentworth, Observations, Agr. Hist. 27:41, 43.
40. Ibid., p. 42.

Chapter 8

1. Bardolph, Agricultural Literature, passim; Ayer & Son's, American Newspaper Annual, 1880, 1891, 1900; Gates, The Farmer's Age, pp. 338-58; Ogilvie, Pioneer Agricultural Journalists, pp. 36-37.

2. Ross, Iowa Agriculture, pp. 35-36, 88-91; Lord, Wallaces of Iowa, pp. 83, 92, 120, 130-31.

3. Prairie Farmer, Jan. 1845, p. 20, Sept. 1850, p. 265, Nov. 1850, p. 349.

4. Hamilton (Iowa) Freeman, July 15, 1858, quoted in Bogue, Prairie to Corn Belt, p. 196; Prairie Farmer, July 1847, p. 201.

5. Prairie Farmer, Dec. 5, 1861, p. 374, Mar. 10, 1877, p. 73, Sept. 1, 1877, p. 273, July 17, 1880, p. 229, Oct. 4, 1884, p. 628.

6. Ibid., Sept. 1851, p. 421, May 1852, pp. 243-44, July 1852, p. 339, May 1853, pp. 176-77, June 1853, pp. 223-34, July 1853, pp. 264, 279, Jan. 10, 1856; Iowa Homestead, Aug. 13, 1869, p. 2; Drovers' Journal, Sept. 12, 1882, p. 2.

7. Prairie Farmer, June 6, 1861, p. 368, Nov. 29, 1862, p. 345, Jan. 26, 1867, p. 56, Oct. 3, 1868, p. 105, July 31, 1869, p. 241, Oct. 7, 1871, p. 313, July 15, 1876, p. 229, Dec. 8, 1877, p. 389, Dec. 15, 1877, p. 397, Dec. 29, 1877, p. 413, Dec. 8, 1883, p. 780, Oct. 4, 1884, p. 628, and passim; National Livestock Journal, Mar. 1885, pp. 110-11.

8. Iowa Homestead, Wallaces' Farmer, Prairie Farmer, passim; Western Agriculturalist, June 1886, p. 4; Prairie Farmer, May 19, 1859, p. 307, Jan. 1845, p. 18.

9. Prairie Farmer, Jan. 1845, p. 18, June 1847, pp. 183, 215, Nov. 1846, p. 337, Dec. 1850, p. 368, Nov. 17, 1883, p. 729; Northwestern Farmer, Feb. 1858, p. 52.

10. Northwestern Farmer, Jan. 1856, p. 9.

11. Prairie Farmer, Aug. 1851, p. 355.

12. Ibid., Dec. 1853, p. 447.

13. Ibid., Oct. 1855, p. 303.

14. Ibid., Aug. 1851, p. 359.

15. Ibid., Apr. 18, 1874, p. 125.

16. For two examples, see ibid., Feb. 15, 1862, p. 100, Apr. 13, 1867, p. 235.

17. Ibid., Dec. 10, 1864.

18. Ibid., Nov. 17, 1866, p. 314, July 4, 1868, p. 2, Feb. 25, 1871, p. 64, Sept. 1, 1877, p. 273.

19. Ibid., Jan. 24, 1862, p. 51, Oct. 29, 1870, p. 340, Feb. 5, 1870, p. 33, Jan. 15, 1887, p. 36; *Field and Farm*, Jan. 15, 1887, p. 5; *Western Agriculturalist*, June 1886, p. 11.

20. See for example: *Prairie Farmer*, Aug. 3, 1872, p. 245, Oct. 17, 1874, p. 333, Jan. 2, 1875, p. 5, July 3, 1875, p. 213, Apr. 14, 1877, p. 117, Sept. 22, 1883, p. 596, Nov. 17, 1883, p. 721, Apr. 19, 1884, p. 244, Sept. 6, 1884, p. 562, Mar. 21, 1885, p. 180; *Western Rural*, Oct. 30, 1886, p. 707; *Iowa Homestead*, Jan. 1, 1890, p. 2, Jan. 31, 1890, p. 4, Feb. 2, 1890, p. 4, Feb. 28, 1890, p. 2; *Wallaces' Farmer*, Jan. 1, 1899, pp. 24, 25.

21. *Prairie Farmer*, Feb. 15, 1884, pp. 100-101, July 26, 1884, p. 468, Aug. 9, 1884, p. 500.

22. *Field and Farm*, Feb. 12, 1887, p. 4, and passim, 1887-1888.

23. *Prairie Farmer*, Jan. 1846, pp. 16-17.

24. Ibid., Jan. 1849, p. 17.

25. Ayer and Son's, *American Newspaper Annual*, 1880, p. 421, 1891, pp. 1182-83, 1900, p. 1356; Bardolph, *Agricultural Literature*, p. 170; circulation figures were guarded items among agricultural papers because they contested among themselves as to which had the most readers.

26. *Prairie Farmer*, Sept. 12, 1863, p. 161.

27. ISAST, I:21; Iowa SASR, 1874, pp. 485 ff. for the founding meetings.

28. ISAST, passim, 1854-1900; Iowa SASR, passim, 1854-1900.

29. Bardolph, *Agricultural Literature*, p. 71.

30. ISAST, II:vii-ix; Iowa SASR, 1857, pp. 448-50; Beinhauer, County, District, and State Agricultural Societies, *Ann. Iowa* 20:50-55; *Prairie Farmer*, Feb. 2, 1860, p. 68; Ross, *Iowa Agriculture*, pp. 84-86; IDAT, XXXVIII:297.

31. *Prairie Farmer*, Nov. 1850, pp. 330, 334.

32. Ibid., Jan. 28, 1858, p. 85; *Northwestern Farmer*, Apr. 1858, pp. 130-31; *Iowa Farmers' Advocate*, Dec. 1847, p. 85 quoted in Throne, "Book Farming" in Iowa, *Iowa J. Hist.* 49:118; Hildreth, *Life and Times*, pp. 180-83; practically any issue of the *Prairie Farmer* in September, October, November, and December up to 1875 will have some reports on fairs in Illinois and Iowa. The state agricultural society reports also carried county summaries, which frequently told of the fairs. After the middle 1870s, the Illinois Transactions contained only a statement of receipts and expenses for

the county societies, but the Iowa Reports continued to have information about local fairs and conditions.

33. *Iowa Homestead*, Feb. 26, 1869, p. 60.

34. *Prairie Farmer*, Dec. 5, 1874, p. 389, Dec. 12, 1874, p. 394, Feb. 24, 1877, p. 60.

35. Ibid., Sept. 19, 1868, p. 89.

36. IDAT, 1878-1900, passim.

37. Ibid., XVI:111.

38. Ibid., pp. 67, 79-82.

39. *Prairie Farmer*, Dec. 14, 1878, p. 397.

40. IDAT, XVII:95, 142-45, 147.

41. *Prairie Farmer*, Mar. 27, 1880, p. 101; IDAT, XVIII:77.

42. IDAT, XVIII:71, XIX:93.

43. Ibid., XIX:5.

44. *Prairie Farmer*, Mar. 24, 1883, p. 180; Illinois Department of Agriculture, *Report on the 8th Fat Stock Show*, pp. 250-53.

45. *Prairie Farmer*, Mar. 24, 1883, p. 180.

46. Ibid., Dec. 8, 1883, p. 785.

47. Ibid., Nov. 22, 1884, p. 737.

48. IDAT, XIX:205-6.

49. Ibid., XXIII:8-9.

50. Ibid., pp. 27-28.

51. *Prairie Farmer*, Nov. 20, 1886, p. 761.

52. *Field and Farm*, Jan. 22, 1887, p. 1.

53. IDAT, XXIV:152, XXV:6.

54. Ibid., XXV:147, 216, XXVI:6; *Prairie Farmer*, Dec. 29, 1888, p. 846.

55. IDAT, XXXV:33-35, 76-77.

56. Union Stock Yards and Transit Company, Review, pp. 10-16.

57. IDAT, XXIII:132, 135-39, 146-51.

58. Ibid., pp. 290, 304.

59. Ibid., XXVII:160. See Van Stuyvenberg, *Margarine*, or Hayter, *Troubled Farmer*, ch. 4.

60. Ritzman, Baby Beef, USDA Circ. 105, pp. 3, 31-32.

61. ISAST, II:317-18.

62. *Prairie Farmer*, Oct. 1853, p. 394; see also ibid., Oct. 1843, p. 217, Nov. 1843, p. 245, Dec. 1851, pp. 561-64, Feb. 1852, p. 96, Aug. 1853, p. 229, Nov. 1853, pp. 424-25, Mar. 1854, pp. 89-90, Aug. 1854, p. 288, Jan. 1855, p. 32, Nov. 1855, p. 350, Oct. 17, 1863, p. 241; ISAST, I:84, 108, 133, II:xiii.

63. USDA, Report, 1873, p. 392.

64. _Prairie Farmer_, Oct. 4, 1873, p. 317.

65. Brinkman, "Historical Geography of Improved Cattle," pp. 106-12, 126-28, 183-204, appendix, pp. 72, 74-78; _Prairie Farmer_, Jan. 15, 1857.

66. Brinkman, "Historical Geography of Improved Cattle," appendix, maps 1, 8.

67. ISAST, V:372.

68. USDA, Report, 1873, p. 392.

69. ISAST, III:301-4.

70. Bogue, _Prairie to Corn Belt_, p. 88; Ross, _Iowa Agriculture_, pp. 45-46; _Northwestern Farmer_, Nov. 1857, p. 420.

71. ISAST, II:xiii, 64-67, III:177, IV:123-28, 254-59, 274, 277, 280, 285, 292, 300, V:111-14, 263-65, 290-92, VI: 26-28, 50, 114-16.

72. Brinkman, "Historical Geography of Improved Cattle," appendix, pp. 91-92, 95, 97-98, 100-105, 107-8, 112-14, maps 14, 15, 26.

73. Prime, _Model Farms_, pp. 154, 280.

74. _Iowa Homestead_, Aug. 13, 1869, p. 2.

75. IDAT, XX:462-67; _Prairie Farmer_, Dec. 2, 1882, p. 4, Feb. 24, 1883, p. 116, May 26, 1883, p. 324.

76. _Prairie Farmer_, May 26, 1883, p. 324.

77. Bogue, _Prairie to Corn Belt_, pp. 90-91.

78. Beinhauer, Early Iowa Agricultural Organizations, _Ann. Iowa_ 34:422-24; _Prairie Farmer_, Jan. 12, 1884, p. 20; _Iowa Homestead_, Dec. 11, 1891, p. 1153.

79. Illinois Farmer's Institute, Report, 1896, p. 176; IDAT, XXXVI:282-83.

80. Bogue, _Prairie to Corn Belt_, pp. 90-91; Eleventh Census, VI, pp. 283-85.

81. _Prairie Farmer_, Apr. 23, 1870, p. 128; _Iowa Homestead_, Jan. 14, 1870, in Hopkins, _Economic History_, p. 212 n62.

82. IDAT, IX:203-4, XV:202, XIX:225, XX:426-29; _Prairie Farmer_, Mar. 28, 1874, p. 104, Feb. 19, 1876, p. 64, Apr. 8, 1876, p. 120, May 12, 1877, p. 152, Nov. 10, 1883, p. 713; Prime, _Model Farms_, pp. 45-47, 154, 429-30, 444-45, 453-54; IDAT, XVII to XXIII, printed a list of all the county fairs and tabulated for each fair the number and kind of purebred stock exhibited.

83. _Prairie Farmer_, Feb. 19, 1876, p. 64.

84. Ibid., Jan. 8, 1876, p. 16.

85. Ibid., May 21, 1881, p. 165.

86. Iowa Homestead, July 9, 1869, p. 210, Apr. 4, 1890, p. 5.

87. Prairie Farmer, May 6, 1876, p. 149, Aug. 27, 1881, p. 277.

88. Prime, Model Farms, pp. 444-45; and for general comments urging improvement see ibid., pp. 154, 280; Prairie Farmer, Apr. 27, 1872, p. 133, May 6, 1876, p. 149, and passim to 1900; IDAT, XVII:202, XX:426-29; Bushnell, Iowa Resources, pp. 59-61.

89. USDA, Report, 1868, pp. 515-17, 1869, p. 542; Hopkins, Economic History, p. 72; Prairie Farmer, May 3, 1879, p. 141.

90 USDA, Report, 1868, pp. 515-16; Prairie Farmer, Dec. 11, 1875, p. 397; Wallaces' Farmer, Sept. 29, 1899, p. 786.

91. Prairie Farmer, May 3, 1879, p. 141, and passim 1880-1890.

92. IDAT, XVI:98.

93. MacDonald, Food from the Far West, pp. 146-48.

94. IDAT, XVII:134.

95. Illinois State Board of Agriculture, Report of the American Fat Stock and Dairy Show, pp. 236-39, 248-53. This report had comparative tables for all eight shows from 1878 to 1885.

96. Ibid., p. 263.

97. Prairie Farmer, Nov. 20, 1880, p. 393, Dec. 4, 1880, p. 389; National Livestock Journal, Mar. 1885, pp. 110-11, Apr. 1885, pp. 155-56, Dec. 1885, p. 500.

98. Breeders' Gazette, Mar. 18, 1886, p. 375.

99. Iowa Homestead, Jan. 9, 1891, p. 25, July 31, 1891, p. 706; Prairie Farmer, Apr. 19, 1890, p. 244.

100. Prairie Farmer, Oct. 4, 1890, p. 637.

101. Twelfth Census, V, pt. 1, p. cliii; Wallaces' Farmer, Jan. 6, 1899, p. 13; Clay Robinson & Co., Report, Jan. 3, 1901, p. 3.

Chapter 9

1. Patent Office, Report, 1841-1861; USDA, Report, 1862-1900; Houck, Bureau of Animal Industry, pp. xi-xii.

2. Patent Office, Report, 1844, pp. 155-56, 387-89, 1012.

3. Ibid., 1861, p. 446.

4. USDA, Report, 1873, pp. 290-91.

5. Ibid., 1880, pp. 90-97.

6. Curtiss, Essentials of Beef Production, USDA Bull. 71, p. 23.

7. BAI, Report, 1884, pp. 245-69, 1885, pp. 362-76, 1887-1888, pp. 39-44, 410-13; Ritzman, Baby Beef, BAI Circ. 105, passim; BAI Circ. 106 contained a list of all bureau publications from 1884 through 1906.

8. Bierer, _Short History of Veterinary Medicine_, pp. 79-85; ISAST, IV:55-57, VI:xi-xv; _Prairie Farmer_, 1860-1861, 1866, passim.

9. Bierer, _Short History of Veterinary Medicine_, pp. 36-45, 79-85; Smithcors, _American Veterinary Profession_, pp. 394-95; _Prairie Farmer_, Mar. 22, 1879, p. 93, Nov. 24, 1883, pp. 737, 740, Jan. 5, 1884, p. 4, Oct. 25, 1884, p. 676, May 9, 1885, p. 292, Oct. 2, 1886, p. 646, June 11, 1887, p. 377; _National Livestock Journal_, Oct. 1885, p. 424.

10. _Field and Farm_, July 16, 1887, p. 4, Mar. 29, 1888, p. 4, Oct. 16, 1888, p. 4, Dec. 8, 1888, p. 8, and passim 1887, 1888; Chew, _Response of Government to Agriculture_, pp. 30-31.

11. USDA, _Yearbook_, 1956, pp. 263-64.

12. Ibid., 1894, pp. 67-80; Houck, _Bureau of Animal Industry_, pp. 25, 256-59; Smithcors, _American Veterinary Profession_, p. 471.

13. There were agricultural experiment stations operating without federal aid in states like Connecticut, Ohio, New York, and Wisconsin before the Hatch Act, but not in Illinois and Iowa. True, History of Agricultural Experimentation, USDA Misc. Publ. 251, pp. 67-164.

14. _Northwestern Farmer_, 1858, p. 5.

15. _Prairie Farmer_, Feb. 12, 1870, p. 41, speech to a farmers' convention, Madison County, Ill.

16. _National Livestock Journal_, pp. 220-22.

17. _Prairie Farmer_, Oct. 17, 1874, p. 333, July 3, 1875, p. 213; True, History of Agricultural Experimentation, pp. 80-82, 107-110; _National Livestock Journal_, Jan. 1885, pp. 20-21.

18. Ross, _Land-Grant Idea_, pp. 71, 90-101; Solberg, _University of Illinois 1867-1894_, pp. 344-48.

19. True and Clark, Agricultural Experiment Stations, USDA, Bull. 80, pp. 202-6, 532-34; University of Illinois, Agr. Exp. Sta. Bulletins, Circulars, and Reports, 1888-1900, especially, Report, 1901, p. 5; ibid., Bulletins 78, 83.

20. True and Clark, Agricultural Experiment Stations, pp. 214, 218-21, 537-40; Iowa Agricultural Experiment Station, Bulletin, 1888-1900.

21. *Butchers' Advocate*, Oct. 24, 1894, pp. 11-12.

22. Hamilton, History of Farmers' Institutes, USDA, Bull. 174, pp. 28-30; *Prairie Farmer*, Mar. 5, 1870, p. 65, Jan. 7, 1871, p. 4, Jan. 18, 1873, p. 20, Dec. 5, 1874, p. 389, Jan. 26, 1878, p. 28, Jan. 18, 1879, p. 20, Jan. 24, 1880, p. 28.

23. Hamilton, History of Farmers' Institutes, pp. 28-30; *Prairie Farmer*, Jan. 27, 1883, p. 56, Feb. 9, 1884, p. 89, Apr. 12, 1884, p. 227; IDAT, XX:426-29, 511-17.

24. Illinois Farmers' Institute, Report, 1895-1900, passim; ibid., 1896, pp. 264, 474.

25. *Prairie Farmer*, Feb. 11, 1871, p. 42, Jan. 13, 1872, p. 12, Jan. 20, 1872, p. 17; Iowa SASR, 1898-1900; Hamilton, History of Farmers' Institutes, pp. 36-38.

Chapter 10

1. *Prairie Farmer*, May 15, 1863, p. 313, June 13, 1863, p. 377, Aug. 29, 1863, p. 130, Apr. 30, 1864, p. 309, Sept. 3, 1864, p. 146, May 6, 1865, p. 352, Jan. 16, 1869, p. 17, Nov. 13, 1869, p. 369; USDA, Report, 1866, pp. 317-18; Throne, Southern Iowa Agriculture, *Iowa J. Hist*. 50:212.

2. *Prairie Farmer*, Oct. 17, 1861, p. 263, Dec. 12, 1861, p. 386, Nov. 1, 1862, p. 275, Oct. 10, 1863, p. 229, Apr. 15, 1865, p. 280, Nov. 17, 1866, p. 314, Jan. 15, 1869, p. 17, May 15, 1869, p. 153; USDA, Report, 1862, pp. 332-33; 1866, pp. 317-18.

3. *Iowa Homestead*, July 30, 1869, p. 4; *Prairie Farmer*, Nov. 29, 1860, p. 341, May 15, 1869, p. 153, May 29, 1869, p. 169, June 5, 1869, p. 177; Iowa SASR, 1868, pp. 216-22.

4. Iowa SASR, 1868, pp. 216-22; *Prairie Farmer*, Jan. 4, 1862, p. 2; *Iowa Homestead*, Aug. 13, 1869, p. 2.

5. *Prairie Farmer*, May 19, 1866, p. 337, Feb. 14, 1864, p. 98, Jan. 28, 1865, p. 51; USDA, Report, 1866, pp. 317-18; Hopkins, *Economic History*, pp. 115-16.

6. *Prairie Farmer*, Mar. 11, 1871, p. 80, May 19, 1877, p. 160.

7. Gates, Cattle Kings, *Miss. Valley Hist. Rev*. 35: 389-405; Hopkins, *Economic History*, pp. 87-95, 103, 128;

Bogue, Pioneer Farmers, Iowa J. Hist. 55:24; Prairie Farmer, Apr. 20, 1878, p. 128, Apr. 27, 1878, p. 136; Hopkins, Passing of the Herds, Palimpsest 11:204-5.

8. Prairie Farmer, Jan. 15, 1869, p. 17; Throne, Southern Iowa Agriculture, p. 212; Hopkins, Economic History, pp. 103, 128.

9. Prairie Farmer, Apr. 30, 1864, p. 309, Apr. 22, 1865, p. 304, May 5, 1865, p. 352, Apr. 29, 1871, p. 136, Sept. 6, 1873, p. 288, Sept. 13, 1873, p. 296; Bogue, Swamp Land Act, Agr. Hist. 25:176-77.

10. Prairie Farmer, Aug. 3, 1872, p. 245, June 30, 1877, p. 205; MacDonald, Food from the Far West, pp. 140-42; ISAST, XVI:95.

11. Prairie Farmer, Jan. 7, 1871, p. 8, Feb. 7, 1874, p. 48, Jan. 8, 1876, p. 16.

12. Illinois Industrial University, Fifth Annual Circular, pp. 220-22; Prairie Farmer, Mar. 23, 1872, p. 89.

13. Prairie Farmer, Aug. 10, 1872, p. 256, Sept. 14, 1872, p. 296, Jan. 11, 1873, p. 13, Feb. 8, 1873, p. 48, Apr. 5, 1873, p. 112, Sept. 1, 1877, p. 273; Chicago Board of Trade, Report, 1901, p. 12.

14. Prairie Farmer, Jan. 4, 1873, p. 8, Mar. 1, 1873, p. 72.

15. MacDonald, Food from the Far West, pp. 120-26, 137-39; Prairie Farmer, Feb. 27, 1876, p. 72.

16. MacDonald, Food from the Far West, pp. 123, 137-39; Western Farm Journal, May 3, 1878, pp. 274-75; Prairie Farmer, Feb. 25, 1871, p. 64, Feb. 1, 1873, p. 40, Feb. 8, 1873, p. 48, Mar. 24, 1873, p. 104, June 14, 1873, p. 189, Mar. 14, 1874, p. 88, Mar. 28, 1874, p. 104, Dec. 5, 1874, p. 392, Jan. 24, 1875, p. 24, Mar. 13, 1875, p. 88, Mar. 20, 1875, p. 96, Mar. 27, 1875, p. 104, Apr. 3, 1875, p. 112, June 5, 1875, p. 184, Nov. 10, 1877, p. 357, Dec. 1, 1877, p. 381, Mar. 27, 1878, p. 93.

17. Prime, Model Farms, pp. 470-72; see also Hopkins, Economic History, pp. 94, 107-8; Field and Farm, June 18, 1887, p. 8; Iowa Register, in Prairie Farmer, Oct. 4, 1884, p. 628.

18. Prime, Model Farms, pp. 17, 27-28, 36, 49-52, 63-64, 82-83, 85-86, 89-90, 92-93, 123, 262-66, 269-71, 282-84.

19. Ibid., 427-30, 443-45, 453-54, 501, 503-4, 519, 484-85.

20. Prairie Farmer, Aug. 14, 1880, p. 261; IDAT, XXI: 185-87.

21. *Prairie Farmer*, Nov. 10, 1883, p. 708.

22. IDAT, XXI:185-86; shorts were a by-product of milling wheat and contained germ, fine bran, and some flour.

23. *Prairie Farmer*, Aug. 15, 1883, p. 516.

24. Ibid., Nov. 14, 1885, p. 744; *National Livestock Journal*, Dec. 1885, p. 501.

25. *Iowa Homestead*, Feb. 13, 1891, p. 149, Feb. 6, 1891, pp. 122, 124, Mar. 13, 1891, p. 241, July 24, 1891, p. 689, July 31, 1891, p. 706, Dec. 11, 1891, p. 1151, Dec. 18, 1891, p. 1174; *Prairie Farmer*, Aug. 2, 1890, p. 484; BAI, Special Report, H. Misc. Doc. 106, pp. 456-61; Hopkins, *Economic History*, pp. 106-18; *Wallaces' Farmer*, Jan. 13, 1899, pp. 24-25, Jan. 20, 1899, p. 42, Feb. 10, 1899, p. 106, July 21, 1899, p. 595.

26. Henry, *Feeds and Feeding*, pp. 382-93.

27. Ibid., pp. 130-31, 152, 196, 204, 207, 225, 257-58, 367-69, 381-84; University of Illinois, Agr. Exp. Sta. Bull. 36, pp. 20, 421-31.

28. Henry, *Feeds and Feeding*, pp. 357-58, 367, 393.

29. *Prairie Farmer*, Jan. 4, 1896, p. 7, Jan. 8, 1898, p. 3, Feb. 19, 1898, p. 1, Mar. 19, 1898, p. 3, Jan. 13, 1900, p. 1; Mumford, *Beef Production*, pp. 57-66; Hopkins, *Economic History*, pp. 129-31.

30. Clay Robinson and Co., *Live Stock Report*, p. 3.

Chapter 11

1. U.S. Census of Agriculture, 1959, General Report, II:975, 985.

BIBLIOGRAPHY

Manuscripts

Flagg Papers, Illinois Historical Survey of the University of Illinois, Urbana.

Hopkins Papers, Illinois State Historical Society, Springfield.

Little Papers, Illinois Historical Survey of the University of Illinois, Urbana.

Scully Papers, Illinois Historical Survey of the University of Illinois, Urbana.

Williams-Woodbury Papers, Illinois Historical Survey of the University of Illinois, Urbana.

Federal and State Documents

Aldrich, Nelson W. Wholesale Prices, Wages, and Transportation. (52nd Cong. 2nd sess., S. Rept. 1394.) Washington, 1893.

Census of Iowa for 1880 . . . with Other Historical and Statistical Data. Des Moines, 1883.

Curtiss, Charles F. Some Essentials of Beef Production. (USDA Farmer's Bulletin 71.) Washington, 1898.

Hamilton, John. History of Farmers' Institutes in the United States. (USDA, Office of the Experiment Stations Bulletin 174.) Washington, 1906.

Illinois Board of Live Stock Commissioners. Annual Report, 1886-1900. Springfield, 1887-1901.

Illinois Department of Agriculture. Report on the 8th Fat Stock Show, 1885. Springfield, 1886.

Illinois Department of Agriculture. Transactions, 1871-1900. Springfield, 1872-1901.

Illinois Farmer's Institute. Annual Report, 1895-1900. Springfield, 1896-1901.

Illinois Revised Statutes, 1874.
Illinois State Agricultural Society. Transactions, 1853-1870. Springfield, 1854-1871.
Illinois State Board of Agriculture. Report of the American Fat Stock and Dairy Show, 1885. Springfield, 1886.
————. Statistical Report for December, 1883-1900. Springfield, 1883-1900.
In Re Barber. 39 Federal Reporter 641.
Iowa State Agricultural Society. Reports, 1853-1899. Des Moines, 1854-1900.
Minnesota v. Barber. 136 U.S. Reports 313.
Nimmo, Joseph, Jr. Report on the Internal Commerce of the United States. (48th Cong. 2nd sess., H. Exec. Doc. 7, pt. 3.) Washington, 1885.
Ritzman, Ernest G. Baby Beef. (USDA, Bureau of Animal Industry Circular 105.) Washington, 1899.
Royce, Charles C., comp. Indian Land Cessions in the United States. (56th Cong. 1st sess., H. Doc. 736, pt. 2.) Washington, 1899.
Swift v. Sutphin. 39 Federal Reporter 630.
Thompson, James Westfall. A History of Livestock Raising in the United States, 1607-1860. (USDA Agricultural History Series 5.) Washington, 1942.
True, Alfred C. A History of Agricultural Experimentation and Research in the United States, 1607-1925. (USDA Miscellaneous Publication 251.) Washington, 1937.
True, Alfred C., and Clark, V. A. The Agricultural Experiment Stations in the United States. (USDA, Office of the Experiment Stations Bulletin 80.) Washington, 1900.
U.S. Census Office. Seventh Census of the United States. Compendium. Washington, 1854.
————. Eighth Census of the United States. II, Agriculture. Washington, 1864.
————. Ninth Census of the United States. III, Industry and Wealth. Washington, 1872.
————. Tenth Census of the United States. III, Agriculture. Washington, 1883.
————. Eleventh Census of the United States. VI, Agriculture. Washington, 1895.
————. Twelfth Census of the United States. I, Population; V, Agriculture; VI, Agriculture; Abstract; Statistical Atlas. Washington, 1901-1903.
————. Thirteenth Census of the United States. V, Agriculture. Washington, 1913.

U.S. Census Office. Seventeenth Census of the United States. I, Population. Washington, 1952.

———. Historical Statistics of the United States, Colonial Times to 1957. Washington, 1960.

U.S. Commissioner of Agriculture. Reports, 1862-1888. Washington, 1863-1889.

U.S. Commissioner of Corporations. Report on the Beef Industry. (58th Cong. 3rd sess., H. Doc. 382.) Washington, 1905.

U.S. Commissioner of the General Land Office. Report, 1838, 1843. Washington, 1838, 1843.

U.S. Commissioner of the Patent Office. Report, II, Agriculture. Washington, 1841-1861.

U.S. Congress. House Executive Document 267. (48th Cong. 2nd sess.) Washington, 1885.

———. House Document 855. (64th Cong. 1st sess.) Washington, 1916.

———. Senate. Report of the Select Committee on the Transportation and Sale of Meat Products. (51st Cong. 1st sess., S. Rept. 829.) Washington, 1890.

U.S. Department of Agriculture, Bureau of Animal Industry. Annual Report, 1884-1900. Washington, 1885-1901.

———. Special Report on Diseases of Cattle and on Cattle Feeding. (52nd Cong. 2nd sess., H. Misc. Doc. 106.) Washington, 1893.

USDA. _Yearbook of Agriculture_, 1940, Washington, 1940.

U.S. Federal Trade Commission. Report of the Federal Trade Commission on Private Car Lines. Washington, 1920.

———. Report on the Meat-Packing Industry, 6 pts. and summary. Washington, 1919-1920.

U.S. Interstate Commerce Commission. Thirteenth Annual Report on the Statistics of Railways in the United States. Washington, 1901.

U.S. Secretary of Agriculture. Report, 1889-1900. Washington, 1889-1900.

U.S. Statutes at Large, II.

Newspapers and Agricultural Periodicals

Butchers' Advocate

Breeders' Gazette

Cattleman

Drovers' Journal

Field and Farm
Iowa Homestead
National Livestock Journal
Northwestern Farmer
Prairie Farmer
Wallaces' Farmer
Western Agriculturalist
Western Farm Journal
Western Rural
"Cole Notes" in the Illinois Historical Survey of the University of Illinois. Thirty-one boxes of newspaper excerpts from 1820 to 1919. Indexed topically.

Primary Works

Allerton, Samuel W. *On Systematic Farming*. Chicago, 1907.
Armour, J. Ogden. *The Packers, the Private Car Lines, and the People*. Philadelphia, 1906.
Ayer, N. W., and Son's. *American Newspaper Annual*, 1880, 1891, 1900. Philadelphia, 1881, 1892, 1901.
Burlend, Rebecca; and Burlend, Edward. *A True Picture of Emigration*, ed. Milo Quaife. New York, 1968.
Bushnell, Horace. *Work and Play: Or Literary Varieties*. New York, 1871.
Bushnell, J. P. *Iowa Resources and Industries: Her Agricultural, Horticultural, Stock-Raising, Dairying, Commercial, Manufacturing and Mining Interests, etc*. Des Moines, 1885.
Chicago Board of Trade. Annual Report, 1858-1901. Chicago, 1859-1902.
Clay Robinson and Company. Live Stock Report. Jan. 3, 1901.
Democratic Press. *The Railroads, History and Commerce of Chicago*, 2nd ed. Chicago, 1854.
Drovers' Journal. *Year Book of Figures of the Live Stock Trade for the Year 1922*, Chicago, 1923.
Flint, Timothy. *A Condensed Geography and History of the Western States, or the Mississippi Valley*, 2 vols. Cincinnati, 1828.
Griffiths' Fifth Annual Review of the Live Stock Trade of Chicago for 1869. Chicago, 1870.
Hildreth, Azro B. F. *The Life and Times of Azro B. F. Hildreth*, 4 pts. Des Moines, 1891.
Henry, W. A. *Feeds and Feeding: A Hand-Book for the Student and Stockman*. Madison, 1898.

Illinois Industrial University. Fifth Annual Circular, 1871-72. Urbana, 1872.
Iowa State College of Agriculture and Mechanic Arts, Agricultural Experiment Station. Bulletin, 1888-1900. Ames, 1888-1900.
Lea, Albert M. Notes on Wisconsin Territory, with a Map. Philadelphia, 1836.
McCoy, Joseph G. Historic Sketches of the Cattle Trade of the West and Southwest, ed. Ralph P. Bieber. Glendale, 1940.
MacDonald, James. Food from the Far West or, American Agriculture with Special Reference to the Beef Production and Importation of Lean Meat From America to Great Britain. London, 1878.
Mumford, Herbert. Beef Production. Urbana, 1907.
National Live Stock Association. Proceedings of the Fifth Annual Convention. Chicago, 1901.
Oliver, William. Eight Months in Illinois: With Information to Immigrants. Chicago, 1924.
Peoria Board of Trade. Annual Report, 1880-1884.
A Pioneer. Northern Iowa, by a Pioneer: Containing Information for Immigrants. Dubuque, 1858.
[Ponting, Tom C.] Life of Tom Candy Ponting: An Autobiography, Introduction and notes by Herbert O. Bayer. Evanston, Illinois, 1952.
Prime, Samuel T. K., ed. The Model Farms and Their Methods. Chicago, 1881.
Schoolcraft, Henry R. Travels in the Central Portions of the Mississippi Valley. New York, 1825.
Shirreff, Patrick. A Tour through North America; Together with a Comprehensive View of the Canadas and United States, as Adapted for Agricultural Emigration. Edinburgh, 1835.
Union Stock Yards and Transit Company. Annual Live Stock Report, 1868-1901. Chicago, 1869-1902.
———. Review of the First International Live Stock Exposition, 1900. Chicago, 1901.
University of Illinois, Agricultural Experiment Station. Annual Report. Urbana, 1888-1910.
———. Bulletin 9, 36, 46. Urbana, 1890, 1894, 1897.
———. Circular. Urbana, 1897-1902.
Wistar, Isaac Jones. Autobiography of Isaac Jones Wistar 1827-1905. New York, 1937.

Secondary Works

Anderson, Oscar E., Jr. Refrigeration in America: A History of a New Technology and Its Impact. Cincinnati, 1953.

Andreas, A. T. History of Chicago, 3 vols. Chicago, 1884.

Angle, Paul M., comp. and ed. Prairie State: Impressions of Illinois, 1673-1967, by Travelers and Other Observers. Chicago, 1968.

Bardolph, Richard C. Agricultural Literature and the Early Illinois Farmer. (University of Illinois Studies in the Social Sciences, 29.) Urbana, 1948.

Bateman, Newton; and Selby, Paul, ed. Historical Encyclopedia of Illinois and History of Morgan County. Chicago, 1906.

Bierer, Bert W. A Short History of Veterinary Medicine in America. East Lansing, 1955.

Bogue, Allan G. From Prairie to Corn Belt: Farming on the Illinois and Iowa Prairies in the Nineteenth Century. Chicago, 1963.

Cavanagh, Helen M. Funk of Funk's Grove: Farmer, Legislator and Cattle King of the Old Northwest 1797-1865. Bloomington, Illinois, 1952.

Chew, Arthur P. The Response of Government to Agriculture. Washington, 1937.

Clark, John G. The Grain Trade in the Old Northwest. Urbana, 1966.

Clemen, Rudolf A. The American Livestock and Meat Industry. New York, 1923.

Cole, Arthur C. The Era of the Civil War 1848-1870. (Centennial History of Illinois, vol. 3.) Springfield, 1919.

Conway, H. M. Cattle Handbook for the Grower and Feeder. Chicago, 1935.

Dale, Edward E. The Range Cattle Industry: Ranching on the Great Plains from 1865 to 1925. Norman, 1960.

Danhof, Clarence H. Change in Agriculture: The Northern United States, 1820-1870. Cambridge, 1969.

Dykstra, Robert R. The Cattle Towns. New York, 1968.

English, William Hayden. Conquest of the Country Northwest of the River Ohio 1778-1783 and the Life of Gen. George Rogers Clark, vol. 1. Indianapolis, 1896.

Fenneman, Nevin M. Physiography of the Eastern United States. New York, 1938.

Garland, John H., ed. The North American Midwest: A Regional Geography. New York, 1955.

Gates, Paul W. The Farmer's Age: Agriculture, 1815-1860. New York, 1960.

Gates, Paul W. The Illinois Central Railroad and Its Colonization Work. (Harvard Economic Studies, 42.) Cambridge, 1934.
Gressley, Gene M. Bankers and Cattlemen. New York, 1966.
Grodinsky, Julius. The Iowa Pool: A Study in Railroad Competition, 1870-1884. Chicago, 1950.
Hayter, Earl W. The Troubled Farmer 1850-1900. De Kalb, Illinois, 1968.
Henlein, Paul C. Cattle Kingdom in the Ohio Valley 1783-1860. Lexington, 1959.
History of Morgan County, Illinois: Its Past and Present, Containing a History of the County; Its Cities, Towns, etc. Chicago, 1878.
Hopkins, John A., Jr. Economic History of the Production of Beef Cattle in Iowa. Iowa City, 1928.
———. A Statistical Study of the Prices and Production of Beef Cattle. (Iowa State College of Agriculture and Mechanic Arts, Agricultural Experiment Station Research Bulletin 101.) Ames, 1926.
Houck, U. G. The Bureau of Animal Industry of the United States Department of Agriculture: Its Establishment, Achievements and Current Activities. Washington, 1924.
Kay, George F., et al. The Pleistocene Geology of Iowa. (Iowa Geological Survey, Special Report.) N.p., n.d., c/ 1944.
Lampard, Eric. The Rise of the Dairy Industry in Wisconsin: A Study in Agricultural Change 1820-1920. Madison, 1963.
Leech, Harper; and Carroll, John Charles. Armour and His Times. New York, 1938.
Lokken, Roscoe L. Iowa Public Land Disposal. Iowa City, 1942.
Lord, Russell. The Wallaces of Iowa. Boston, 1944.
Morris, Ira Nelson. Heritage from My Father: An Autobiography. New York, 1947.
Odell, R. T., et al. Soils of the North Central Region of the United States: Their Characteristics, Classification, Distribution, and Related Management Problems. (University of Wisconsin, Agricultural Experiment Station Bulletin 601.) Madison, 1960.
Ogilvie, William E. Pioneer Agricultural Journalists. Chicago, 1927.
Pierce, Bessie Louise. A History of Chicago, 3 vols. Chicago, 1937-1957.
Pooley, William Vipond. The Settlement of Illinois from 1830 to 1850. (University of Wisconsin Bulletin 220.) Madison, 1908.

Power, Richard L. Planting Corn Belt Culture: The Impress of the Upland Southerner and Yankee in the Old Northwest. Indianapolis, 1953.

Putnam, James W. The Illinois and Michigan Canal: A Study in Economic History. Chicago, 1918.

Rogin, Leo. The Introduction of Farm Machinery in Its Relation to the Productivity of Labor in the Agriculture of the United States during the Nineteenth Century. Berkeley, 1931.

Ross, Earle D. Iowa Agriculture: An Historical Survey. Iowa City, 1951.

———. The Land-Grant Idea at Iowa State College. Ames, 1958.

Ross, R. C.; and Case, H. C. M. Types of Farming in Illinois: An Analysis of Difference by Areas. (University of Illinois, Agricultural Experiment Station Bulletin 601.) Urbana, 1956.

Shannon, Fred A. The Farmer's Last Frontier: Agriculture, 1860-1897. New York, 1945.

Smithcors, J. F. The American Veterinary Profession: Its Background and Development. Ames, 1963.

Solberg, Winton U. The University of Illinois 1867-1894: An Intellectual and Cultural History. Urbana, 1968.

Swift, Louis F., with Arthur Van Vlissingen, Jr. The Yankee of the Yards: The Biography of Gustavus Franklin Swift. Chicago, 1927.

Taylor, George Rodgers; and Neu, Irene. The American Railroad Network, 1861-1890. Cambridge, 1956.

Van Stuyvenberg, J. H., ed. Margarine: An Economic, Social and Scientific History 1869-1969. Toronto, 1969.

Wallace, Henry A.; and Bressman, Earl N. Corn and Corn Growing, 5th ed., rev. New York, 1949.

Articles

Agnew, Dwight L. The Rock Island Railroad in Iowa, Iowa Journal of History 52(July 1954):203-22.

Baldwin, W. W., ed. Driving Cattle from Texas to Iowa, 1866, Annals of Iowa 14(April 1924):243-62.

Beinhauer, Myrtle. The County, District, and State Agricultural Societies of Iowa, Annals of Iowa 20(July 1935): 50-69.

———. Some Early Iowa Agricultural Organizations, Annals of Iowa 34(October 1958):422-33.

Bogue, Allan G. Pioneer Farmers and Innovation, Iowa Journal of History 55(January 1958):1-36.

Bogue, Margaret B. The Swamp Land Act and Wet Land Utilization in Illinois, 1850-1890, Agricultural History 25(October 1951):169-80.

Buck, Solon J. Pioneer Letters of Gershom Flagg, Transactions of the Illinois State Historical Society, 1910, pp. 139-83.

Buckingham, Clyde E. Early Settlers of the Rock River Valley, Journal of the Illinois State Historical Society 35(September 1942):236-59.

Carman, Harry J. English Views of Middle Western Agriculture, 1850-1870, Agricultural History 8(January 1934):3-19.

Chandler, Alfred D., Jr. The Beginnings of "Big Business" in American Industry, Business History Review 33(Spring 1959):1-31.

Dunbar, Gary S. Colonial Carolina Cowpens, Agricultural History 35(July 1961):125-31.

East, Ernest E. The Distillers' and Cattle Feeders' Trust, 1887-1895, Journal of the Illinois State Historical Society 45(Summer 1952):101-23.

Gates, Paul W. Cattle Kings in the Prairies, Mississippi Valley Historical Review 35(December 1948):379-412.

———. Frontier Landlords and Pioneer Tenants, Journal of the Illinois State Historical Society 38(June 1945): 143-206.

Goodwin, Cardinal. The American Occupation of Iowa, 1833 to 1860, The Iowa Journal of History and Politics 17(January 1919):83-102.

Harper, C. A. The Railroad and the Prairie, Transactions of the Illinois State Historical Society, 1923, pp. 102-10.

Harris, Mary V., ed. The Autobiography of Benjamin Franklin Harris, Transactions of the Illinois State Historical Society, 1923, pp. 72-101.

Hopkins, John A., Jr. The Passing of the Herds, Palimpsest 11(July 1930):282-91.

Johnson, Curtis L. E. S. Ellsworth: Iowa Land Baron, Annals of Iowa 35(July 1959):1-35.

Johnston, William J. Sketches of the History of Stephenson County, Illinois, Transactions of the Illinois State Historical Society, 1923, pp. 217-320.

Kerrick, L. H. Life and Character of Honorable Isaac Funk, Transactions of the Illinois State Historical Society, 1902, pp. 159-70.

Kujovich, Mary Y. The Refrigeration Car and the Growth of

the American Dressed Beef Industry, Business History Review 44(Winter 1970):460-82.

Le Duc, Thomas. Public Policy, Private Investment, and Land Use in American Agriculture, 1825-1875, Agricultural History 37(January 1963):3-9.

Lee, Judson F. Transportation—A Factor in the Development of Northern Illinois Previous to 1860, Journal of the Illinois State Historical Society 10(April 1917):17-85.

McClelland, Clarence P. Jacob Strawn and John T. Alexander: Central Illinois Stockmen, Journal of the Illinois State Historical Society 34(June 1941):177-208.

Murray, Janette S. Shipping the Fat Cattle, Palimpsest 28 (March 1947):85-94.

Newcombe, Alfred W. Alson J. Streeter—An Agrarian Liberal, Journal of the Illinois State Historical Society 38(December 1945):414-45; 39(March 1946):68-95.

Perrin, Richard. The North American Beef and Cattle Trade with Great Britain 1870-1914, The Economic History Review, 2nd Ser., 24(August 1971):430-44.

Porter, John L. History of Harmon Township, Lee County, Illinois, Journal of the Illinois State Historical Society 10(January 1918):593-638.

Rammelkamp, C. H., ed. The Memoirs of John Henry: A Pioneer of Morgan County, Journal of the Illinois State Historical Society 18(April 1925):39-75.

Sanders, Alvin H. James N. Brown as a Breeder and Importer of Live Stock, Transactions of the Illinois State Historical Society, 1912, pp. 170-75.

Throne, Mildred. "Book Farming" in Iowa, 1840-1870, Iowa Journal of History 49(April 1951):117-42.

———. Southern Iowa Agriculture, 1865-1870, Iowa Journal of History 50(July 1952):212.

———. The Burlington and Missouri Railroad in Iowa, Palimpsest 33(January 1952):1-32.

———. A Population Study of an Iowa County in 1850, Iowa Journal of History 57(October 1959):317-19, 321.

Wentworth, Edward N. Some Observations on Constructive Agricultural Movements, Agricultural History 27(April 1953): 41-48.

Yungmeyer, D. W. An Excursion into the Early History of the Chicago and Alton Railroad, Journal of the Illinois State Historical Society 38(March 1945):7-37.

Zimmerman, William David. Live Cattle Export Trade between United States and Great Britain, 1868-1885, Agricultural History 36(January 1962):46-52.

Theses and Dissertations

Anderson, Russell H. "Agriculture in Illinois during the Civil War Period, 1850-1870." Ph.D. dissertation, University of Illinois, 1929.

Brinkman, Leonard. "Historical Geography of Improved Cattle in the United States to 1870." Ph.D. dissertation, University of Wisconsin, 1964.

Harding, Jennie L. "Economic and Social Conditions in Iowa to 1880 as Reflected in the Observations of Travelers." M.S. thesis, Iowa State College of Agriculture and Mechanic Arts, 1942.

Hawk, Duane C. "Iowa Farming Types: 1850-1880." M.A. thesis, State University of Iowa, 1957.

Massey, Robert E. "British Travelers' Impressions of Agriculture in Illinois, 1800-1860." M.A. thesis, University of Wisconsin, 1963.

Nelson, Peter. "A History of Agriculture in Illinois with Special Reference to Types of Farming." Ph.D. dissertation, University of Illinois, 1930.

Newlove, George H. "Economic History of Illinois Agriculture." Ph.D. dissertation, University of Illinois, 1917.

Smith, Marjorie C. "Historical Geography of Champaign County, Illinois," Ph.D. dissertation, University of Illinois, 1957.

S U B J E C T   I N D E X

[County Index follows Subject Index]

A

Abilene, Kans., 58-59
Agricultural experiment station. <u>See also</u> Illinois, Iowa, Wisconsin
  beginnings, 110
  reports, 120
Agricultural regions
  Illinois, 16
  Iowa, 15
Agricultural societies, 92-93
Albany, N.Y., 46
Alexander, John T., 19, 24-26, 58-60, 65, 78
Allerton, Samuel, 52, 78
Allerton stockyards (New York City), 22
Alton, Ill., 36
American Fat Stock Show, 90, 94-99, 103-4, 117, 119
American Hereford Cattle Breeders' Association, 101
American Short-Horn Breeders' Association, 101
Armour and Company, 42, 47, 51-52, 110
Arthur, A. D., 82

B

Baby beef, 68-69, 99, 104, 108
Baltimore
  beef consumption, 40
  cattle market, 18
Bardolph, Richard, 91-92
Barns. <u>See</u> Shelter
Barreled beef, 34, 37, 40, 123
Beef packing, 37-38, 40-41, 43. <u>See also</u> Armour and Company; Morris, Nelson; Swift and Company
<u>Beef Production</u>, 113
"Big Five," 51-53
Bloomington, Ill., 35-36
Bluegrass pasture, 116-17
Board of Trade
  Chicago, 35-36, 40, 97
  Elgin, Ill. (butter market), 98
Borden, Gail, 13
Boston, 34, 45-46, 53
<u>Breeders' Gazette</u>, 84, 86, 90-91, 105
Brighton, Mass., 34
British Order-in-Council (1879), 108-9
Brockway, E. T., 83
Brown, James N., 23, 26, 32, 71, 78
  and Sons, 94, 111
Bryant, Arthur, 93
Buel Institute, 93

Bull's Head Yards (Chicago), 41
Burch, John, 79
Bureau of Animal Industry, 49, 61, 71, 106
  disease control, 108-10
  publications, 107-8
Burlington, Iowa, 39
Bushnell, Horace, 7
Butchers' Advocate, 49, 113
Butchers' National Protective Association, 49
Butchers, object to dressed beef, 47-49
Butterine, 97-98

C

Cairo, Ill., 58-59
Capital needs in cattle feeding, 23, 65, 83, 125
Capron, Horace, 12
Case, Morris, 115
Cash-grain crops, 9, 14, 16
Cattle. *See also* Baby beef
  age and weight, 26-27, 30-32, 38, 68-69, 94-95, 104-5, 118-19, 121
  areal distribution, 74-84, 124
  census numbers and value, 126
  contract sales to packers, 30
  cost of production, 95
  early maturity, 94-95, 104-5, 107
  English, 100
    exports, 49-50
    foreign, 57, 95, 125
    importations, 100
    improvement, 22-23, 58, 64, 94-95, 104-5, 107
  numbers, 8, 11-12, 14, 21, 74-77, 81-82
Chester, E. E., 71
Chicago
  Board of Trade, 35-36, 40, 97
  as cattle market, 20, 25, 30, 33, 35, 45, 56-59, 66-67, 79, 108, 126
  farm periodicals, 85
  growth and development, 35-44, 77
  machinery, 14
  meat packers, 24, 34, 53, 71-72
  railroads, 15, 57
Chicago Cattle Loan Company, 43
Chicago Tribune, 38
Cincinnati, Ohio, 36, 41
Civil War, 13, 22, 56
Clark, George Rogers, 7
Clarkson, Coker F., 56
Cleary, William, 79
Climate, 3, 6, 8
Colorado, 63
Commissioner of Agriculture, 56, 84
Commissioner of Patents, 20
Commission firms, 67-68, 73, 105, 121
Corn
  concentration, 15-16
  feeding, 18, 20, 28, 64, 73, 80-81, 86, 87, 115-18
  growing conditions, 3, 7-9
  production, 11-13, 55-56, 77
Cottonseed oil, 53
Council Bluffs, Iowa, 39, 57
Cudahay, Patrick, 51
Culbertson, C. M., 96
Cultivator, 25

Cultural impact on agriculture, 80
Curtiss, Charles F., 113

D

Dairy, 9, 13, 15-16, 97-98
Dakota, 57, 63
Deere, John, 14
Denver, Colo., 53
Department of Agriculture, 63, 106-7. See also Patent Office
Depression (1890s), 55, 57
Disease, cattle. See also Bureau of Animal Industry
  blackleg, 110
  control, 61-62, 108-10
  effect on foreign market, 50, 108-9
  "mad itch," 28-29
  pleuropneumonia, 50-51, 89, 108-10
  Texas or "tick" fever, 59-63, 108
Dole, George W., 36
Dressed beef, 34-35, 44-54, 123
  butcher objections, 46-47
  exports, 50-51
  foreign criticism, 110
  G. F. Swift, 46
  "monopoly," 52-53
  prices, 48
  railroad discrimination, 46-48
Drovers' Journal, 67, 86, 91
Drovers' National Bank, 43
Droving, 18-24, 30-31
Duffield, George, 58
Dyer, Thomas, 37

E

Eastman, Timothy, 50
East St. Louis, 16, 46, 53-54
Education, agricultural, 93, 110-13
Ellsworth, Eugene S., 57
English farmers, 80
English market, 49-51
Erie Canal, 7
Exports. See Cattle, Dressed beef

F

Fairs, agricultural, 23, 84, 88, 91-92, 97, 101
Farmers' clubs, 92-93
Farmers' institutes, 71, 102, 113-14
Farms
  number, 8, 11, 15
  size, 11
Fat Stock Show. See American Fat Stock Show
Federal Trade Commission, 52
Feeder cattle supply, 8, 19-22, 56-57
Feeding, cattle
  advice of press, 87
  agricultural experiment station, 112-13
  capital requirements, 24, 64-65, 83, 125
  contract, 33
  cooked feed, 25, 89, 107
  distillery mash, 53
  after 1865, 55-74
  before 1865, 18, 22, 24-29
  English feeds, 106
  experiments, 111

Feeding, cattle (<u>continued</u>)
methods, 24-25, 73, 88-89, 115-21
profits, 86-87
root crops, 25, 88-89
"stall-feeding," 23, 25, 79, 118
<u>Feeds and Feeding</u>, 113
<u>Field and Farm</u>, 109
Flagg, Gershom, 13
Flax, 9
Flint, Timothy, 10
Ft. Worth, Tex., 53-54
Funk, Absalom, 30
Funk, Isaac, 19, 21, 24, 30, 33, 38, 65, 71, 75, 79
Funk, LaFayette, 96

G

Gamgee, John, 61
Gillett, John D., 56, 78-79, 83, 94-95, 104, 119, 121
Great Plains, 55, 57-58, 123
Green, George F., 82
Groat, A. B., 72

H

Hammond, George, 45-46, 52
Hann, William G., 81
Harris, Benjamin F., 19, 20, 24, 27, 75, 79, 83
Hatch Act (1887), 110, 112
Henry, John, 79
Henry, W. A., 102, 113, 120-21
Hopkins, Joel W., 69-70
Hops, 9
Horticulture, 13, 16
Hostetter, C. L., 101
Hubbard, Gurdon Saltonstall, 37

I

Illinois
agricultural experiment station, 110-112
Breeding Association, 100
Cattle Breeders' Association, 72
Central Railroad, 13, 27, 39-40, 58
Department of Agriculture, 64, 67
Industrial University, 111-13, 119
Livestock Breeders' Association, 102
-Michigan Canal, 6, 12, 35-37
State Agricultural Society 60, 92, 108-9
State Board of Agriculture 92-95, 97, 113
State Fair, 12, 97, 101
State Railroad Commission, 43
Stock Importing Association 100
University of, 107
Immigrants, 79-80, 82
Immigration, 7
Indian Territory, 8, 57
International Livestock Exposition, 97
Iowa
Agricultural College, 111, 114
agricultural experiment station, 110-13
Improved Stock Breeders' Association, 102
Shorthorn Association, 101
State Agricultural Society 92, 114

State Board of Agriculture, 92
State College of Agriculture and Mechanic Arts, 111-12, 114
State Fair, 84
Iowa Homestead, 57, 69, 71, 101, 103, 109
founding and development, 86, 90, 93

J

John, Alexander, 81
Jungle, The, 110

K

Kansas, 55, 57, 62
Kansas City, 41, 43, 46, 53-54, 57, 126
as market, 56-57, 73
Kellogg, O. W., 30
Kentuckians, 79
Kentucky, cattle feeding, 18, 79
Kerrick, L. H., 71
Kilbourne, F. L., 61

L

Land
drainage, 77
in farms, 15
surveys, 8
values, 20, 77, 83
Lawrence, E. S., 111
Lea, Albert Miller, 7
Le Duc, William G., 109
Libby, McNeil, and Libby, 49
Liverpool, England, 51
Loan companies, cattle, 65
Logan, John A., Jr., 97

M

McCoy, Joseph, 22, 59, 60, 62, 83
Machinery, farm, 14-15
Mahanah, Dennis, 82
Market, cattle. *See also* Baltimore, Boston, Chicago, Kansas City, New Orleans, New York City, Omaha, Philadelphia
conditions, 30-33
grades, 68
numbers, 43
reports, 31, 67, 89-90
Mason, Charles, 21
Meat. *See also* Dressed beef
canned, 49
distribution, 47-48
inspection, 106, 110
packers. *See also* Armour and Company; Morris, Nelson; Swift and Company
as cattle feeders, 52
decentralization, 53-54
influence on rail rates, 48
price fixing, 52-53, 71-72
processing. *See* Barreled beef, Dressed beef, Pork packing
Michaels, P. E., 77
Miller, Mark, 86
Miller, T. L., 96
Mills, Oliver, 81, 99-100
Minnesota, 55, 63
Missouri, 57-59, 62
Montana, 57
Morrill Act (1862), 110-11
Morris, Nelson, 47, 50, 52-53, 109

Morrow, George, 108
Mumford, Herbert W., 113
Myrick, Willard, 41

N

National Cattle Growers' Association, 48
National Livestock Bank, 43
National Livestock Journal, 90, 105, 111, 119
Nebraska, 55, 57, 62
New Genesee Farmer, 25
New Orleans, 11, 19, 30, 79
New York City, 34, 53, 66, 79
  beef consumption, 40
  markets, 18, 21, 31
New York Tribune, 20, 31
Nofsinger and Company, 46

O

Oakwood, J. H., 84
Oats, 15-16
Ohio
  cattle feeding, 18-19, 33, 55
  Dairymen's Protective Association, 98
  Stock Raising Company, 101
Oklahoma City, 53
Oleomargarine, 97-98
Omaha, 43, 46, 53-54, 57, 126
Ottumwa, Iowa, 39

P

Patent Office
  advice on feeding, 25
  agricultural reports, 106-7
Periodicals, agricultural, 52, 85-91, 99, 101. *See also Breeders' Gazette*, *Iowa Homestead*, *Prairie Farmer*
  abuse of Bureau of Animal Industry, 109
Philadelphia, 18, 27, 40, 46
Ponting, Tom C., 22, 24, 119
Pool. *See* Price fixing
Population, 7-8, 11, 15-17, 122
Pork packing, 36, 41
Prairie, 3-5, 7-8, 10-11, 19, 21, 116
Prairie Farmer
  cattle raising, 23, 32, 56-57, 65, 94-96, 102-4, 111
  commission firms, 67-68
  cooked food, 25
  correspondents, 20, 40
  dressed beef, 38, 40, 46-48, 51
  founding and development, 85-90
  live trade, 50
  market reports, 72, 117
  shelter, 28-29
Price fixing, 52-53
Prices
  agricultural, 125
  beef, investigation, 71-72
  cattle, 30-33, 42-43, 55, 62-64, 70-72, 115
  corn, 20
  meat, 52
Prince, Frederick, 42
Profits, cattle, 27-28, 32-33, 78, 86-87, 126
Purebreds, 78, 81-83, 91-94, 100-102

Angus, 71, 91, 97, 104
capital for, 25, 99
Devon, 94, 100
Hereford, 64, 80, 91, 94, 96-97, 101, 104
impact of, 23, 102-5
Shorthorn, 23, 26, 32, 56, 64, 80, 91, 94, 96-97, 99, 100-104, 107, 117

Q

Quarantine, cattle, 61-62

R

Railroads, 12, 14-15, 35-41, 48, 57
damage to live cattle, 30, 66
dressed beef, 47-48
impact on cattle feeding, 25, 39-40, 75
national market, 122-23
rates, 66
refrigerator cars, 45-46, 53
Ranching, 55, 57-58
Range
competition, 63-64, 99
feeding cattle, 21
source of feeder cattle, 22
transition to grain production, 20
Refrigeration, 17, 34-35, 45-46
Renick, George, 18
Restraint of trade, 49
Robinson, Solon, 20, 31-32
Rock Island, Ill., 36, 39
Royal Agricultural Improvement Society of Ireland, 87

S

St. Joseph, Mo., 46, 53-54
St. Louis, Mo., 12, 27-28, 30, 46, 53-54, 56-57
St. Paul, Minn., 43, 46, 53-54
Salmon, D. E., 109
Schoolcraft, Henry R., 10
Schwarzchild and Sulzberger, 52
Scottish farmers, 80
Scottish market, 51
Scott, John, 100
Scully, Thomas, 78
Sheep, 9, 15
Shelter
cattle, 29-30, 115-16, 118
lack, 11, 89
Shipping, live cattle, 34-35, 40-41, 50-51. _See also_ Railroads
Shirreff, Patrick, 7, 11, 19
Sioux City, Iowa, 53, 54
Smith, Shaw, 24
Smith, Theobald, 61
Soils, 3-7, 77, 80
Sorghum, 9, 10
Springfield, Ill., 19
Stahl, John M., 119
Steele, J. W., 71
Stevensen, William, 67
Stowe, Fred, 71
Strawn, Eli, 21
Strawn, Jacob, 19, 24, 27-28, 31, 65, 78-79, 83
Sugar beets, 9, 10
Swift and Co., 46-47, 52-53

Swine, 3, 10-12, 14-16, 18, 26, 28, 41, 115-16, 126

T

Tenancy, 78, 83
Texas cattle, 22, 57-59, 63, 73, 117
  excluded by Illinois, 62
Timothy grass, 19
Tobacco, 9
Tolono, Ill., 58
Topography, 3-7, 77, 80
Transportation, 3, 6, 13, 15, 36. See also Railroads
"Trusts," meat packing, 125
"Twenty-Eight Hour Law" (1873), 66

U

Union Stock Yards, 41-43, 52, 67, 72, 97
United States Senate investigation of meat prices, 71-72
United States Supreme Court, 49
Urban market, 17, 34-35, 123

V

Veeder, Henry, 52
Veterinarians, 60-61

W

Wabash River valley, 75
Wadsworth, Dyer, and Chapin, 37
Wallace, Henry C., 86, 102
Wallaces' Farmer, 86, 120
Warfield, Benjamin, 23
Wentworth, Edward Norris, 84
Wentworth, John, 100
West Liberty, Iowa, 80
Wheat, 9, 11-12, 14-16, 20, 36, 55-56, 77, 80-81
Whitson, W., 71
Wichita, Kans., 53
Willits, Dudley, 30
Wilson and Company, 52
Wilson, George, 65
Wilson, James, "Tama Jim," 56, 67, 84, 112
Wilson, William J., 49
Wisconsin, 9, 57, 62
Wisconsin agricultural experiment station, 113
Wood, Samuel, 79
Wool, 55
Wright, John S., 85
Wyoming, 57, 62

COUNTY INDEX

Illinois
Adams, 5, 12, 32
Boone, 13
Bureau, 4, 8
Carroll, 76
Cass, 12, 76, 78, 100
Champaign, 4, 8, 19, 56, 58-59, 62, 71, 77, 83, 100
Christian, 5, 76
Clark, 77
Coles, 5, 75
Cook, 13, 77
Crawford, 12
De Kalb, 13
De Witt, 12, 38, 76
Du Page, 13
Edgar, 12, 75
Effingham, 40
Ford, 4-5, 8, 12, 59
Fulton, 12
Greene, 12
Grundy, 59
Hancock, 75
Henry, 4, 78, 118
Iroquois, 4-5, 8, 116
Jackson, 77
Jersey, 77, 100
Kane, 13, 75
Kankakee, 4-5, 8, 13, 116
Knox, 11-12, 102
Lake, 13
La Salle, 22
Lee, 4, 8, 116
Livingston, 4, 8, 10
Logan, 75, 78, 83
McDonough, 12
McHenry, 13
McLean, 12, 19, 21, 30, 62, 71, 75-76, 78, 83
Macoupin, 12
Madison, 12-13, 19, 100
Mason, 4, 75, 77, 81
Menard, 12, 76, 100
Mercer, 30, 80
Morgan, 12, 19, 23, 27-28, 30, 59, 62, 64, 74-76, 78-79, 83, 100
Peoria, 23, 30, 36, 53
Perry, 62
Piatt, 8, 22, 44, 78
Pike, 100
Putnam, 19, 69
Randolph, 12, 75
St. Clair, 100
Sangamon, 10, 12, 21, 23, 27, 30, 57, 62, 75-76, 78-79, 81, 100
Schuyler, 75
Scott, 12, 74, 76, 78
Shelby, 10, 13, 16
Tazewell, 12, 100
Vermilion, 4, 12, 75, 78, 84
Wabash, 77
Warren, 12, 80

Illinois (continued)
Whiteside, 4
Will, 100
Williamson, 12
Winnebago, 13
Indiana
Benton, 116
Lake, 116
Newton, 116
Iowa
Adair, 81
Allamakee, 5, 14, 77
Appanoose, 75
Benton, 14
Bremer, 93
Buena Vista, 77
Butler, 101
Calhoun, 118
Cass, 40, 81, 99-100
Cedar, 10, 76, 80
Cherokee, 77
Clarke, 76
Clinton, 14, 23, 75-76, 80-81
Dallas, 5
Decatur, 14
Delaware, 14, 76
Des Moines, 14
Dickinson, 5
Dubuque, 75
Emmet, 57
Franklin, 6, 57, 77
Fremont, 14
Hamilton, 81
Hancock, 57
Iowa, 14
Jasper, 76, 80, 82
Jefferson, 40
Johnson, 14
Jones, 76, 80
Keokuk, 14
Kossuth, 77
Lee, 14, 23, 100
Linn, 76
Louisa, 14, 76
Lyon, 77
Madison, 14
Mahaska, 14
Marshall, 12, 14, 76
Monona, 71
Muscatine, 14, 76, 80, 101
O'Brien, 57
Osceola, 77
Page, 14
Palo Alto, 77
Pocahontas, 77
Polk, 80
Pottawattamie, 71
Poweshiek, 102
Ringgold, 76
Scott, 23, 93
Story, 23
Tama, 67
Taylor, 14, 81
Wapello, 14, 39
Warren, 76
Washington, 14, 75, 83
Woodberry, 82
Worth, 81
Wright, 57, 82

FEEDLOT EMPIRE: Beef Cattle Feeding in Illinois and Iowa, 1840-1900

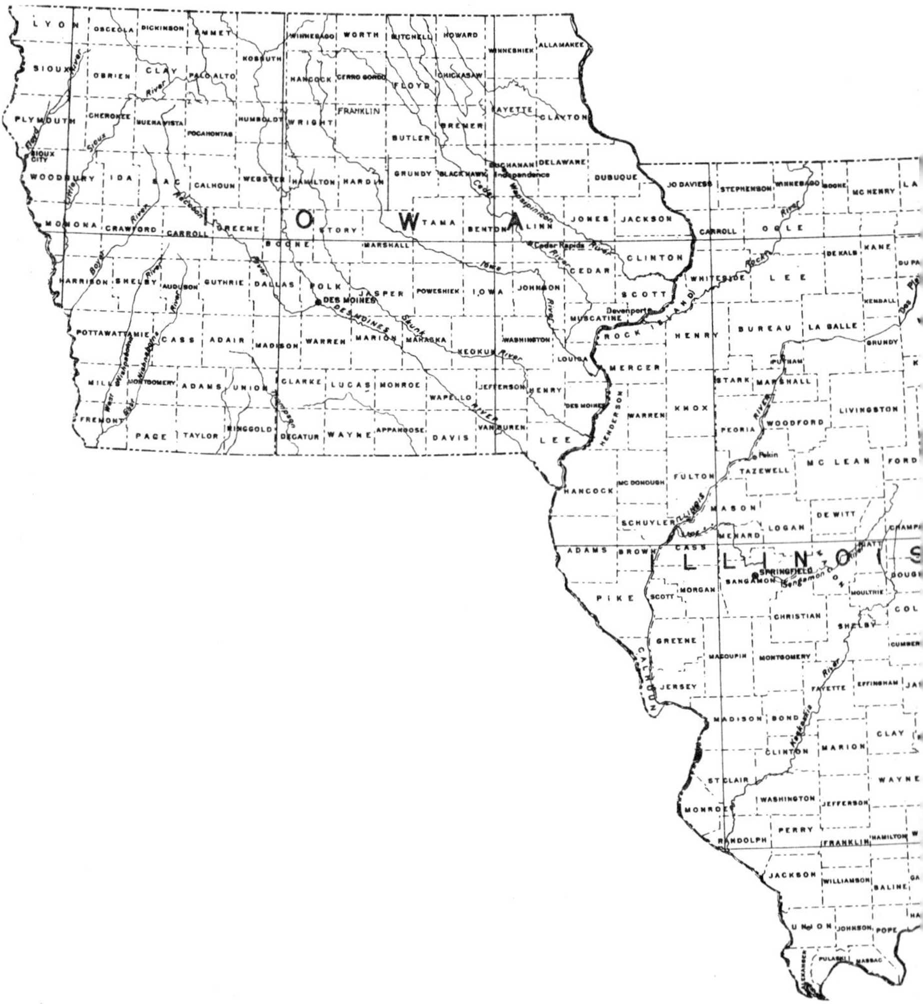
IOWA
ILLINOIS
LYON
OSCEOLA
DICKINSON
EMMET
WINNEBAGO
WORTH
MITCHELL
HOWARD
WINNESHIEK
ALLAMAKEE
SIOUX
OBRIEN
CLAY
PALO ALTO
KOSSUTH
HANCOCK
CERRO GORDO
FLOYD
CHICKASAW
PLYMOUTH
CHEROKEE
BUENA VISTA
POCAHONTAS
HUMBOLDT
WRIGHT
FRANKLIN
BUTLER
BREMER
FAYETTE
CLAYTON
SIOUX CITY
WOODBURY
IDA
SAC
CALHOUN
WEBSTER
HAMILTON
HARDIN
GRUNDY
BLACK HAWK
BUCHANAN
Independence
DELAWARE
DUBUQUE
MONONA
CRAWFORD
CARROLL
GREENE
BOONE
STORY
MARSHALL
TAMA
BENTON
LINN
JONES
JACKSON
Cedar Rapids
HARRISON
SHELBY
AUDUBON
GUTHRIE
DALLAS
POLK
DES MOINES
JASPER
POWESHIEK
IOWA
JOHNSON
CEDAR
CLINTON
SCOTT
Davenport
MUSCATINE
POTTAWATTAMIE
CASS
ADAIR
MADISON
WARREN
MARION
MAHASKA
KEOKUK
WASHINGTON
LOUISA
MILLS
MONTGOMERY
ADAMS
UNION
CLARKE
LUCAS
MONROE
WAPELLO
JEFFERSON
HENRY
DES MOINES
FREMONT
PAGE
TAYLOR
RINGGOLD
DECATUR
WAYNE
APPANOOSE
DAVIS
VAN BUREN
LEE
Floyd
Little Sioux River
Sioux
River
Boyer
River
Raccoon
Nishnabotna
West Nishnabotna
East Nishnabotna
Cedar
Wapsipinicon
Iowa
Skunk
DES MOINES RIVER
Keokuk River
Thompson
JO DAVIESS
STEPHENSON
WINNEBAGO
BOONE
MC HENRY
CARROLL
OGLE
Rock River
DE KALB
KANE
WHITESIDE
LEE
ROCK ISLAND
HENRY
BUREAU
LA SALLE
KENDALL
GRUNDY
Des Plaines
MERCER
PUTNAM
STARK
MARSHALL
KNOX
WARREN
HENDERSON
PEORIA
WOODFORD
LIVINGSTON
Pekin
TAZEWELL
MC LEAN
HANCOCK
MC DONOUGH
FULTON
MASON
SCHUYLER
Illinois
DE WITT
LOGAN
MENARD
CASS
ADAMS
BROWN
SPRINGFIELD
SANGAMON
Sangamon
PIKE
SCOTT
MORGAN
MOULTRIE
CHRISTIAN
SHELBY
GREENE
MACOUPIN
MONTGOMERY
CALHOUN
JERSEY
FAYETTE
EFFINGHAM
MADISON
BOND
CLAY
CLINTON
MARION
Kaskaskia
ST CLAIR
WAYNE
WASHINGTON
JEFFERSON
MONROE
PERRY
RANDOLPH
FRANKLIN
HAMILTON
JACKSON
WILLIAMSON
SALINE
UNION
JOHNSON
POPE
PULASKI
MASSAC
ALEXANDER

The Rose Exterminator

OTHER BOOKS BY WILLIAM CARNEY

A Year in a Closet
The Real Thing

The Rose Exterminator

A NOVEL BY

William Carney

EVEREST HOUSE
Publishers
New York

Published by Everest House, New York
Published simultaneously in Canada by
Beaverbooks, Don Mills, Ontario
Manufactured in the United States of America
Designed by Judith Lerner
First Edition *RRD982*

Library of Congress Cataloging in Publication Data:

Carney, William.
The rose exterminator.

I. Title.
PS3553.A758R6 1982 813'.54 82-9260
ISBN 0-89696-177-X AACR2

The Rose Exterminator

S'il importe d'être sublime en
quelque genre, c'est surtout en mal.

DIDEROT
Le Neveu de Rameau

Where you come from is gone, where
you thought you were going to never
was there, and where you are is no
good unless you can get away from it.

FLANNERY O'CONNOR
Wise Blood

Saturday

1

IT WAS in the afternoon after the night that Eugene disappeared when Pinkerton called to apprise me of what had happened. He telephoned the museum. The switchboard was closed, as were the offices, and I remember wondering at the time how he had gotten hold of my direct line. He did not bother with any preliminary salutation; just, "Have you seen Eugene?"—assuming, no doubt, that I would instantly recognize that gravelly drawl wrapped in a lazy, mendacious grin even when it was filtered through the clamor of what could only be a bar swinging in the background.

"Not recently," I told him.

"You heard about Symonds?"

"All there is to know, I expect."

I waited, and so did he. I broke first.

"All right, what do you know that's new?"

"He's dead, chum."

The way he said it had the salacious smack of a gossip columnist's revelation. I repeated the word, "Dead?" And let it hang there for a few seconds while the din of a jukebox thumped in the receiver. Then I asked him what had happened.

"Who knows? I ain't The Shadow. Eugene's mother called me just before I came to work. She's been tryin' to get a-hold of you. I didn't give her your number—at the museum, I mean."

There was no use in asking how he guessed I would be working in the lab on a Saturday afternoon, and even less in inquiring as to how he knew my off-duty number. Outside the administrative staff, nobody but Eugene knew it, and although he might possibly have let it drop in an unguarded moment, it was not like Eugene to do such a thing.

I closed my notebook and started putting the desk in order with one hand, knowing I was finished there for the day—possibly forever. What, I asked him, did Eugene's mother want with me? "I barely know the woman."

"She's lookin' to find out where her wanderin' boy's gone to, I reckon. An' so's the police. They found him in Eugene's trailer."

"Symonds?"

I could almost see him nodding in that sententious, Texas way of his. "Dead." He pronounced it *day-ud.*

"Dead how?"

"Not sure. It'll probably be in the paper tonight. It's been on the radio already."

"Any details?"

"Nope. Just who it was and where they found 'im. You got a radio in that place?"

There was one, but it wasn't on. I had already heard it anyway. I stood up, unbuttoning my lab coat. "Are you at Jo-Do's?"

"How'd you guess?"

"Stay there. I'll be down as soon as I lock up here."

"You gonna call her first—Eugene's ole lady?"

"I hadn't thought about it."

Mrs. Carmichael was the last person I wanted to talk to at that moment, but as I grabbed my jacket and headed for the door, the telephone began ringing again. I hesitated. It could have been Eugene, but the prospect of finding his mother on the other end of the line, questioning me in that sing-song whine that was the likeness of her son's, only raised an octave, was something I didn't want to chance. I left it ringing, locked the door on whose frosted glass was stencilled *Life Sciences,* and headed for the service stairs. I wanted to see what Pinkerton knew before I talked to anybody else. Pinkerton had Special Sources.

San Diego was enjoying one of those bright fall days that would pass for late spring anywhere else in the country. We had gotten a touch of rain, and the grass-covered hills of the golf course across the canyon were turning green in places; hibiscus and flamboyants against the building were in full bloom. The parking lot in back of the museum, open to anyone on weekends, was full of cars, and I had to take my time straddle-walking my bike across the sidewalk and pavement crowded with people strolling back from the zoo and the art museum as closing time neared. It was the kind of family crowd that part of Balboa Park attracts on weekends. Clear of them, I took the drive that curves south and west past the Naval

Hospital, and curled onto the freeway, cutting north across downtown. The wind foiling against my back and through my hair felt almost balmy, but a promise of fog lay on the horizon where a gray, hard-edged fringe of clouds rose up out of the ocean like a torn strip of cardboard against the yellowing sky.

The freeway only lasted for about a mile at that particular section. Where it paralleled Lindbergh Field (now called the International Airport) it gave out in a tangled heap of grading and construction equipment left where the four o'clock whistle had halted work on Friday. From there the old Pacific Highway took over along the big cut of right-of-way where they were still trying to connect up this missing link in what was changing U.S. 99 into something called Interstate 5. At Grand Avenue I cut west again and tried to compose my jittery nerves to at least the degree where I might have a chance of learning from Pinkerton more than what he was counting on learning from me.

His announcement about Symonds was no news to me. It was what he had not said that had brought me out. I was testing what he had to give me as a foretaste of what might lay ahead. A familiar, empty feeling had been expanding in my chest all afternoon, beginning when I had happened to hear the news broadcast a few hours before this after coming in from a swim up the coast. Glenn Symonds and all that his presence entailed had been a Damoclean sword hanging over me for a long time, and the thread had shown signs of fraying badly during the previous few weeks. Ever since the Tuesday of this particular week when I had had dinner with him and listened through most of it to his oblique, self-satisfied references to some masterpiece of deviltry he was about to consummate, a feeling of uneasiness had dogged me. It was a need to get away for a while more than anything else that had prompted me to accept an invitation to weekend up at Crystal Cove. But even there Symonds had managed to intrude. And now, this.

I felt as if I ought to breathe into a paper bag; nervous depression was no condition in which to be maneuvering in traffic on a motorcycle down there.

Inquiries as to Eugene Carmichael's whereabouts might logically be directed at me with no significance in the asking beyond what everybody knew: I was his friend. But this was not a very reassuring line of argument to take under the circumstances. The last I had heard from him was a telephone call the day before when, by chance, he caught me at home just as I was leaving for the weekend. He said he was in L.A., calling from a post office near Silver Lake, and that he had Symonds in tow. Eugene—I could see the outlines of it taking shape—had to be the prime suspect in this latest absurdity. The fact that it had occured in that singularly furnished and festooned trailer would draw from most the reasonable assumption that he did it, once the sordidness of the particulars were known. But for those familiar with the workings of Symonds' and Eugene's minds, and for whom the circumstances would seem no more than fitting (a group at once smaller and more knowledgeable where these kinds of things are concerned), a caveat might be raised, for it must have come as a surprise to them that it was Symonds, and not Eugene, who was the victim. Everybody who knew anything at all about Eugene had said, or at least thought, at one time or another, that he would sooner or later end up being carried feet first out of that house-on-wheels he called a home.

But not Symonds.

Knowing both of them better than anyone down here, I savored the prospect ahead of me the way a helmsman tastes a rising wind, and was prepared by the time I got to the beach to keep my responses down and my guard up.

At Jo-Do's I found Pinkerton sitting on a high stool with his back to the wall just inside the front door, filtering the clientele for ABC bait. The place was already so packed with weekend traffic there didn't appear to be space for any more, but in they streamed, from the beach and the street. He grabbed my arm.

"Stand around here in back of me so's I can hear you, chum," he shouted above the noise.

I wedged myself in between the crush of bodies against the wall by his stool while he went on checking the IDs of a trio of surfers who were wearing nothing but soaked pajama shorts, their bodies still wet from the last crest. On a quiet evening during the week you could hear the rollers a hundred yards away through the back door. That was closed now, and everybody was being channelled through the front. I bent down close to his ear and asked what else he had found out.

"No more'n what I told you."

It was no more than a standard rhetorical denial, one of several stock replies Pinkerton employed when pressed by anybody about something he might know and they might not. He carried information around with him the way most men carry money: a lot of small change readily available, but the important stuff folded away and more difficult to get at. I waited him out while he ticked through some more arrivals, then he twisted around to look at me.

"The police went in an' carried 'im out a little after noon."

"How did it break?"

"The trailer park manager called 'em."

"Was he the one that found him?"

Pinkerton nodded, then broke off again to check some others coming in. While he was doing that, I looked around to see if I knew anybody. Jo-Do's was not my idea of a good bar, particularly during a weekend Happy Hour. Most places don't have a Happy Hour on weekends, but Jo-Do's did. Falling down drunk for two bits was the style here, and dead drunk in an hour or less, since you could down all you could hold for free as long as you had hold of the glass the first one came in. Beer didn't count. Bars that followed the custom got together around town and staggered the thing. It generally started at a dyke joint over near the rollercoaster on Mission Bay around twelve o'clock, then at one, some other place would take over, and so on down the line. It kept everybody on the move and spread the profit around, but the casualty count mounted rapidly after the second hour; it was now after five. Given the native greed of Americans in general, and the uncommon dissipative tendencies of Californians in particular, it was not surprising to hear of someone every week who never recovered from an attempt to accomplish the entire circuit. The San Diego city council had been making noises for some time about declaring the practice illegal, but nothing ever came of it, any more than with the issue of bars serving free suppers on Sunday nights, which already was illegal.

I could recognize nobody through the smoke and racket, nobody familiar enough for me to want to speak to at least. Pinkerton reached around and pulled me to his side, resting a big hand on my shoulder, leaving his left one free to adjust the flow. He picked up where he had left off.

"Eugene's dog was raisin' Cain all night long, and I reckon the manager got tired of the other trailers complainin'. He went down to see about it and found the door unlocked and Eugene gone. The dog was shut up in the bathroom."

"And Symonds?"

"About what you'd expect. Hey, you!"

He half rose off the stool to collar a teenager who had pushed by him with some others older than himself. Pinkerton held out a bony, long-fingered hand as if he were waiting for a tip while the kid dug in his billfold for identification. While he waited, he went on talking.

"He was in the back room, the bedroom—if that's what you call it."

"Did you get to talk to the manager?"

"Not exactly."

"What does that mean?"

"I heard, that's all."

I did not have to ask how he heard. Instead, I wondered aloud if there had been any sign of Eugene yet. He shook his head and cocked a brown, slightly bloodshot eye at me.

"You don't know where he's at?"

"I haven't the slightest idea." It was the truth, but I did not expect him to believe it.

He grinned. "Not your lover's keeper, is that it?"

"I'm not Eugene's lover *or* his keeper."

"Neither is Symonds, any more." His grin relaxed into a sardonic, lantern-jawed look of amusement as he glanced once more at the kid who had finally come up with something. Pinkerton hardly gave the card any notice before flipping it back to him and motioning the trio in.

"Look at that!" he said. "Navy. And he ain't a day over eighteen, or I'll fuck your mother."

"My mother's dead, and you wouldn't have gotten to first base with her. And what do you let him in here for if you think his ID's a fake?"

"I don't think, I *know!* They make 'em up in a couple of minutes for a couple of bucks in Tijuana, an' for a little more than that right here in town. I was just makin' an observation, that's all. I don't give a damn. The police don't expect me to be no forgery expert. I call 'em like I see 'em."

I started to ask him how he knew the kid was in the Navy, but let it drop. Dale Pinkerton had served a stint in a tin can himself during the Korean affair, and it was probably a case of taking one to know one. Besides, San Diego is the headquarters of the Eleventh Naval District, and you had a better than even chance of guessing any stranger of draftable age to be on shore leave. On the other hand, I said I would be interested sometime in learning exactly what the police did expect from him—in bars as well as out. "Meanwhile, how about telling me what else you know?"

He shrugged. "That's about it."

"Didn't that manager tell you what he found, what he saw when he went in the trailer? Did he hear anything the night before? How did he say Symonds was killed? You must've asked him that."

"I wasn't all that innerested, to tell you the truth. But he didn't get a close look at him. He was scared at what he did see, though. He backed out the way he come in as soon as he got a peek at that back room. As I understand, he didn't even let the damned dog out. He at least had sense enough not to touch nothin'. Went right on back to the office and called the police."

I waited, and after a moment he added grudgingly, as if he'd been caught shortchanging me, "He did say it was a terrible sight. But that don't tell you much."

I agreed that it did not.

"Eugene must've really lost his mind," he went on, shaking his head. "It was just like the goddam fool though, to go hellin' around and get hisself into this kinda fix."

"Eugene wasn't a heller, Dale, and you know it. What have the cops been telling you?"

He snorted in disgust. "The police never tell me nothin'!" He pronounced it *po*-lice.

"They questioned you, I'll bet."

"Of course they questioned me!" He adopted a wounded, put-upon expression, like something he had learned as a child from his mother—if he ever knew his mother. "I knew the dumb sonofabitch, and they know where to find me. They got small, one-track minds, and they're too lazy to get off their butts and find anybody else to talk to who ain't as easy to get a-hold of. They'll most likely be wantin' to talk to you about it too, chum." He nodded with self-righteous emphasis. ". . . when they get around to it."

"Why me?"

"You was his friend."

"So are a half dozen others around this town."

He nodded again, this time in sympathetic agreement. "They'll get around to them too, don't you worry."

"Why should I worry?" In spite of my resolve to keep cool, I felt the resentment rising along with my voice. "I don't know what was going on in that nutty trailer. Symonds was a fool, and so was Eugene."

Again he nodded. "Couldn't agree with you more. The point is, they'll be beatin' the bushes for anybody who was in spittin' distance of either one of 'em over the past few days—or longer maybe."

"Well," I said, pulling my voice down as low as I could and still be heard above the babble and music, "I ought not be very high on their list."

It did not sound very convincing, and he caught the affectation in my voice, looking around at me with sly amusement deepening the hatchet lines in his face.

"Shit!" He drawled it out. "Somebody who knows the one who got it *and* the one they figured give it to 'im? . . . Stick around!"

"I wonder," I said, pulling his spade-shaped chin around so I could look into those lying, shit-colored eyes. "I wonder how they know all that?"

He jerked his head away. "Hell, everybody knows that."

At the moment I could think of a goodly number of people, down there at least, who knew nothing of the kind, but whom the police might want to talk to. On the other hand, from his and certain others' point of view, it *was* common knowledge, knowledge that would weigh heavily in determining which direction the investigation was going to point.

Where Pinkerton's path now crossed mine was an overgrown side road for me, but it constituted the main highway of his activities. It had long been rumored that he had an informal arrangement with the local constabulary that benefitted him with special privileges denied the general run of the population. It was a symbiotic relationship dating back to his swabee days when he ran shore patrol out of Coronado before the Navy had for some reason or other found him unfit for the service and booted him out. This discredit seemed to have in some way cemented, rather than severed, whatever ties he had with the police. It at least served to explain the peculiar invulnerability he enjoyed in areas where everyone else had to watch his step. Dale gave the impression, to those who knew him, of a man who did what he pleased with impunity, not because he personally had any push, but because some kind of power and influence shielded him. At what price was not known. Moving around as he did, he knew a lot of people, but how much of what he knew he told, or might tell under pressure, depended on the nature of this special association he enjoyed.

These were imponderables, of course, but Pinkerton moved along the fringes of disaster with a kind of cussed skill and unpredictableness, like a homeless tomcat stalking backyard fences, always landing on his feet. He performed a number of odd, often very odd, jobs around town (of which doorman at Jo-Do's was merely one of the more obvious and respectable), and, as a consequence, knew something worth repeating about nearly anybody the police might at any moment be interested in. It did not simply happen that way, as he so often protested; it was a calling, a compulsion like the tomcat's mania for mauling rivals and mounting queens in heat. He was, whether he admitted it or not, at the law's beck. Not to put too fine a point on it, Dale was an informer. But he loved to complain in a lugubrious, soft twang about the way the fuzz picked on him every time some dumbbell ran to riot and made the headlines.

"You must have really gotten your nose dirty once, Dale," I told him.

"You hang in there long enough, doc, and you'll get yours dirty too," he countered with a knowing smirk.

I reflected at the time that he was closer to the truth than he knew. "There," for him, expressed an unabashedly jaundiced view of society. For me it meant "here," here being this southernmost pocket of southern California which seemed to harbor a peculiarly virulent strain of just about every kind of lunacy and viciousness in the book, plus a few that had not yet been identified and catalogued. Not that one could not find the same derangements elsewhere in the state and the nation. But what in San Francisco, or even in L.A., was allowed a certain tolerance, or at least room, to work itself out in public was hedged in here in a way that made it fester in the dark, so to speak, and so was liable to come to a head like an ugly boil, causing inordinate pain and leaving unsightly scars. I hear it is all different down there now, but then, a lot of things have changed all over the country in the past fifteen years.

It probably did not do to stay too long in an environment like that. Even if you had nothing to hide to begin with, you would soon find something that needed covering up. Normal people newly arrived tended to fall apart fairly quickly. The young seemed not to mature, and the old got older at a more rapid rate than elsewhere—this last contrary to the Chamber of Commerce's claims on behalf of the salubrious climate. It was the very climate that may have been to blame for the stagnation and early rot. The climate and certain regional attitudes transplanted from the Middle West had resulted in a rank growth of expanded prejudice that provided, at the same time that it stifled, the flowering and fruition of its own baneful progeny. A very physically aggressive and emotionally repressive community withal.

That climate and those attitudes got to you in funny ways, and in other ways that were not so funny. The air of the place carried, on its mild, palm-bending breezes, a scent of impending and foolish undoing. There was the perpetual feel of earthquake weather to the place, although the tremors felt were usually of a moral or felonious nature, and had nothing to do with the geological instability of the terrain. Save for the old, the retired, or those born and still growing up in the city and its outlying communities, most of the inhabitants found themselves down there by accident or absentmindedness, and stayed on through inertia. I had settled there as a last chance and been looking for a way out ever since.

Like the animals in the celebrated zoo near where I worked, most of the population lived in an environment that did not appear to be a cage, but whose restrictions were unmistakable and inescapable.

Pinkerton didn't see it that way at all though. He had never had it so good. "It's a wonderful town!" he declaimed to me once when I had disparaged it in some way or other. "An' there's somethin' here for ever'body!" Now he was barking about me getting my nose dirtied in the local mire, and the prospect of it appeared to please him. Not that Dale enjoyed the prospect of another's undoing. Neither malice nor bitterness did he harbor, as far as I was able to tell, and unprovoked violence was not his game either. He simply liked his work, such as it was, and if he saw things from a somewhat seamy angle, it was no more than to be expected, considering what he had been through and the people he came in contact with in his line of employment. The same, of course, might be said of almost anyone. Point of view is everything.

He was smiling at me now, revealing long, tobacco-stained tusks and the slight white welt of a scar across his near cheek, the memento of a slice received as the parting gesture of a random nut on his way out of some other bar a couple of years earlier. Incensed, no doubt, at the sass this displaced Texan had handed him at the door, he missed in his haste by a mere half centimeter the critical soft underpinnings of that overhanging jaw for which he had aimed.

"Kinda turned the tables on ol' Symonds, that sort of thing, don't you think?"

"Symonds wasn't old," I said. His remark, I knew, was made as an expression of teasing admiration, the "old" implying a familiarity I took to be spurious, and the content of which constituted bait to which I rose without thinking.

"You oughta know." He turned away to pass another through the door. "You lived with him."

I tried to get him back on the track. "Did anybody get a look at him when they took him out?"

"Oh, I expect they covered it up like they always do." He paused to light a cigarette, then commented further. "I get the impression, though, that he wasn't the kind of sight you'd care to come home to after a hard day at the office. He didn't go gently into that last goodbye."

"'Goodnight,'" I corrected him.

"You going already?"

"The line is 'Go gently into that last good night,' friend. And yes, I've got to get home."

"Somebody waitin'?"

I hoped there would not be. I told him I was tired, and added, "It's a good thing you called—aside from what occasioned it. I was getting groggy."

"Workin' overtime at the museum?"

I nodded.

"Where d'you reckon Eugene's hidin'?"

He pronounced Eugene the same way he did museum.

"What makes you think he's hiding?" I said. "For all I know, he's home with his mother in bed. And that's just where I'm going."

"Thought you said your mother was dead."

2

IT WAS fortunate that she was. Had she still been alive, she would have wished us both in our graves before the night was finished. My parents had adhered to the standard notions that used to form the mainstay of family life, even on the West Coast. Los Angeles was an overgrown joke to my mother's way of thinking, and San Diego hardly more than a name on the map. Ventura, where I was born and grew up, satisfied her as a fine enough place to live and die in—at least until the war brought about political and demographic changes which ran counter to her prejudices. But her prejudices, at least, had a common-sense logic about them that allowed her a certain composure in the rising tide of restlessness that began to surround her in her declining, and me in my developing, years.

California, to be sure, is no better than any other region, but those who live here for very long get to thinking that it is, and the thinking is what makes it so, and is sufficient reason for their staying here if they come from out of state. Having been born here, and having seldom left it (and then only to spend some time in other countries), I had little to compare it with nationally. The only objection I ever found to it was having to live in that southernmost corner of it where I found myself now.

Not that it was in any way un-Californian. On the contrary, by some peculiar antipodean alchemy it seemed to have combined the best and worst the state had to offer in geography and social attitudes and transmuted them into a bright deceptive bane. Here change and stability worked together in mysterious, often irrational ways that left large stripes for small offenses. To be fingered by the law was tantamount to guilt in the eyes of this community, and my depravity by casual association with half a dozen whose delinquency was on the books several times over, to say nothing of a more intimate connection with Eugene Carmichael and Glenn Symonds, would make me suspect to a degree that was anything but minor.

I knew about what to expect from the outside, but my concern was enriched by a mixture of interior unease and a growing sense of loss which I was only now beginning to contend with. Because I was not entirely clean in this matter.

Eugene's call to me from L.A. (and I had only his word for it that that was where he was dialing from) had left me worried after I hung up simply because he sounded worried. He was not as given to fanciful hyperbole as were most that I knew, and the restrained fear that cloaked his voice put me on my guard. He was unable or unwilling, and at any rate too much in a hurry, to go into details. He said only that he was bringing Symonds down to his place, and that he—Symonds—ought not to be traveling. He wanted me to call and check with him later in the evening. I told him I was going away for the weekend, but that I would try to phone him if I got a free moment, or on Saturday for sure. That was all. Our conversation could not have taken up more than a minute's time, but I stood staring into space for a long while after I hung up, thinking over what was said, unable to fathom its implications. That Eugene was up in L.A. with Symonds at all was somewhat puzzling, since they had not been seeing much of each other for the past four months.

It came as no surprise, but almost as a relief, when I saw them waiting for me, parked down at the turnoff to the gravel lane that led halfway up the hill to where my house stood screened by a big pepper tree in the back and by a clump of tall Canary Island pines in front. They were in an unmarked car, and their headlights went on as I passed them. The full glare of the beams in the rear-view mirrors nearly blinded me as they pulled out and followed close behind.

It was not the first time a patrol car had tailed me, or had followed me home for that matter. They did such things for sport down there. After dark was open season, and you could not live in the city for very long without experiencing the inconvenience of the boys in gray sprinkling salt on your tail whether you were a protected species or not.

I was not, although I was gainfully employed, a property owner and taxpayer who took a moderate interest in civic issues, and violated no more minor traffic ordinances than the next person. None of that mattered much. Sitting in a parked car was a loitering offence, driving around the block would get you questioned, doing it slowly might find you booked, and walking, anywhere, after dark was vagrancy. This last incrimination had been thrown out as unconstitutional as far as the state was concerned, but the city of San Diego, as the local boodlers and the rags that catered to them never tired of reminding the amorphous electorate of displaced persons who encouraged this kind of high-handedness, was not the state. The local peace force was a law unto itself, and the climate of opinion that supported such usurpation was fully as stable as the celebrated *douceur* of the weather. Perhaps there was some compensatory mechanism at work here. Along the smiling thoroughfares might be seen billboards extolling the virtues of both, illustrating clean beaches and the opportunities for family fun side by side with exhortations fifteen feet high to impeach the nation's Chief Justice, a native son these self-made natives wanted no part of. Bumper stickers directed fellow citizens to support their local police (as if it would have made any difference whether they did so or not), and J. Edgar Hoover spent his free time, when he had any, with his buddy right down here in our well-guarded midst. The Navy security office really ran the town, a fact few found fault with. After all, the Navy was San Diego's *raison d'être,* along with its notional three-hundred-and-fifty-plus sunny days a year and the nearness of exotic Mexico in the form of that smoldering trash heap called Tijuana—these three factors being not unrelated. It was a way of life not quite like any other, but one got used to it if one elected, or was forced, to stay around for long.

Although forced, for economic and professional reasons, I had come to terms with the situation and felt relatively at ease. It helps if your work interests you, and it was my work that had, up until very recently, kept me on an even keel in the face of emotional buffetings few, if any, of my few acquaintances knew anything about, and that made the petty harassment that went with the local lifestyle pale by comparison. Now, however, the storm warnings were going up, and I would not be able to sit this one out the way I had so many others that had led up to it. This time there was a killing to contend with, and they were at my door.

"Doctor Scott?"

They were out of their car before I had the kick stand down and was off the saddle. I turned to see them coming toward me, past my car which blocked the shallow driveway, crossing the semblance of lawn I had spent two winters nursing out of the adobe crust. Plainclothes. The dog was barking, putting on his glad-to-see-me act, trying to tear the small, glassed-in patio to pieces in his affectionate frenzy. I had to quiet him and dole out some affection in return before I could answer them. They flashed their badges—National City—and said they would like to ask me some questions.

"Come on in," I said, letting the dog loose outside and closing the door. "And you can call me mister."

"Isn't that what they call you where you work?" one of them asked me once we were in the house. He was kind of big and beefy, but not bad looking, and he already had a notebook out. "Doctor?"

"The office staff and the public do," I explained. "I'm curator of invertebrate zoology, not an M.D."

"Doctor of Science," the other one offered, sure of himself.

"Philosophy," I said, wondering for a moment if he were going to question me as to the distinction. Discussions of the subtleties of academic titles and the history of the gradual indiscrimination among the seven liberal arts was one of the less interesting confrontations I was liable to. But such trivia was not what they were after.

"You have a friend who's an M.D., don't you, Mr. Scott?"

"I have several. Why don't you sit down?"

They did, on a couch against a bookcase. I switched on a heater imbedded in the wall near the front door, then excused myself to go outside and retrieve Thief. After quieting him again and shutting him in the patio, I took a stand with my back to the Panel-Ray, facing the heat.

"You were a friend of Dr. Glenn Symonds," the beefy one with the notebook led off.

"Was and am."

I waited, knowing I had made an initial misstep. The chances were good that Pinkerton had called them to say I was going—or had already been—to see him. They did not blink.

"Do you know he was found dead this afternoon?"

"I heard."

"How did you hear?"

"It was on the radio."

"Is that how you first heard it?"

"Not at first, no. Someone. . . . a friend of mine . . . told me about an hour ago. He heard it on the radio, I suppose."

They did not pursue that by asking about the "friend." Instead, they both pulled out cigarettes. The one with the pad started to put the pack back in his coat after he had lipped one, then, as an afterthought, held it out to me. I shook my head.

"I quit," I said. Then, "What exactly happened?"

"That's hard to say right now." He went through the business of striking a match, lighting his and his partner's cigarette, sucking in the first long drag, and the initial gust of smoke. Then he stared at me, relaxed, expressionless, as if waiting for me to comment. Then he went on. "The preliminary report indicates the primary cause of death was suffocation, but the coroner's doing an autopsy."

I absorbed this, trying to match their noncommittal air and keep my voice steady. "What did you want to see me about?"

He waggled his hand and pursed his lips as if to indicate that it was no big thing. "We're trying to find another friend of yours at the moment, Mr. Scott. And by the way . . ." He looked at me sharply, the cigarette stuck in his mouth. ". . . I suppose you can account for your whereabouts last night and this morning."

I told them I could, and the topic seemed to hold no further interest for him. He returned to the subject he had been about to bring up. "There's a fella by the name of Eugene Carmichael. He's a friend of yours, isn't he?"

I nodded. "I know Eugene all right, but I haven't seen him for maybe a week. Have you tried his mother?"

They ignored that. The one doing the questioning consulted his pad, then said, "That would be last weekend sometime?"

"I think so."

"No other contact with him? Telephone?"

"Yes," I said. "He called me yesterday afternoon—around four o'clock, I believe."

"Where were you when he called?"

"Here."

"Where was he?"

"He said was calling from Los Angeles."

"What time do you get off work, Mr. Scott?"

"Five, usually. But I left the museum early yesterday."

"Where does Carmichael work?"

He was beating all around the bush, and it was making me nervous. They must have known all about Eugene by now, but I told them. "He works down in the shipyards. He's a welder."

"Why did you leave work early yesterday?"

"The director of the museum and his wife invited me up to their place at Crystal Cove for the weekend. I wanted to get ready, avoid the traffic, maybe try to get in some swimming before dark . . ."

"What did Carmichael call you about? What did he say?"

I let them have a smile—a condescending one. "Talking with Eugene," I explained, "never is very much of a conversation. He tends to run off at the mouth, and I lose interest pretty soon."

"I've never spoken with him," he said, very cold. "What did he talk about? Why did he call?"

"To tell me he was worried." I paused a split second, but then decided I had better tell them. "About Dr. Symonds."

"Did he say why?"

"No, he didn't. He wasn't exactly incoherent, but I got the impression he was either in a hurry or scared."

"How was that?"

"Just the way he talked. He didn't ramble the way he usually does, but he didn't make any sense either."

"Where was he in Los Angeles?"

"He didn't say."

"Do you think he was at Dr. Symonds' house?"

"No."

"At his office?"

"No. It was in some public place though, a phone booth, I'd guess. I think he mentioned a post office."

"All right. He said he was worried about Dr. Symonds. Anything else?"

"Yes. He said he was bringing him down here—to his trailer in National City."

"Why was he telling you this?"

"I don't know. It sounded as if he felt he had to talk to someone about it, and I know him pretty well."

"Did you ask him what was going on?"

"No. That's the thing. I was in a hurry, and it was lucky he caught me when he did. He must've called the museum first, and then tried me here. At the time I didn't think much about it. Eugene is always complaining about something."

"What specifically was he complaining about? Did he want you to do something? Help him with Dr. Symonds maybe?"

"No, nothing specific. But he did ask me to call him after he got down here. He wanted me to check on him later."

"Did you?"

I nodded. "Late last night. I was having a good time up at the

Silvers' and it wasn't until time to go to bed that I remembered I said I'd call him. There wasn't any answer."

"What did you do then?"

I shrugged. "I forgot about it."

My questioner relaxed a little, leaned back, and cocked his head as if he were prepared to be friendly now if I would like it that way.

"Now how about telling us what you know about Dr. Symonds, Mr. Scott."

Thief chose that moment to start barking. There was the sound of a bike pulling up the hill with some difficulty. From the blast of it, I knew it was a Harley, and at first I thought it might be one of the law's helpers, although they don't usually tackle these unpaved roads on two wheels. I waited until it went past, spewing up gravel on the grade. When the sound died away across the hill in the direction of Reynolds' place, I guessed who it must be, and settled back, prepared to engage in some evasive action.

"How do you mean, 'what I know'?"

"Did you, for instance, ever notice, or ever hear, any peculiar things about him?"

"Peculiar?" I snorted in faked amusement. "There's something peculiar about anybody if you look hard enough for it or listen to enough gossip."

He disregarded that for the paste it was. "Let me put it this way then," he said. "What have *you* seen or heard that you'd consider out of the ordinary about Dr. Symonds?"

"Not a damn thing."

It was a little strong, but my mind is less untidy than most, and this vague kind of probing for information they could better obtain with a plain question tried my patience. "What exactly do you want to know?" I said, then added, toning down the sharpness that had crept into my voice, "How can I help you?"

It came out sounding prim.

"He was a close friend of yours, wasn't he?"

"He wasn't a close friend."

And that was the truth, I think. We had been close, God knows, but from point of time, we no longer were.

"You knew him for a long time, didn't you?"

"Eight, nine years, I guess."

It was going on ten, but I was beginning to see the wisdom of employing vagueness.

"Didn't you live with him at one time, Mr. Scott?"

For this one I switched back to the letter of the truth. "We shared a house in L.A. for about a year."

He nodded and glanced at his sidekick who had done nothing up to now but sit at the other end of the couch like a chaperone. He might have been a deaf mute for all the response he showed to what was said between his partner and me. Beefy picked it up again. "Then you must have known he was a homosexual."

What to say?

"You can live in the same house with somebody and lead a completely separate life." It sounded sententious. I added, "It happens all the time."

His response was to ask me again if I knew Symonds to be a homosexual. So I lied again with a simple "No."

"He wasn't married."

I agreed that he was not.

"Did he go out with women?"

"Sure." I had to smile again. You don't get to be a very successful plastic surgeon in Beverly Hills without a certain amount of social life, one usually as varied as your practice. Symonds went out with everything, but what did that mean? Watching them curling smoke out of their mouths, I wanted a cigarette in the worst way; it would have given me something to do with my hands besides keeping them stuck in my pockets.

"Look," I said, unable to dampen the mounting exasperation in my voice. "We were roommates. What he, or anybody else, did or does with their lives is none of my business so long as nobody gets hurt."

He reminded me that somebody just got killed. I said I was profoundly sorry about that.

"And as far as you know, Dr. Symonds wasn't a homosexual?"

"As far as I can say, no. I don't think I'd say anything about him that I didn't know for a fact, and I've never seen him doing anything of that sort with anybody."

My distinctions were overly fine, and they were properly unimpressed. I shrugged and went on. "I've heard things. You hear that sort of thing sooner or later about almost anybody that isn't married today. It's something people conjecture about . . . about bachelors, I mean."

"That's interesting, Mr. Scott." He looked over at his silent partner, wide-eyed with mock surprise. A real ham. "I never thought about it in quite that way. Pete here is a bachelor. Are people talking about you, Pete?"

Pete had his own pad out now, as if to emulate his obvious superior. He grunted and looked away in disgust while the other indulged in a gutty little laugh at my expense. Then Don (for that was my interrogator's name) settled down, shifting his heavy frame and his line of inquiry, taking up the matter of my movements over the past few days.

The accounting I gave pretty well jibed with the truth. On Thursday the director of the museum, Morris Silver, had invited me to spend the weekend with him, his wife Louise, and his two children by a previous marriage. They had a summer place up at Crystal Cove, and would be closing it for the season soon. I accepted even though I did not particularly like the man and considered him a failure as the director of even so modest an institution as the San Diego Natural History Museum. It was politically advantageous to stay on the good side of him; he had hired me at a time when nobody would, and he liked me. It was the proper thing to do, I found Louise a charming woman, intelligent and easy to be around, and I liked the surf. The two on the couch listened, expressionless.

"Do you go up there often?"

"No, this was the first time." Silver was a funny kind of man who liked to keep his distance and make sudden forays into your department to ask foolish questions and pat you on the head for doing a good job. He had inherited a lot of money and made a lot more of it in oil. Having spread some of it around in scholarships and grants to nonprofit organizations like the one I was in, he believed himself to be a scientist of sorts. He was an engineer by training and a natural history buff with excellent fund-raising abilities. He had conned the local and regional magnates on the board into appointing him to the directorship a few years before I arrived. Several years before that he had married Louise after his first wife died of cancer. Louise was a widow with two sons of her own, both of whom would be grown now, but one had committed suicide pretty far back. I did not go into all this with the two on the couch, however.

"When did you leave here to go to Crystal Cove?"

"That would be right after Eugene called. I was ready to go when the phone rang. About four o'clock. I got up there before six, but too late for any swimming."

"And you called Carmichael from there?"

"I phoned his place . . . the trailer . . . oh, it must've been around nine o'clock. There was another guest up there, and we played bridge after dinner. We turned in rather early."

"There wasn't any answer when you called?"

"There wasn't any answer."

"Did you think anything was wrong about that?"

"Wrong?"

"You said Carmichael asked you to call him, and when you did there was no answer."

"No, I didn't think it odd, if that's what you mean. I wasn't very interested. I said I'd call, and I called."

"Now, was your invitation for the weekend, or just for Friday night?"

"It was for the weekend. I came back early because I was bored, and I'm working on some interesting specimens I've been collecting over at Anza-Borrega."

"When did you come back here?"

"I left the Silvers' place around one o'clock, right after lunch, and I didn't come back here. I went directly to the museum."

"What about your dog? What happens to him when you're away?"

"If I'm going out in the field I usually take him with me. When I can't, then a neighbor over the hill takes care of him for me."

"Did you just come from the museum?"

"No. I told you, a friend called me at the office and told me about Dr. Symonds. I rode over to Pacific Beach to a bar where he works, and we talked about it. His name is Dale Pinkerton."

He nodded, and turned a page in his notebook.

"You lived with Eugene Carmichael too, didn't you, Mr. Scott?"

"Not recently, no. He rented the spare room here last winter for a few months while he was between jobs. He moved out in March, I think."

"Was that when he got the trailer?"

"Yes. It was one he'd owned before. He bought it back from the person he sold it to about a year ago."

"Do you know who that was?"

I did not. Eugene had originally bought the thing for his mother, but she would have nothing to do with living on wheels, so he took it over himself and did a remodeling job in it that included complicated changes in decor, soundproofing the walls with cork, and equipping the back room with bars, block-and-tackle, ring bolts, and other paraphernalia which, to the uninstructed eye, might have passed for the knocked-together furnishings of a home gymnasium. There he settled in and pursued his own distinctive lifestyle undisturbed by the blue-collar couples and senior citizens who populated that spread down on National Boulevard, and who remained, up to this weekend, unaware of what went on of an evening in that nice young man's Breezeway Town Home. National City is like that, a bedroom community bordering the south of San Diego whose inhabitants have learned to look the other way in the interests of cheap rent and a tolerant give-and-take where their seamier activities are concerned. It used to be known as the whorehouse of San Diego, although Tijuana probably had as valid a claim to that distinction, with jai alai and the bullfights thrown in besides for the benefit of those who liked to sublimate their rut in gambling and violence at a safe remove. As a city it isn't much: a dusty spread of unredeemed squalor paralleling an inlet of Mission Bay full of ship-repair and mothballing facilities.

"How long have Carmichael and Dr. Symonds known each other?"

That was easy. "Since last December."

"Have they been together a lot, would you say?"

"I'm not sure what you mean by a lot. Living apart the way they did, they had to wait for the weekends in order to see each other. But I'd say they were together as often as they could be."

"Was Carmichael a homosexual?"

"He never told me he was."

"I see. You'd just say they were pretty close then?"

"Yes, I guess you could say that."

"Now, you knew them both, Mr. Scott. What would you say they had in common? I mean, Dr. Symonds was a pretty important man up in Los Angeles, well known, respected, and then there's this welder who lives in a trailer who he had for a very close buddy."

I waited.

"If neither one of them was queer, what was the connection, do you think?"

"Bikes," I said.

"Bikes?"

"Motorcycles. They were both of them bike freaks. That's what they did when they got together: ride their bikes. Some men go boating or hot-rodding; they went biking."

The two of them exchanged glances, then Don went on.

"Did they ever have any difficulties that you know of? Disagreements? Fights?"

"If they did, they never told me about it."

"Okay." He turned another page and asked me when I had last seen Symonds.

"We had dinner together Tuesday night."

"Last Tuesday?"

I nodded. Symonds had been attending a convention of the American Association of Cosmetic Plastic Surgeons in town, and called me at the museum to meet him at the Grotto, one of the better seafood places down on the waterfront.

"Did he mention Carmichael when you were together?"

"Eugene never came up in the conversation. We talked mostly about my work, the convention; it was just an ordinary . . ." I was at a loss to say what kind of an ordinary thing it was, because nothing done with Symonds was ever ordinary. His presence, welcome or not, made it an occasion. Only—and I did not tell them this—on Tuesday I had found Symonds less extraordinary than usual in that he seemed somehow subdued. Not depressed or bored, but quiet in a way that indicated his mind was on something else. It was something he wanted to tell me, but would not. I knew him well enough to read the signs. I did not press him to reveal what obviously excited him; he came out with enough of it to turn me off completely. I had had to listen to too many of his intricately planned arrangements before and after the fact to want to sit through another one. The only curious thing about it was that he should have started that sort of thing again—with me, that is. Even before I left him, I had stopped listening to his boasting and his plots. It was as if he were sending me signals. "This is going to be a good one, something extraordinary, fabulous. Ask me!" Well, I did not ask him. Symonds had always exhibited a clearly defined tropism towards teasing, and I had had to learn, early on, how to parry this kind of bullying with feigned indifference. Our dinner broke up earlier than either of us expected it to.

The interrogation, like that dinner, came to an end as abruptly as it began. My interlocutor closed his notebook, put it in his jacket, and they both stood up, as if by a prearranged signal. Moving to the door, through which I preceded them in order to let the dog out of the patio first, the beefy one said they might want to talk to me again, and that I was to let them know if I left town for any reason. Turning at the patio door, he handed me his card.

"If you think of anything else that might help us, or if you hear anything, *anything* . . . give me a ring at that number. And if you hear from Carmichael, don't waste any time, call me immediately."

Outside, I followed them to their car. The other one got in the driver's seat. Beefy leaned out the window and said good night, then added, "That's a nice dog you've got there."

3

HE WAS a very good dog, one of a pair Symonds and I had picked out of a litter of German Shepherd pups in a San Fernando kennel three years before when I had finally come to live with him, and a year before I left him. "Come on, Thief!" I called him in and put his food down in the kitchen, then went back out into the night air to think while he ate; I was too rattled to think about food for myself.

Those cops knew I was holding back, but they probably did not guess, in this first encounter, that I was doing any more than skirting the boundaries of the truth. They could not have failed to sense how shaken I was, but they failed to ask the question I feared they would, for, in the shape I was, I doubt that I could have lied in answer to it. They would talk to the Silvers, who would confirm that I made a call, but nobody there knew that I had tried again, much later, to reach Eugene, since there was no answer that time either. And, as far as I knew, no one up at the Cove was aware of my having slipped out around midnight when I rode down to National City to see for myself what was going on and, having seen enough to determine not to see any more, returned before dawn, puzzled at what I thought I had witnessed and depressed at what I believed it implied.

How far those two detectives would push their inquiries would depend on how long it took them to track down Eugene. It was obvious to me that they found their assignment distasteful. The mess that was taking shape was something they would usually have expended little effort in investigating, since the police and courts considered that the suborders of humankind did not warrant the expenditure of much time and energy in the search for truth and justice. After all, few of them were inclined to stand up and demand their rights, and fewer still had the money to take a stand had they wanted to.

The confining social boundaries of color and craft, fractious borderlands already erupting elsewhere in the country, had not yet really been breached in San Diego, but the threat was there, a worrisome possibility that nagged at the public mind, informed as it was by the cant of popular prejudice dispensed by the media and the brokers of a limited job market. Sexual nonconformity was conveniently, if arbitrarily, lumped together with this leven of economic depression so that all three (niggers, bums, and queers) were regarded as a single, conspiring menace to good order and established mores. And if this made for strange bedfellows, none of the agencies fostering such a point of view saw any need for a reassessment of the muddled thinking that brought them to such conclusions. They had, after all, the votes.

The deviant sector's threat to the establishment was particularly troublesome, I suppose, because it was recognized as a kind of fifth column, veining as it did the entire strata of the community. And Symonds was a case in point. He straddled the bounds in such a way that neither the media nor the law could ignore his demise. The violent dispatch of a successful and moderately well known plastic surgeon is no everyday occurrence. It called for a more thorough looking into than might, in the regular course of things, have been the case. The circumstances surrounding his death would give rise, among those out there who did not know him, to feelings of fear and outrage, and the morbid professionalism of the newspapers and TV commentators would play up these reactions in lieu of any other sensational items coming to hand.

To those who were in one way acquainted with him, however, the manner of Glenn's passing would constitute a source of ironic amusement and idle, if not totally unfeeling, curiosity. It was to this element that both the police and I had to turn, each in our own way, in order to make sense out of what had happened. They did it because it was their job, I because I no longer had sense or courage enough to keep out of it. Of the two, they were the better equipped to undertake the search. But I, like Pinkerton, had special sources.

He had set the dogs on me, of that I was sure. But whatever he had told them earlier, it was, I was also sure, no more than he had to. The functioning of an informer is somewhat like plea bargaining in that he measures what he receives against what he has to drop, and as a matter of policy Dale always left something aside in case he was crowded later on.

With a numbness that was the aftereffect of shock, I had managed to keep from coming unstuck through the interview just concluded, but now I felt the cracks beginning to appear. Even as I sat there on a wooden bench near the patio door, the screen I had built around myself to hide behind was beginning to crumble and flake away like the stucco molded to look like carved stone on the old exhibition buildings in Balboa Park. For two years I had been trying to fade into the background and cultivate a certain peace of mind, but now this little cornered existence was about to be highlighted in a territory limned in violence and colored with the garish strokes of scandal. All fingers, it seemed, were pointing to Eugene—by way of me.

Some might say it was no more than the merited consequence of the company I kept, but I was not willing yet to assent to such automatic liability. At the time of our coalescing I was in no condition to estimate with that foresight granted those calm in heart and secure of income just what the future of our getting together might hold in store for me in the way of surprises. I had, after all, come across Eugene Carmichael in a time of emptiness, having picked him up much as one lays hold of a makeshift tool for lack of something better at hand. I can remember saying to myself the first time I woke up and felt his tough, peasant hulk asleep beside me in the foredawn, "He'll do . . . for a while."

He might well have said the same of me had he been the sort to speculate about these things. He was not that sort though. There was nothing very introspective, clever, or even particularly bright about Eugene. He was a plodder who got to where he was going in slow, determined stages. An unapparent shrewdness was the only astringency in his character. He was predictable to such an extent that whenever a sudden or erratic swerve in his behavior occurred, it was a sure sign that some other hand was at work on him.

I suppose it was my hand that directed him for a while. It was what he was looking for at the time—someone to take charge. When he first hove into view, he offered the kind of uncomplicated company I needed when I got too lonely to stand myself any more. Gradually he became a habit, as that sort will, like Saturday night at the movies, or Friday night at the bars, or Sunday afternoon on the beach. It might be said that before it was over we even became something like lovers. Everybody who knew either or both of us assumed as much without ever saying so, although there must have been more than the usual amount of head-shaking over such an unlikely combination. Politeness, rather than a lack of curiosity, probably inhibits the questioning of participants in this kind of comedy—at least while it is still going on.

As it was, it never dawned on me how much I had gotten used to having Eugene around until he moved out. But the shock of seeing him turn up at the front door one afternoon towards the end of December riding a new bike, of having him tromp in and skin down his Levis to proudly exhibit the Symonds *paraphe,* a rose and a whip etched in a still raw second-degree burn on the hairless right cheek of his ass, followed by the announcement in an unwontedly direct and enthusiastic manner that he had been taken over all but took the wind out of me.

I didn't let him know this, and as a matter of fact I took a while to let myself in on how I really felt, and why. It was not until a few days after that, when they had both been parading around town on a weekend like conquering heroes, showing themselves off to the goggle-eyed admiration of those interested in the kind of flashy display that often accompanies such changes in ownership, that what I had begun to take for indifference on my part was revealed to me as no more than a lid that was coming unfastened over a pot of rage achieving a slow boil inside me. And what scared me was that the rage was directed not at Eugene, but at Symonds, whom I had imagined winnowed out of my life forever in the wind of my escape down here a year and more before.

Yet there are some wounds that neither time nor absence heal; they only blur the hurtful aspect of them with a scar that itches when the emotional weather changes, forever serving as a disquieting reminder of something that cut too deep ever to be forgotten, or wanted, again. And it was not necessary for Eugene to tell me that it was Symonds' doing. A Harley trash-wagon; seared, personally embossed meat; unquestioning obedience to the spoken word and the barely discernable gesture: His Mark.

Still, I never desired him dead, for that or for anything that went before it. Nor did I ever imagine Symonds fellable until I glimpsed him hanging by the heels in that trailer and fled the sight of it, embarrassed for him as much as I was worried for myself whom I saw at that moment only as an intruding witness to what I took to be the mastering of the only human being I had ever loved.

This was still my secret, but for how long I could not even conjecture. Yet of one thing I was certain: the other person in that trailer, whom in my haste to get away I did not get a good look at, could not have been Eugene. But Eugene might well be the key to finding out. The law was after him and so, I knew now, was I. And it occurred to me, as it must already have to them, that there might be others who were seeking him as well; others who might already have reached him. Standing now in the rowen of their questioning, and awash in feelings of guilt at having run when by staying I might have saved Glenn Symonds from the death I doubted he deserved, I cast about for what I ought to do. For to do nothing and wait was the last thing I was capable of at the moment. I felt like a driver on a coastal road at night when the fog moves in. There is choice of stopping and sitting it out until the weather lifts, or of creeping ahead with the lights on low. I went ahead, but I did not creep. And I was in the dark almost all the way.

I decided to go over and see my neighbor, Ray, but first I went back in the house and dialed his number. He must have been sitting on the telephone; one ring and he answered it.

"Have you got company?" I said.

"Yes." A rich, creamy tenor, camping the syllable with an upward inflection. "And you?"

"They're gone. Is yours who I think it is?"

"Ye-es." Same inflection.

"We could hear him going past outside."

"The whole world heard him. Are you coming over here?"

"I'm on my way. Keep him there."

"*Keep* him! He wouldn't dream of leaving now. You don't think he came up here just to see me, do you?"

I locked the front door, called Thief, who had not touched his dinner, and we set out along a path that cut across the hill to the west through tall grass.

Ray Reynolds was a schoolteacher, and probably still is. He lived, and maybe still does, on the other side of the knoll we shared with a couple of other houses. It was the highest of three rounded hills that formed a sparsely settled enclave in Encanto we called Radio Heights because of the twin antennae of a local station that rose from the tops of the other two which were bare of habitation.

The area, while within city limits, was not yet built up; city-close but country-clean. And quiet. The freeway into town was only a couple of minutes distance by way of the Federal Avenue turnoff and a road called Weaver which wove up nearly to my house, but one of the hills to the north blocked out the noise of traffic traveling between the downtown and Lemon Grove, the closest suburb, which also happened to be where Eugene's mother lived. A rural somnolence enveloped this urban pocket. Cocks crowed in Chicano shanty yards down in the gullies, and the occasional sound of tires on gravel was no more than a pleasant interruption of the constant whisper of wind moving across the house and singing through the pines and eucalyptus and the clumps of bamboo grown wild out of somebody's yard. Red-tailed hawks rode the prevailing westerlies during the day scanning the slopes for mice and snakes and stray kittens; barn owls took over the territory at night when the scent of wild sage saturated the air. The scattered inhabitants of the area went their way and left each other alone for the most part. Cannabis grew in window pots among the ubiquitous geraniums or was nurtured in wild waste places in the sunny tucks of the hills screened by pampas grass and unseen from the main roads. From the tops of any of the three rises you could see Mexico, and on clear days the Coronados Islands stood out like ghost ships, gray against the far western horizon.

Ray's house was a little larger than mine, better built, and commanded a view of the ocean. He had gone to considerable expense over the years to do things to it even though he did not own it. I owned mine and had done as little as was necessary to keep it from falling apart. There was a high board fence around his, and in back a flagstone and dichondra patio hid behind a corrugated fiberboard windscreen. The house was maybe fifty years old, gabled and trimmed in a manner atypical of the architecture found this far south: somebody's summer house back before the First World War. Ray used a glassed-in side porch to paint in, and the other five rooms he kept as neat and clean as might an old maid aunt—which is one way of describing him.

The house looked dark now because the heavy curtains were drawn. I took a moment out to check the big white Hog parked near the gate, tilted over on its kick stand under a gnarled pepper tree. The seat was already dew covered, but the pipes were still warm. Thief let out a little whine of anticipation; Ray and he were good friends. So were Ray and I, but there were limits to our sodality, limits prescribed by me, and which I was about to breach. I waited at the foot of the back steps, patting Thief on the head to quiet him while I weighed contingencies.

Answering Pinkerton's summons was wise, and could in no way be construed as an unusual action on my part. The police were something to be expected too; I had no choice with them; they would have treed me sooner or later and would undoubtedly be around to see me again. Coming over here, though, was something else. Not because there was anything out of the ordinary in my walking over to visit with Ray, but because of the visitor he was already harboring in there and what that visitor might, in turn, portend.

At this point my reason for getting out of the house was not curiosity so much as the apprehension I felt at the prospect of being left alone with my own thoughts. To do something—anything—seemed at this moment preferable to waiting with no one but myself to talk to. I compromised by telling myself that after I had consulted what was waiting inside for me, I would return home and go to bed.

Ray had heard the dog. The kitchen door opened and the screen door swung out revealing his pudgy outline in the opening. He turned on the light and indulged in a wrestling welcome with Thief. An extended grin triangulated his moon face as he fended him off and motioned me inside. Closing the door behind us, he indicated the front of the house without a word, and I went on ahead of him through the darkened dining room. Past the archway that had once housed sliding doors was a lighted living room. A pair of well-formed legs encased in frayed Levis and black bike boots were derricked out in view on the faded Wilton. Their owner sat concealed, but I knew it was Lawson even before the legs were retracted and he levered himself into view, peering around the edge of the archway, scowling.

"I thought that had to be you plowing up the gravel outside my place," I said, taking a seat in a wicker settee across the room from him. Wicker was a thing with Ray right then. It was a cheap commodity at the National City swap-meet and he had furnished most of the house with it, anticipating the fad by over half a decade. He followed me in and stood for a moment in the archway, holding a can of beer, and with the expectant grin still plastered across his face.

"Did you set out to visit Ray," I said, "or did the fuzz scare you away from my place?"

"Fuzz always scares me away," Lawson grunted, and stretched his legs out again. He was lapped in wicker too; it and the leather jacket he wore creaked in chafing colloquy whenever he moved. Ray came over and sat on the opposite end of the settee, handing me the beer. Lawson shot him a sidelong glance that showed how bloodshot his eyes were. He had either been drinking or was recently in a session. I wondered if he had fallen off the wagon or reneged on his vow, or both, since, unlike Eugene, he was wont to mix the two reforms as he did the two failings.

"What was on their minds?"

Devious. To look at him one would have thought he did some kind of productive work. He might have been a mechanic, a plumber, a laborer. He had hands as rough and hard as a carpenter's, except that they were well-kept. Lawson never failed to amuse me, and I took my time looking him over while he waited, nursing a cigarette balled up in an adequate fist held close to his face, a map of interesting ruins charting the early ravages of teenage acne and the more recent detriments accomplished on it by outside agencies. I had not seen him since he had sworn off, and that would have been about a month before when there had come this big announcement to the effect that he was Giving It All Up. He handed this out one evening at a bar the way a prominent public figure might have distributed fliers announcing his candidacy for elective office.

"What the hell are you talking about?" I said at the time. "You *can't* give it up."

"Yes I can." He shook his head like a bull stuck full of picks. "I'm two days into it already. Givin' away all my equipment too. I went to confession for the first time in ten years last Saturday, and now I gotta get rid of it. You want anything? It's all got to go; every bit of it."

I told him no. There was nothing he had that I had any use for, but I questioned him further about his new resolve. It turned out that the priest had ordered him to disencumber himself of all the accoutrements of his folly lest the sight of them lead him to relapse into his former errant ways, and to report back regularly on the progress he was making along the way to salvation. I told him he had to be kidding. But no. . . .

"Not this time." He shook his head again, doleful and oh-so-sincere. "And neither is he."

"You're jaded, Chuck," I chided him. "There just isn't anything around this town that interests you any more—or that's interested in you."

He conceded that I was not far wrong, but added that the path of grace might lead through boredom as well as zeal. The corollary to this, I reminded him, was that the road to hell was paved with good intentions.

"It goes deeper than intentions," he persisted. "I'm fed up with this life, and that's for sure."

At the time I considered putting him to the test, but it hardly seemed worth even the mild efforts I would bring to such a trial. Lawson never could resist an assault. Looking at him now, I wondered again if he was still on this renunciation kick. He had certainly not changed his way of dressing. Probably the obscurity of the shriving booth did not allow his confessor a good enough look at him to judge that aspect of his reform.

"Didn't your spiritual director say anything about changing your public image as well?" I asked him.

His frown deepened, indicating he wanted a translation.

"What about all that junk you're wearing?"

"That ain't what's important." There was a sanctimonious note of irritation in his pronouncement. "It's the act that counts. Besides, I ain't got no other clothes."

Which was easy to believe, up to a point. Certainly he had no others he could wear in public, but I happened to know that he possessed several outfits that would have landed him behind bars if he had worn them on the streets. The full bloom of the flower children was still a good year away in other parts of the land, and down here bizarre costuming was looked on as a social outrage and a threat to public order on any occasion but Halloween and maybe

New Year's Eve. And never mind the ACLU—I doubt if a local chapter existed at the time.

"You keep this up, Lawson, and you'll turn into a tease."

"Keep what up?" That dumb, porcine squint almost dissolved into a smile. He enjoyed talking about himself.

"This yes-I-am, no-I'm-not game you're playing."

"I ain't out to tempt nobody," he said with a self-righteous jerk of his all-but-shorn blond head.

And he was unwittingly right in this too, being too much at cross-purposes with himself ever to be much of a tease or a temptation to anybody in any calculated maneuver. Fussy drag was his secret vice; heavy leather his undisguised worship. The two tended at times to clash. He had once belonged to a bike club of sorts, a local effort he undertook to organize, only to be thrown out of it as an affront to the image he had sought to cultivate in founding it. Lawson was a would-be shepherd fleeced by his flock; a perpetual yearner who never realized what he had going for him, because he sometimes scared the pants off people who did not know him. Literally.

Chuck Lawson was essentially a harmless soul with enough problems to cripple any three men for life, but he looked competent and mean as hell. And, he *was* tough. I mean physically tough. Solid. He was known as "Iron Ass" to those who had worked on him. He had to be durable to swing with the crowd he tried to hang out with, although they generally only used him for warmups.

"There's other things in life, you know."

It was now the backwoods schoolmarm speaking out from under all that scowl and brawn. He was full of platitudes, and dropped them like a dog relieving himself on the sidewalk. I had stared at him as I was doing now, until he lowered his glance to study the hair on the back of his hands. Ray-the-Obscure finally broke the silence by asking the obvious.

"Did the police want to know about Symonds?"

"Among others," I said, not taking my eyes off the hulk across the room. "They were also under the impression that I might know where Eugene Carmichael is—or was."

Lawson looked up at that and leaned forward. The ash from his cigarette dropped on the carpet, and I knew without looking that a slight crispation must have distorted the hospitable smile on Ray's face. Like everything else in the house, the carpet was old and worn, but clean, clean.

"Who else did they want to know about?"

"The talk never got around to you, Chuck. Disappointed?"

He mumbled something and gave Ray, and then me, a look that indicated he wished we were alone—he and I, that is. I caught it, but I doubt if Ray did. Ray was not obtuse; it was simply that his subtle social sense was no match for the edacious curiosity he had never outgrown from the time when he was the good boy of his block excluded from the fun the other kids were having out behind the garage.

"What d'you know about what happened?" I said.

"Me?" Lawson mimed surprise. "What the hell would I know about it? I don't go around strangling people."

"Neither does Eugene. And who told you Symonds was strangled? I understand he was beaten to death."

A gasp of distaste stirred the couch beside me, but I kept my eyes on Lawson.

"Do they think Eugene did it?" he said.

"I imagine they think it could've been Eugene. They don't know Eugene. When was the last time you saw Symonds?"

Lawson rolled his reddened eyes all around in their sockets while he considered this, or gave the appearance of considering it. At last he shook his head and said he did not remember exactly. "Not recently anyway. What about you?" And just to be fair, he glanced at Ray. ". . . or you?"

"Me?" Ray put a hand to his chest, shocked. "That man scared me to death just to look at him. I used to run the other way when I saw him."

"What're you talkin' about? He was a good lookin' man! There wasn't nothin' wrong with him."

"Who's talking about his looks?" Ray was enjoying this. He held Symonds in awe, and Symonds, for his part, had hardly acknowledged Ray's existence beyond the fact that he was someone I knew. "His looks only made it worse! All that beauty, and so *ee-vil!*" He giggled, then stopped abruptly out of deference to me, I suppose.

Chuck dismissed him with a grunt of disgust and turned to me again. "When did you see him last?"

"I had dinner with him Tuesday, I think."

I did not "think," I knew. But the encounter with the two police inspectors had left me with a tendency to be almost as imprecise as Lawson when questioned. But, "Yes," I added, "Tuesday."

"Does the police know that?"

I nodded. I was getting a little irritated with him, and with myself. My idea in coming over there—if I had any idea beyond simply getting out of the house and talking with someone—was to see if he had any information of a clarifying nature. Instead, he seemed to be asking all the questions. It did not do to let a person like Lawson take the lead in things; you lost leverage with him that way. As if reading my thoughts, he let go with another one.

"D'you know where Eugene is?" he said.

"I was going to ask you the same thing."

"Why me? I don't hang around with him?"

"Cut the shit, Lawson! When did you see him last?"

"Take it easy, doc." He suddenly settled back into the role of a deferent bumpkin, just trying, in that nasal twang of his, to set things in their proper perspective. "I'm as upset about the whole thing as you are. I was just wondering—you and Eugene being friends and all. . . ."

I started to comment, but Ray interrupted.

"Was he really . . . beaten to death?"

"I don't know. There's going to be an autopsy."

"But wouldn't you think *somebody* down in that awful trailer park would've heard what was going on?"

"They don't never hear nothin'," Lawson commented. "It's soundproofed."

"Well," Ray sighed, smoothing the creases in the corduroy pegbottoms that were too tight for him across the thighs, "They're out of their minds to think Eugene would do a thing like that—and particularly to Symonds."

"Oh, you never can tell," Lawson said. "Sometimes the worm turns."

"There!" I had to laugh. "You've heard it straight from the worm."

Lawson shrugged this off and looked at me. "Did you hear anything about Symonds havin' a new slave?"

"No, I didn't. Where did you hear that?"

"I do remember Eugene mentionin' somethin' about it, but I can't recall when. It might've been somebody else." He stood up and stretched his narrow-hipped, barrel-chested frame with much creaking of hide and jingling of metal. In a few more years he would have a pot on him if he didn't lay off the booze.

"Where are you going?" I said.

"I thought I'd go see what's goin' on in town."

"What about this slave?" I said. "What's his name?"

"All I heard was that he had somethin' new, and that him and Eugene was just friends now. There wasn't no names given to me."

I remarked the red eyes again as he settled a white-billed cap down carefully on his head, shading them. It occurred to me that he might be keeping his resolution and drinking more to compensate for it. His attention, such as it was, was still directed at me as he stood there getting ready to leave. Lawson's way of paying attention consisted of studiously avoiding any direct confrontation of his gaze with that of whomever he was interested in, and was no more than a stylized invitation to make him shape up. It was the way he was trained, or had trained himself. Devious.

"I'm glad the fuzz didn't crowd you much," he said.

"How did you come to that conclusion?" I was trying to find some excuse to keep him there, but I knew better than to let him know how badly I desired it. And anyway, I was at a loss as to how to proceed even if he did stick around. "I got the distinct impression they were crowding me plenty. And where in town are you going?"

He shook his head, and the hardware encrusting the cap trembled and glittered from all the rally and club buttons pinned to it like tiny *ex-votos*. "Oh, here and there; maybe look in at The Club, I guess. Nothin' much else to do."

Nor was there anything more I could do with him at the moment, so I nodded, which in Lawson's system of etiquette was an indication that he might withdraw and not be missed. Ray saw him to the back of the house and returned. We listened while he stomped his machine into life and set every dog in the Heights barking except Thief, who was used to bikes. After the noise of his going died away down the hill, I said, "What do you make of it? What did he want?"

"He wanted to see you, of course. Were the police very bad?"

"No, they weren't bad at all."

"I thought maybe they were; you were so out of sorts. I've never seen you mean to Chuck like that."

"I wasn't as mean as I should've been. He's playing one of his goddamned games. Either that, or he's worried about something. I'm not in any mood to put up with his foolishness now. Probably if you hadn't been around he'd have spit it out. He knows something. Or at least he thinks he does."

Ray sat down in the wicker armchair Lawson had vacated and lit a cigarette, prepared to hear my mumblings, without comment if need be. In no way my confidant, he was nevertheless a good and trusted listener, able, in the presence of stress, to remain low key and patient, helpful even. It was a condition that stemmed from, if it was not the reason for, his job as a special teacher in the school district. He oversaw a couple of dozen crazies abandoned by the primary administration as unteachables. Some were criminally delinquent; a few were absolute mental burnouts. But Ray maintained that most of them were no more than slow learners inhibited for a variety of reasons in their ability to achieve the goals other people set for them, to say nothing of those they might be induced to aim for for themselves. He had a fair percentage of successes in this salvage operation, and was left with pretty much of a free hand by an administration interested in body counts and balance sheets, and by parents or other surrogates who wanted, as much as the law and conscience would allow, to forget their existence. These little vestiges—or at least that part of them who recognized him from one day to the next—liked Ray. He had twice brought them in a group to tour the museum, and I acted as docent. I was charmed by their docility and the sudden, unexpected flashes of interest the exhibits touched off in them. Some were very beautiful, and likely to be more so in a few years. I asked Ray once if he ever had to wrestle with the desire to take any of them to bed. The idea appeared to stir no interest in him. His tastes, I knew, ran to the more prosaically mature, and were possibly more sick than would have been any rut stirred up by the proximity of the dependent, willing, and often provocative flesh he handled every day. Some deformed inspiration of brute manhood was what preoccupied Ray's mind in heat—which is not to say that he went in for the kind of thing that Eugene, Lawson, and Symonds did. He was known to get pukey if the action got rough. What he sought was the ideal of the self-rejected: the Real Man. And, such being the way with the faint of heart, he usually had to come by it second hand—literally.

The agent by which he realized these dear fantasies was a young woman of more than average proportions and weight called Linda. Linda taught school too: English in a junior high. She had a husband everyone called Dump, a somewhat subdued housepainter who had spent time undergoing therapy in the state hospital at Napa to loosen him up from a severe case of the blues that came on after their child, Claire, was born. He had always been moody, to hear Linda tell it, but she became so herself after the birth of the little girl. She had no idea of why this should be, but the sight of Dump lavishing care and affection on little Claire at first depressed her and then lashed up her rage to where she took to inventing reasons for not going to bed with him while searching for other men to fill the exorbitant demands of a sexual appetite that threatened to surpass those of the uncontrollable bulimia with which she had been afflicted all her life.

Within a year she had pussy-whipped him into a state of befuddled abnegation, a semi-comatose state from which he could only be roused by the sight of their child. He stopped work and laid about the house until she finally had him committed. After about nine months on electricity or insulin, he was returned to her, a quiet, dutiful cipher. He went back to painting houses and, aside from Claire, appeared indifferent to anything outside the routine of work, sleep, and the frozen foods that constituted his existence. He showed no interest in what Linda did, and looked benignly on her friendship with Ray, an association which had grown up during his removal to Napa. This was Dump's way of coping with the connubial morass which had enveloped him. Linda's was to eat herself into a state of adiposity repugnant to all but a few men who like to get it on with pigs. To facilitate this last, she and Ray had schooled themselves into a kind of conjunctive symbiosis—of which more anon.

Besides Linda, Ray had a loose arrangement of long standing with a young man who taught art at Fresno State College and got down to see him once a month through the academic year and spent most of the summer with him. I found Ray soothing at times like this, and at others, when our comic moods happened to mesh, amusing to be with. He was like a steer among bulls.

"All right," I said. "Lawson saw me. So what?"

"He knows where Eugene is—or was."

He said it in that quiet way of his, the almost buffoonish grin on his face making mock of all the concern Chuck and I had been exhibiting for the last quarter-hour or so, stalking around the subject like two cranes in the mating season.

"He told you that?" I said, on the verge now of venting a little spleen his way.

"Not in so many words, no. I think that's why he came up here though. Seeing that police car down at your place probably scared him off."

"How could he knew it was a police car? They were plainclothesmen, and the thing was unmarked."

"He knew somebody was there, and he's not too thick-headed to guess."

He was right about that. "Well, why didn't he spit it out after I came up here if he knows something?"

Ray considered this a moment. "Maybe he doesn't want to get mixed up in it now that he knows the police are bothering you. I know I wouldn't."

"You're different and, besides, you don't know anything. But if he knows where Eugene is, and isn't talking, then he *is* mixed up in it."

"It'll probably take a little while for that kind of logic to percolate through his pea brain. What do you suppose he wanted to tell you for, assuming he does know where Eugene is?"

Thinking about it, I reasoned that Ray might have the cart before the horse. "It would be like Lawson to come up here to compare notes in the event that I also knew where Eugene was hiding—if he is hiding—or to keep his mouth shut if I didn't. Or, what if he's just curious because he doesn't know a damned thing?"

"That's the trouble with second-guessing dumbbells." Ray stubbed the cigarette out and picked up the ashtray from the coffee table. He went off to the kitchen with it, his voice trailing behind him. "It's hard to know what direction their stupidity's taking them."

I did not believe that Chuck Lawson was stupid. Slow perhaps, and self-defeating in most of the things he did, but not stupid. Something was bothering him, though; something he wanted to tell me about but had shied away from at the last minute because of something I said, or did not say. Or Ray may have said something that changed his mind for him before I arrived. When Ray returned, carrying the ashtray wiped and shining, I asked him if there was something he was not telling me. He looked at me, another cigarette halfway to his mouth, hurt amazement slackening his features.

"Mr. Scientist-Man, I don't know anything about this idiotic business. Chuck Lawson came up here and said he wanted to talk to you about something, but that the police were over at your place and could he wait until they were gone. He sat here going on about what a terrible thing it was, asking me what I thought about it, and what you thought about it—as if I ever know what you think about anything!"

"But you think he knows something."

"Of course he knows something! Whether it's worth telling is something else. But he's aching to spill it."

I stood up. "All right. I think I'll go find him and relieve that ache."

"There *is* one thing," he said, looking at his cigarette as if it had suddenly tasted peculiar. I waited. "He asked me if you knew this new person Symonds was seeing."

"That's interesting," I said, "but not very. I don't know him."

"Neither do I. I didn't even know Symonds was fooling around with somebody new. I thought he and Eugene were still lovers."

"I doubt if either one of them ever gave that name to their relationship, but your point's well taken."

"Do you think it's got anything to do with what happened?"

"I mean to find Old Red-Eye and get to the bottom of it, if I have to get to his bottom to do it."

Ray was grinning again. "He'd like that."

"Don't be too sure."

"Can I come too?"

"Thought you didn't want to get mixed up in it."

"I just want to watch."

His style.

"This isn't fun and games, Sucky. And no, you can't come."

Then, at the kitchen door, I changed my mind. Ray and I often went out together for a beer or two. He was good at fading into the background when that was called for, and it would be the most natural thing in the world for us to stop off at any of several places together on a Saturday night—even on the Saturday night after Symonds got killed. The people I would run into were no less unfeeling than those anywhere else in the world.

"Come on if you want to," I said. "You'll make a good witness."

"A witness?" Ray hesitated on the steps.

"You never know what you may be getting into. Come on."

"I don't have to *do* anything, do I?"

"No. I'll do what's necessary. You just stand by, maybe keep score and run errands."

"And witness."

"If it comes to that." He was joking, but I meant it. I had no idea of what I was going to do.

We stepped out into the cool night with Thief bounding all around us. "Drop me off at my place and I'll pick up my car," I said. "Then follow me. I don't want to be burdened."

"Yes, massah!" he said, and we headed for the shed where he parked his Volkswagon. Thief followed behind and beside the car over the short lane that rounded the hill to my house where I locked him inside again. The poor beast had not had a good run in two days. I mentioned that he had not eaten his food, and Ray said he had fed him before I got home. Ray was helpful like that.

4

MY REMARK to the effect that Symonds might have been beaten to death was no more than a check run on Lawson to catch his reaction. If he knew anything for certain, then he would think I knew nothing; if he knew nothing, then there was no harm in further muddying the waters, for him at least. I did not reason it out this way at the moment of speaking to him. Rather was it like much of what I had done in the past few hours, and would do in those to come: quick shots on impulse followed by a survey of what I meant and what others were liable to construe from it after the smoke had cleared and there was time to think about it.

Almost all of what I had said on the subject of Symonds since that call from Pinkerton was in the nature of evasive action. Take Lawson's mention of Glenn's new connection. The name of the candidate in question was Owen—Owen Somebody. That much I knew because Symonds had talked about him to me on Tuesday night. He did not tell me much; most of what he said was a gratuitous retailing of the feats he was about to accomplish on this new one. I wasn't interested, which he well knew, but he had to have somebody to whom he could crow, and since he was down there for the convention I was the only one available to whom he would deign to tell it. The only reason the name and situation stuck with me was that I wondered at the time how Eugene was reacting to this new alignment—assuming he knew that his successor had been chosen. There was no reason either for me to go into all this with Ray, particularly when I was not too sure about the details of it myself.

The truth was I was in no shape to think very clearly about Symonds or about what had happened to him, but I was trying to convey an impression of, if not indifference, at least distance where my relationship to him was concerned. It was a stance I had adopted ever since arriving in San Diego, a way of avoiding the issue.

As I wove my way down Weaver towards the freeway with Ray

tailing me in his Bug, I tried to put Chuck Lawson more clearly into the picture. His awe of Symonds was in the nature of that homage fraud pays to integrity. He knew Glenn by reputation long before he came to know me, and Symonds' refusal to have anything to do with him after a single session was a putdown, the next best thing to an enduring acquaintance in Lawson's view of things. By the same token it colored his perception of me.

I did not meet Chuck until I had been in San Diego for several months, and my familiarity with him, once met, was limited to brief and infrequent conversation. Lawson's interest in me, beyond that conjectural stance he always adopted when faced with anyone who might, if provoked enough, put him to the test, stemmed from the aura still surrounding me from my—how shall I put it—legendary? Yes, let's put it that way—legendary alliance with Glenn Symonds.

Some unknown harbinger of my arrival in those parts had let it out among the lunatic fringe that I had been, or was still (notions and rumors varied), Symonds' boy. Either way was enough to establish instant credit in my account for those transactions it was assumed I had an interest in negotiating in this new territory. That I did not immediately or subsequently avail myself of their respect and confidence was cause for speculation. The consensus, I suppose, held that after having experienced Symonds a person could go no farther, and so naturally I had, (as Chuck so often claimed to have) given it all up, with the difference that Lawson approached renunciation through despair at having dragged bottom while I (all fairy tales are logically sound) *must* have viewed as futile any possible future encounter after the prolonged bliss I had unquestionably enjoyed with the *ne plus ultra* of almost everybody's ideal, everybody in this case comprising the restricted crowd I was dealing with.

I listened to them and said nothing. But to hear the admiring, even laudatory comments of others about someone you no longer admit as a part of your life, recapitulates in a disheartening way the earlier, joyful anxieties you suffered while in the process of forming a linkage with that person, and such crippled recollections were the last thing I needed at the time. I retreated before these unwanted attentions and tended to keep to myself. One of the reasons Ray Reynolds and I became friends was that he knew nothing of my former existence and had small acquaintance and no patience with those who did.

Eugene Carmichael was the one who, not quite a year after my arrival, spoiled the pretty vision these watchers and talkers nourished of me and my ways. Taking on so used a product as Gene made me, after all, human in their eyes. In their purblind view I was like some Phaeton flung back to earth after a daring but unmerited ride with the Sun. I viewed with amusement tempered by irritation the glimmers of pity and admiration that lit their eyes in the darkened purlieus of their gathering places after Gene and I began to frequent them. The reception I now received, tinged as it still was with residual respect, was the kind usually accorded down there to used-up movie actors, like someone who once went over big in juvenile roles but now manages a Buick franchise in Azusa.

Yet at first, as I said, I pretty much kept to myself, staying at home, absorbed in my work, seeing almost no one. Symonds used to call during the initial stages of our separation. In the beginning the telephone rang at least once a week and there he would be, like a spider on the line waiting at the end of his webbed tunnel. Then a month might pass without his dialing San Diego. There was a period when an entire season slipped by before I heard from him again. Occasionally a letter or a postcard would arrive: clippings about him, an article he had just published, scribblings in that large aggressive script resembling a copybook exercise in ovals—his touch: a finger on a bruise. For he was, I knew, fingering me the way a surgeon palps a locus of infection; carefully, deliberately gauging the degree of inflammation and the amount of hurt that might still be there. He was, after all, an expert in his field, but his field lay not in cures, or even in the reparations of damage done by accident. With him the healing art was ancillary to the marvels he accomplished on disfeaturements and the counterfeiting of vigor in those undone by time and their own intemperance.

Of course I would talk to him whenever he called, all the while trying to keep my own utterances as noncommittal and yet as unequivocal as possible in the face of what was nothing but probative aggression wrapped in the guise of blandishments.

Naturally I never initiated or returned any calls. Nor did I ever respond to the mail. He must have known that I was fighting for my life, and I suppose he knew, knowing me (and I have no doubt he read me like a surgical atlas), that there was no fundamental strength in any of the withdrawal I practiced. My only strength lay in a passive refusal to respond, for the mechanism of discourse and response between us that had once sprung so quickly and fully to life was run down. We both knew this without any need to admit it formally to each other. True, it could be made to whir briefly into motion whenever he cranked it up by the bullying affrontery of his seemingly limitless initiative. But, unsustained by any help from me, it quickly petered out and we were left frozen in a stale tableau of tired and untrusting strangers who had once meant the world to each other.

Glenn may have chosen to regard me as some kind of a challenge, although all I really was was a piece of him he missed. God knows there were pieces of him still rotting away in me where they had become lodged in the process of my tearing away. No matter. I looked on him as my enemy. His importunate testings I regarded as attacks, and my response to them was a defense in depth by means of time and distance. Given more time and greater distance it might have worked, but his tactics, I realized too late, were simply flanking actions in a grander strategy that had preoccupied him for a long time and for which Eugene and I served as screens. And now it looked as if these grand maneuvers of his had somehow gotten out of control.

Ray pulled up beside me in the parking lot behind The Club, a cinder-block storefront building set down on the corner of a residential street that had suffered a downgrading by the proximity of the new freeway. I had rented a bungalow a block from there when I first arrived, before I sank all my cash in a down payment on the hovel out on Radio Heights to get clear of the bay-blown mists and the neuralgia-inducing noise from a wind tunnel down on the airfield operated by a firm aptly named Rohr Industries.

Together we went inside, but there was no Lawson.

After waiting through half a beer, we left and drove over to Pacific Beach. It was still into the late dog watch of dinner time and not many people were out. Jo-Do's, in contrast to a few hours earlier, was almost deserted. Pinkerton was gone and the bartender was plugging the jukebox to keep himself company. He wanted to talk about Symonds' death, but we excused ourselves and left the free beers he offered untouched.

Driving up to La Jolla, we began to work our way down again, checking on a half a dozen other bars Lawson might have hit. Ray would go in alone to case the crowd, if there was one, and question the bartender since, having known Symonds, I would immediately be surrounded and quizzed about what had happened. I didn't want that kind of attention, nor did I want it known that I was looking for Lawson. Not that it mattered, as far as I could see. The law had not discouraged me from seeking out and talking to whom I pleased, but if Lawson got wind that I was after him he might become hard to find, particularly if he had a lead on Eugene. All the way through this search, as one thing led to another, I kept establishing and revamping certain limited goals, each leading, I hoped, to Eugene. Lawson was now my immediate concern, and I did not want to be sidetracked by sensation-hungry queens.

We found him where he said he would be: back at The Club. We had not given him time enough to get there, that was all. Lawson never took a direct route to anything if he could help it. Anticipation and foreplay were more than half the game with him. It was a way he had of driving prospective masters up the wall before they nailed him to it.

"Have a beer with me," I said to Ray after my eyes became accustomed to the dimness inside and I had spotted the quarry, "then get lost."

"Can I hang around here?" he asked.

"I don't care what you do as long as you stay away from me." Then I pulled him back as he started to move off into what had now grown into a sizeable Saturday crowd. "Let me know, though, if you go out with anybody. I may need you for something."

"Messenger service or towel boy?" he sighed.

"Maybe both."

Ray understood that I wanted to be free to move. Which I did, away from the entrance to the far end of the horseshoe bar from where I could keep an eye on Lawson who stood directly across the room past the other horn of the bar from me. Finding a vacant stool, I sat down and leaned over the naugahyde-padded forward edge of the bar top. (It was that kind of place.) Chris, the owner, had just come in and was presiding in person while his barmen did the real work. Seeing me, he waddled over, a squeal like a steam leak escaping from him as he expressed gleeful distress.

"Jesus! Isn't that a terrible thing about Symonds?"

Instantly every head was turned our way. He leaned forward, fat pale hair-tufted hands on the bar in front of me.

"I heard about it a little while ago and I couldn't finish my dinner. It's sickening!"

I agreed that it was and asked him to give me a beer.

"Sure, Scotty. This one's on the house, hear? I can just imagine how you must feel, being a friend of his and all. You wonder what the world's coming to sometimes."

I wondered what it was coming to all the time. Right now I wanted him off my back so I could maneuver.

"Chris," I said when he came up from below the counter with a bottle of beer. "Who's the guy over there talking to Chuck Lawson?"

He turned around to look.

"You mean Pinkerton?"

"Clean your glasses, for Chrissake! I know Pinkerton when I see him. The other one—on the left."

He actually adjusted the black-rimmed spectacles which had faintly tinted lenses, then turned back to me in wide-eyed bewilderment. He acted everything out in dumb show before howling the message in a tenor that could carry over the loudest gathering.

"Why," he said, as if someone had accused him of hiding something, "I don't know! I never saw him before in my life. Didn't even see him come in. Of course, I just got here, and I've been so busy, and so upset too, like I told you. . . ."

He was off again like some virilescent biddy, and I had to sit through more lugubrious cluckings over the day's events before he waggled off around the curve of the bar to cater to other arrivals he recognized.

The influx was keeping his two barmen busy, but another hireling, a dyke named Wendy who worked the floor with a tray on busy nights, was nowhere in sight. I looked at my watch. It was only ten o'clock. Left alone now, I nursed the beer in peace and looked the room over to see who else of interest was there. Most were regulars. Ray, I noticed, had latched onto a more or less straight-looking stranger in coat and tie. He kept his eye on me, with glances every once in a while at the trio across the room while he conversed with his potential trick.

I returned my attention to what I had come after. The three in the corner made a picture to contrast with the surroundings. All in leather, but not all in the same degree, they were very much apart. The Club was not what would be called even an approach to a leather bar, not even at that time, but it often served as a starting point for warmups and displays. Chris indulged this alien influx as long as it was sporadic and well-behaved, but was too gaga in his old-time candy-coated way to know what to do with it, or how to foster it into any kind of a consistent draw for others who might be interested in the flavor it imparted to the usual fruity mix. For, to use one of his favorite expressions, why bother? He was making a bundle the way it was. Besides, it takes Leather to know Leather, as leathermen never tire of repeating, and Chris was only vaguely titillated by the idea, responding to it with a delicious camp shiver which last night's occurrence had converted into a shudder, equally camp.

Even so, the accommodating atmosphere of The Club and the atypical pickup that could sometimes be found there, made it a well known dropping-off place for out-of-town stuff—Western and Leather from The Gauntlet and The Explorer up in L.A., top and bottom men from as far away as San Francisco's Tool Box, and occasional numbers from Chicago and New York who found their way down to San Diego as a sidelight to other business that brought them to the West Coast. Symonds had never come here until after he took charge of Eugene, but Gene, who had for a long time been a target for expeditions this far south by the so-called Big Boys, used it as a steady base. Eugene, however, had never donned leather in public until he got tapped by Symonds.

Whenever these exotics descended on The Club, they usually came in clusters. Strangers, easy to spot, quieter than the frenetic local cats, they dressed in a more thought-out way, whatever they wore. Their manners were more studied and less sociable than the giddy stream around them, and their reckoning glances had about them the covert attention bidders at an auction give to items on the block. In short, they acted like men with something on their minds, in contrast to the hometown team who behaved as if they hadn't a brain in their heads. This impression was not totally deceiving.

One of the reasons San Diego used to get these off-squirtings from the larger centers of salacity was that people with big names could prowl this sector without any great likelihood of being recognized if they didn't hang around too long and were careful not to make any noise. This was before The Great Awakening, remember, when figures of prominence were careful about what they did and where they were seen in public.

These transients were not interested in chicken runs, or flashing in the bushes, or action in tea rooms, monopolies enjoyed by the local closet set, and which supported the pettifogging trade of entrapment and exposure that heightened the sensibilities and reset the moral tone of the solid majority who ran the town, or supported those who did. No, they were seeking other pleasures. One came to recognize their provenance not only from their dress, the words they used, and by their more restrained gestures, but by an undefinable something else that was the sum of more than these parts. The one who was talking to Pinkerton and Lawson I judged, as it turned out correctly, to be from L.A.

None of the three of them appeared to have noticed me, but this meant nothing. Lawson had remarked and acknowledged my presence from the moment I entered, and without any overt sign on his part. Pinkerton appeared too busy talking to have noticed anyone but his listeners, but you never could tell with that flat-assed string bean. The other one would not have known me anyway, unless told who I was.

Chuck, as befitted his lowly station, kept silent except when addressed by one or the other of the two flanking him, but his gaze was busy in the crosstalk, canvassing everything that went on around him. He had the multifaceted eye of the wary barfly. He might be sloshed, but if his life depended on his having to know what was going one, he would have been in no danger. No grouping, no pairing- or tripling-off or subtle shift in position among the evening's company got by him without his noting it and tabbing the hidden meaning of it, if it had any. He knew everybody, or found out about whom he did not know. He could estimate the degree of drunkenness or sobriety any of them might be afflicted with at any given moment. He knew who was on what. No one worth clocking went to the head but that he was aware of how long they remained there, or of who followed them in. He not only noted who left the place with whom, but he made it his business to find out what went on afterwards—again, if it was worth knowing. And all this without seeming to do so, as if it came to him through some process of atmospheric osmosis or peculiar informational magnetism that he alone possessed. With a little prodding he would usually disgorge what he knew; he had a well-hung mouth. And it dawned on me there for the first time, seeing him with Pinkerton, that he fulfilled perfectly the role of an informer's informer.

So, for that matter, might Eugene. But the three of them were quite different in this knowing aspect of their personalities. Eugene, for instance, did not drink, and the only reason he paid attention to someone was for the working-over he might receive from them in return. Pinkerton, on the other hand, had no quarrel with alcohol, but he played the get-acquainted game as a sidelight for protection, and, except for a minor, albeit distinctive aberrancy, he took no part in the violence practiced on and submitted to by the other two—although he was wont to watch if the opportunity presented itself and he had nothing better to do.

Eugene and Dale exhibited singleness of purpose with different ends in view; aims obvious in the one and hidden beneath a veneer of good-natured insouciance in the other. Lawson, however, had small aims and practically no control over the workings of his omnivorous talent for observation. It came to him, if not naturally, at least out of necessity. As he told it, he had sharpened these faculties during the lonely Catholic childhood he was supposed to have spent in a fundamentalist Protestant mining hamlet hidden away in an Alleghany glen where he was forever on guard against the daily ritual of pursuit and abuse at the hands of papist-baiting bullies. Yet after the onset of manhood and deliverence from that poor white hell, he could find no better use for his circumspect abilities than in the stalking of a stud who would consent to use him as a piece of ass for an hour or so and then maybe knock the shit out of him afterwards as thanks for the favors given and received.

They might have stood him in good stead in the work he did as a bartender—for that was his trade, if he could be said to have had one. When I first arrived in town he was tending bar for Chris. It was not called The Club then, but carried some fatuous appellation like The Shogi Room and was decorated with paper screens and hangings which the fire marshal soon got around to condemning. According to Chris, it was the steadiest employment Chuck had ever had since he ran away from home. Chris would not dignify the four-year stint he served in the Marines by calling it work.

The Marine Corps at least gave Lawson an honorable discharge, but his tour of duty with Chris ended the way all his subsequent jobs terminated, in disgrace. Chris fired him. Sadly, but with an unbending, righteous air of having been taken advantage of too long, he told him to clear out one morning after he had failed to show up for work the night before. He was too hung-over or too beat-up to get out of bed; he forgot which.

"Chris," he would explain without rancor, "don't understand these things."

Which was true enough, as far as it went. Chris made sincere efforts to ignore "these things," even when they took place right under his nose. A pudgy fairy with a broad motherly streak running through him as wide as his butt, he knew what he wanted for love or money, and what it was worth to him to get and hold onto it. He owned The Club, another bar on the beach that catered to straights, an apartment building, and a scattered, hillside colony of quaintly remodeled shacks which he rented out (both the apartments and the shacks) at less than market rates, mostly to men whom he referred to collectively as "my boys."

His boys were single (or at least separated from wives by writ or abandonment), and most were gainfully employed. Chris was known to tenant youthful indigents and find them work if, in his estimation, they merited this consideration. It was not too difficult to get on his merit list; looks and loyalty were what he cherished in those he would help. There were other obligations too. These varied according to the degree of downness a prospect for his aleatory missions of selfish mercy might be going through and the converse high their situation and appearance produced in him.

For the rest, Chris knew where he stood with the law, paid dues to public service organizations, gave to publicized charities, and was ruthless in business matters. "I'm so *bourgeois!*" was how he characterized himself in a creamy chortle. "Upright and soft all at the same time. See if you can figure *that* one out!" Followed by the comedian's braying cachination. He could have made a fortune on the stage or the tube, piping the garrulity of aging vice.

As it was he was loaded with dough and dripping with sentiment, like a sticky Greek pastry. He surrounded himself with what he judged to be wit and good looks, neither of which Lawson possessed. But then, as he never tired of admitting to patrons who seemed never wearied of hearing it, "I always give an A for a good bod or a nice basket." It was a rule of thumb left over from the time, years before, when he had been a high school principal somewhere in the Midwest. Enlightenment and not scandal had driven him to these greener pastures where he could mother the young, the beautiful, the homeless, and those others endowed with little or nothing of the world's good and stranded in the retrogressionist mire of Orange and San Diego Counties.

Chuck Lawson must have been a trial for him though, and the only reason he put up with his antics as long as he did was because Chuck occasionally let him watch the more conventional performances he was able to get himself into. Once they happened to get their signals crossed, however, and Chris found himself witnessing a piece of heavy action that left him blanched and indignant for days afterward. He apparently could not tear himself away from what he saw going on, but he vowed he never wanted to see anything like it again. Yet he was continually bringing it up. Hearing him recount the experience, I could not buy the attitude of innocent abhorrence he adopted in the face of his admitted tarrying on the scene after he sized up the situation for what it was. After all, the voyeur is as valid a role as any other in the total madness all this carrying-on amounted to. I told him as much.

"Hell, Chris! You brought yourself to the feast, you might as well have enjoyed it since you stayed through all the servings."

"It was horrible!" he quivered, his jowls shaking, "simply horrible!"

"Of course it's horrible!" I laughed. "But what's that got to do with it?"

We were both kidding, and we both knew it.

For his part, Chuck never held it against his benefactor for canning him. "I let him down," he admitted with a facsimile of contrition. Lawson was known to bite the hand that fed him even as he was moved to kiss that same hand when it became a fist to correct his breaches of gratitude and good conduct. Neither of them bore a grudge, and Lawson still enjoyed visiting privileges which included the use of a basement room in the apartment building when he did not have any other place in town to stay, or needed a hole to be taken to. I wondered if the priest's injunctions had included clearing out the fixtures and doodads that embellished that subterranean chamber.

5

THESE RUMINATIONS were interrupted by a hand clapped on my shoulder. I turned around to behold Cusped the Harpist standing behind me, predictably attired in black.

"I haven't seen you in a coon's age, man!" he yowled, bringing me once again to the attention of most of that end of the room, now packed with standees. He leaned down over the sweatered occupant of the stool next to mine and requested him in a voice of dire friendliness to give it up to one who had need of it for some private matter he wished to discuss with his neighbor. The sweater looked at me with a view of refusing, but his resolve appeared to cave in when he got a look at what had made the request. He rose without comment and melted into the welter behind us as this old acquaintance from L.A. days took his place.

Cusped was wearing a black curl-brimmed buckboard with a chain hatband, black broadcloth shirt with black satin piping and pearl buttons, black silk neckerchief, black silver-stitched belt and silver buckle with a longhorn's head in bold relief, black Frisco jeans tailored to a snug fit, black-dyed sheepskin vest with the short-clipped wool on the inside and the outside decorated with Mexican silver conchos and finely embroidered silver appliqué all over it, tooled black-on-black Tony Lama stovepipes, and (I looked twice just to please him) Mexican tooled silver spurs with crimped chain laps. These last I had brought him back as a present from Guadalajara several years before.

"What would you know about Symonds' getting it?" he said, nursing a bottle of beer between pale, narrow hands with long, nail-chewed fingers.

"About as much as you'd know about a coon's age."

"Tilly been talking to you?"

He grinned. It was more like a sneer, but that was due to the almost skeletal contours of his face rather than the contemptuous turn of his mind. And his use of "Tilly" to indicate the law was an earmark of his age and underlying style which, despite the veneer of leather and the frontier crusting, betrayed formative years spent at no other corral than the rail of The Jade Room on Hollywood Boulevard in the '40s.

"What makes you think they'd be interested in talking to me about what happened to Glenn Symonds?"

"Just wondering," he said. "No offense. And by the way, what did happen in that trailer? Nobody seems to know exactly."

"He was drowned."

"Drowned?" The look he gave me indicated surprise and not incredulity. Drowning was a possibility Cusped had obviously not considered. The mechanics of drowning somebody in Eugene's trailer, offering as it did the limited resources of a stall shower, sink, lavatory, and toilet bowl as means to this end, represented a challenge the solving of which in the dispatch of one as unlikely as Symonds to submit meekly to such a perishing aroused his admiration. "Is that what they told you?"

"They who?"

"Come on! I know the heat's been at you."

"It was what I was given to understand."

"What else did they say?"

"I can't talk."

"Well, I never thought Eugene had that much imagination."

It was a point to be considered. Eugene, to be sure, was not endowed with much originality, but he had picked up what he could use in the way of knowledge and technique with a shrewd retentiveness lacking in mere dabblers and undervalued by most of those who got close enough to him to test his abilities in his specialized line of pleasure. The world is full of people who have made a name for themselves by copying from their betters and calling the results their own, and Eugene had passed through so many hands, assimilating what he wanted to use from what they had to teach him, that what he now was was really something else, an original.

"And," Cusped was rambling on, "I never would've imagined Symonds going out like that."

"You mean by drowning?"

"No. I mean getting taken. You know Symonds. . . . "

There was nothing I had to say to this implication of a shared knowledge. Cusped had maintained a loose kind of acquaintanceship with Glenn for many years, one that almost amounted to friendship, I suppose. He satisfied certain cultural pretentions Glenn had, affectations often found in successful doctors, and which points to a left-out feeling they experienced when exposed to esoteric fields of interest other than their own. This seems to arise from their having sacrificed so much of the formative time of their schooling to the studies and demands of their particular discipline. By the time they are certified in their specialties, they have become so expert in one branch of knowledge that to hear them talk about anything else can leave the listener with the not totally false impression that they are gifted ignoramuses, if not actual idiot savants. Glenn once opined that anyone would be a fool to trust his cure to a doctor who enjoyed a well-rounded life of varied interests. On the other hand, he also expatiated on the dangers awaiting the patient with complicated dysfunctionings who fell into the hands of an eminent specialist too nearsighted and assured of his own expertise ever to entertain the notion that his diagnosis or treatment might have a hole in it somewhere. It was the kind of Scylla and Charybdis present-day medicine had to get past somehow, and he was content that his patients came to him for their looks and not their lives—although their livelihoods, as well as their looks, might depend upon his skill.

Anyway, what Cusped and I knew of Symonds in his night-time avatar might well be two different things. I believed him capable of almost anything he thought he could get away with, and this was what had worried me all the time I had been his lover (not that he practiced any of his foolishness on me), and it continued to worry me when he took up the chase down here where I had thought to have gotten clear of him. But as to Cusped's perplexity over his death, I could not but agree: it was not like him. I changed the subject.

"How long have you been in town? I read somewhere you had a concert tonight up in L.A."

"That was this afternoon," he said, always glad to talk about himself.

"Did you play dressed like that?" I looked down again at his spurs, and he laughed. But it was no joke. Once a year he performed at a semi-private recital given by and for the Southern California Harpist's Association in white tie, tails, and black boots much like the ones he was sporting tonight. Minus the spurs though; they interfered with the pedals.

"I changed after it was over. Three of us came down. We were supposed to meet Symonds." He shook his head as if disheartened by it all. "Isn't that something?"

"Tonight?" I said. "Here?"

"Tonight, but not here." His face suddenly brightened. "Hey! Would you believe there's going to be a party at Chris'?"

It was not at all hard to believe; Chris had a party every Saturday night.

"What's that got to do with what we're talking about? Where were you going to meet Symonds and who arranged this thing?"

"As I understand, it was supposed to be some sort of a coming-out party, if you know what I mean. *And*—get this—in Carmichael's trailer."

What he was saying made no sense to me. Pinkerton was moving towards the door and I half rose from the barstool, but lost sight of him in the crush of people there. I noticed now that Ray had disappeared too, but I could still see Lawson, across the other bar with the one I did not recognize now all to himself.

"Who was coming out?" I said, keeping my eyes on the remaining twosome.

"Don't know. Eugene wasn't very precise. But whoever he is, somebody else'll have to show him off now—if he shows up at all."

"Are you telling me Eugene put this thing together?"

"Not on his own he didn't. He said Symonds said do this and that, and we all did it, but now it looks like we're going to have to find something else to do."

I was still watching across the room. Lawson was now talking, his back to me. He swiveled this way and that, checking the action while his jaw barely moved. His listener never took his eyes off him, staring at him as if he were furious at whatever he was saying. It was beginning to look as if I might be shortstopped in getting to Chuck if I didn't make my move soon. On the other hand, I wanted to hear more about what Cusped had just touched on.

His news of this arrangement meshed with what Glenn had said to me on Tuesday night, but at that time I hadn't gotten the impression that he was so far into his latest chattel that he was letting him out for use by others. It had not sounded to me as if he had gotten into him at all up to that time, now that I thought about it. He had said, in so many words, that he was *about* to consummate something. But why would he call everybody down here to do it?

"Is this new one from San Diego?"

"Depends on which one you mean. I know one of them is from L.A., so the other one must be from around here; otherwise why would Symonds plug in in a place like this. San Francisco I could see, but . . . San Diego?"

"Back up, Cusped. What do you mean, 'which one'? Do you mean there were two of them lined up for tonight?"

"Why not?"

I could think of several reasons. Symonds never had more than one leash in his hand on any occasion, and he was careful in the allotment of time he gave to this kind of thing. His energy was not at the high level that it had been when I first knew him. And there were other, logistical considerations: in Eugene's trailer it just would not rhyme.

"Did he or didn't he?" I said. "And who's the one from L.A.?"

"Well, I'm not really sure about that either."

"In other words, you don't know a damned thing."

"I know it was planned and I and a half dozen others were invited."

"But why here?"

"Maybe it was because of that convention he was attending, and he couldn't wait."

"The conference was over on Tuesday," I said, "and Glenn could always wait. I don't believe he set this up at all."

"Well then, who did?"

"How did you find out about it?"

"I told you! Carmichael called me yesterday afternoon. He called all of us."

I let out a long sigh, then asked him, "Who knows about this—about Eugene doing all this calling?"

"Hell! Everybody does." He took time to blow his nose, then added, "If you're wondering if the police know it, you can relax. They do."

It did not relax me. I asked him if he knew the one who was talking to Lawson.

"His name's Rublatz. He was invited down for this arrangement too."

"Was Chuck included on that invitation list?"

"He wishes he was. He's over there begging to be included right now, I'll bet."

"I wouldn't take that bet," I said. "And right now you ought not to turn up your nose at a sure thing. If the chief candidate for this evening's honors doesn't check in after what's happened down in National City, then you may have to make do with him. Did you ever work on Lawson?"

"Not really. Never considered him worth the trouble. Oh, I jacked-off in his face one time out of boredom, I think. You might say I had him eating out of my hand."

Followed by a snicker.

I looked away, suddenly tired, worn down with a feeling of being hemmed in by a pack of wayward schemers who, absorbed in their own extravagant displays, were forcing me to take circuitous paths to find out something that I desperately wanted to know. Perhaps the answer did not lie with them, but I had no way of determining this without getting the answers to some detailed questions. Yet I hesitated to alert anyone to how deeply what had happened was affecting me. It never paid to put yourself in a client position with them—or with most of them at any rate.

From my association with Symonds I had had nearly a decade of familiarity with these people, but Glenn, while revealing their world to me, had also acted as a screen through which I viewed their antics but remained untouched by their influence—or so I thought. These people, as I referred to them, wrapped themselves in the trappings of a cult, but the rituals they performed tended to overwhelm the content of their beliefs. The rewards they sought in what they did were transcendental experiences achieved through an emotional control aimed at freeing themselves from passion. Few of them would have put it in so many words. Symonds was the only theorist I ever met among them, and given a room to rant in and devoted listeners to fill it, he could hold forth by the hour on the fundamental strength this activity conferred upon its devotees. Their successes, to hear them parse it out, preserved them from the dangers of the rational strangulation which ordinary sexual dependency too often entails. Perhaps it was so for a limited few, but they overlooked other, greater, and more subtle dangers, or, if they acknowledged them, they dismissed them as being of no consequence in their quest for the mastery of others and the management of themselves.

Much of what they said and did I could find amusing from time to time because, except for some head-sprung cycle freaks and a few sickies who had no real need to impinge on this area of social warp to get their bloody kicks, there were none of them as unbridled, vicious, or abandoned as they would have it believed. Again, there were exceptions. The real danger for anyone hanging around them lay in the rare but ever-present possibility of some very wolf wandering undetected into their folds to practice among them in the guise of a shepherd. I had been coming around to considering this possibility in connection with what happened to Symonds.

"Who's the third party in your group?" I said.

"What do you mean?"

"The other one you said you came down here with?"

"Nelson Faulkner."

"Faulkner!"

"That's right."

He grinned in triumph at having surprised me, then swallowed hard with an audible "gunk." Cusped was afflicted with the clicking nasal twang of the chronic sinus sufferer. He was continually blowing his nose and looking at the residue before tossing away the Kleenex. He did it now, letting the wadded tissue drop to the floor at his feet after inspecting the mess he had made in it.

"You see?" he went on. "It promised to be a first-class arrangement. But then . . ." He wiped his nose again with inane solemnity. ". . . anything with Symonds or Faulkner involved in it couldn't help but be."

I said nothing for a while, pondering the presence of this new element in what had become a senseless jumble of disparate parts. Nelson Faulkner had been an issue between Glenn and me long before we broke up. It had to do with pride on Symonds' part and an expression of independence on mine. These two masters knew each other by reputation; other than that there was only what might be called a nodding acquaintance between them. Faulkner, as far as I ever knew, had no interest in Symonds, and Glenn's desire to know, and possibly to test, Faulkner had come about from learning that I knew him. Had an arrangement been made in which the two of them were to have a hand, and Symonds knew it on Tuesday night, then I could not imagine his not revealing it to me over dinner if for no other reason than to see just how I would take it. Had he done so at that time I'm sure I would have received the information with the disinterest I exhibited for the rest of his revelations. Yet here, now, with what had happened since then, I found Cusped's news disturbing to a degree I did not want him to be aware of. I changed the subject.

"Who told Lawson about this, and when?"

Cusped shrugged. "That one can smell a session shaping up clear across state lines."

"Somebody had to tell him about it."

"What does it matter?" He looked at me, amused at my persistent interest. "Maybe Carmichael told him."

"Gene would never tell him anything. They barely speak to each other."

"Two negative charges that repel each other. Isn't that typical."

I only wished at that moment that I could see something typical, familiar, recognizable for what it was in what I was being told. Eugene used to trust me to keep tabs on him when he went off for some heavy action where he couldn't be reached, and I followed up on him because I cared enough about him not to want him really hurt, even though I was otherwise complaisant about his activities. The same, in a different and more intense way, was how I managed with Symonds for over seven years, although, of course, I never imagined Glenn ever getting hurt, much less killed. But I had given over paying out time and good sense on their folly. The result of this indifference (if that was what it could be called) was that when I turned around again, so to speak, my attention caught by something out of the ordinary, I misjudged what was going on and did the wrong thing. Criminally wrong, as it might turn out. For I was possibly the only person who could have headed this off or stopped it at the moment it was nearing completion, but I had done nothing—or rather, I had turned tail and run, not wanting to get involved. As a result, it looked as if Eugene was being sewed up stitch by stitch in a tight sack of circumstances that could get him gassed.

And here was I sitting on a barstool with a finger up my ass casting about for a ray of enlightenment in the fog of implausibility thickening around me. I told Cusped I did not believe a word of what he had said about Eugene making an arrangement.

"No? Well, here we are!"

He gestured to Lawson and the L.A. type whose name I had already forgotten, still wrapped up in each other across the room, as if that were proof enough for anyone except a niggling fault-finder. He reached under his vest and produced a pack of special, medicated cigarettes and smiled at me, pleased at having gotten my goat. All I could do was shake my head. One were a fool to count on the crippled for support, or the blind for guidance, or the hopeless for cheer, yet I found myself reduced to such succor at every turn. I made a mental note to change my way of living and sow a whole new crop of acquaintances to help in the project if I came out of this scrape in one piece.

Meanwhile I waited for Cusped to continue his comments after he lit his odoriferous cigarette, but, from the silence that followed, it appeared his last expression was meant to sum up his thoughts on the matter. I picked it up again.

"Well, it looks like you came down here for nothing."

"Oh, we're not going to let a little thing like what's happened to Symonds stand in the way of things. The session's still on."

"No night of mourning for the esteemed dead?"

He shrugged. "What's done's done. We'll miss Symonds, no doubt about it. But he should've been more careful."

"You don't think it might be wiser to cancel this arrangement under the circumstances?"

He looked at me with those soft black eyes set in mauve pockets of tired flesh. Symonds had done canthoplasty on him a few years before this, and from a distance, or in the light of a bar, he was still good looking; the sag was not too apparent. But when he relaxed, letting his face go blank, unanimated by the habitual sneer, there was a slackness there that exposed the gathering middle age behind all that sardonic façade. I had seen some of his publicity scrapbooks. They chronicled his professional life from as far back as before the Second World War when he had been some kind of child prodigy. An aging, sentimental top man now, he dropped his inquiring gaze and said, "What the hell for?"

"If not out of respect for the dead, then because of the police. The law's going to be on the lookout for a while down here for any disorderly gatherings."

"It's not going to be disorderly!" He looked genuinely shocked. "You know better than that!"

"I was using the word in a legal sense, Cusped, since we're talking about the law."

"Shit!" He looked away in disgust. "They know it goes on all the time."

"It's not a question of what they know. It's what they don't know that'll put a burr under their tail. Glenn wasn't just anybody. He was a prominent figure in circles other than what you mean to move in tonight. The way the papers and the radio and TV will feed it to the public, they're going to get jittery about something as messy as this happening to someone as solid as Symonds. They don't understand that it's really just good clean fun, with harmless playacting on the part of all concerned."

The look he gave me this time was reward enough for the jibe I'd aimed at his pastime. To have an outsider, even one privy to the inner counsels of their extravagant foolery as I was demean their lodge rites with facetiation of the sort that was all-too-often employed by lawyers in court on behalf of some hapless defendant charged with ogrish assault on a willing partner was enough to leave his mouth gaping in indignation. Such straight-faced pleasantries aimed at their shenanigans rankled, as I intended it should.

"What happened to Symonds wasn't playacting," he grumbled. "And the public doesn't care. It's just something they read about over their morning coffee."

As he was uttering this non sequitur Chris leaned over the bar and said there was a telephone call for me.

"You can take it in the office, hon," he beamed. A special favor for me.

I got up and went into the back, behind a partition that screened off the barroom from the toilets and a little cubbyhole containing a sink, a hotplate on a bracket shelf, and a small desk where sat the phone. Wendy the waitress looked up as I entered; she was stubbing out a final cigarette before taking up tray and swab to go out and work for tips. I said hello to her, a greeting she returned with a graceless nod as she made room for me. Picking up the phone, I waited until the extension at the bar clicked off, then found myself saying hello to Ray Reynolds.

"When did you leave and why didn't you tell me you were going?" I said.

"You were too busy talking to that crazy harp player. I found something that wanted to go home, so I left with him. You're not mad, are you?"

"Enjoy," I told him, then asked what was on his mind.

"I ran into Pinkerton on my way out. He says he's got something important he wants to see you about."

"Why couldn't he talk to me here?'

"That's what I asked him, but he said he had to see you on a private matter, and asked me to tell you to get hold of him. I meant to come back in and tell you, but this guy I'm with got to talking with some friends of his and I was afraid of losing him. Anyway, I forgot to go back in and tell you."

"Where are you now? And where did Pinkerton go?"

"I'm back home. Dale's probably at the Hollywood. He said he had to take that child of theirs home and then bring her mother down there for the first show. He'll be there for a while. Ask for him at the stage door. He said he'd leave word for them to let you in."

Suddenly I heard a shout through the door Wendy had left ajar, a cry of anger vaulting above the rumble of voices. It was joined by others, and above it all came the high-pitched squeal of Chris, protesting.

"I've got to go!" I said, and hung up.

The narrow corridor leading from the bar to the little-used rear entrance was clogged with queens trying to push their way through in both directions at once, and it took a long minute to get out of the squeeze and up front to where things were happening.

Instinctively I went around the other way from where I had come so that I was on the side of the horseshoe where Lawson had last been standing. I knew, also instinctively, that the ruckus had something to do with him, but at first I could make nothing of the confusion all around me. Everybody was yelling or laughing, and a tight cluster of people were knotted up at the entrance, blocking any near approach. What had caused the uproar was now outside. The majority of drinkers stood back against the walls, watching, scandalized. Fat Chris emerged from the impacted mass at the front, bawling.

"Now will you please go!" he whined, distraught. "Just go!—all of you!"

Whom he meant was not clear. Nobody moved to leave, and he began shoving flustered clientele on either side of him with no apparent method in his selection. Then he stopped, spotting me.

"And you too!" he screamed.

"What the hell have I done?" I said. "I wasn't even out here."

"You're part of it!" he harrumphed, brushing past me to take his place behind the bar again. He faced me as from behind breastworks and from there continued with a harangue addressed to the now quieting room.

"I won't stand for any violence in this bar. I've said that before, and I mean it. I'm not going to play host for one moment to rude and ungrateful scum!

He looked to be on the verge of tears. I let the remark about scum pass, and leaned over the bar to speak as quietly as I could above the renewed hubbub his speech called forth.

"Chris," I said, "would you mind telling me what happened?"

"Go!" he squalled. *"Out!"*

I stepped back and looked around me, the universal object of disapproving stares. It was no use attempting to reason with Chris who was now receiving comfort from Wendy and his two barmen, so I turned and left.

Last one out gets the hardest lumps. Going out through the nearly cleared front entrance, I passed a tall squirrel in a Hawaiian print who said to his companion, "I just wish that L.A. trash would stay up where they belong and leave us in peace." To which the other, studiously looking the other way, replied, "Umm . . . " I could only agree with that, but it was a little like poor whites putting down the "niggers."

Outside on the street the cool night air was rent by the sounds of a modest cavalcade getting under way. Three bikes, two of which I recognized as belonging to Lawson and Cusped, were stuttering to life. An unknown riding Lawson's white Harley with Chuck, his nose running blood, buddy-seated behind him, led off followed by the one who had been talking to Pinkerton straddling a BMW. Cusped, mounted on his own green Indian, brought up the rear. He paused at the curb beside me, displaying a long-toothed leer, very pleased with himself.

"Wanna come along?" he shouted above the racket of his machine. "Take your mind off things."

"No thanks," I said. "You people have given me enough trouble for one night. Now you better get out of here before the cops raid this place."

Cusped pulled the buckboard's string tie down under his chin, turned up the brim, and tilted off in a blast of exhaust to join the others already two blocks ahead of him, tooling north towards Washington Street canyon.

I watched their ruby lights wink out of sight as they took the turn, then went back along the side of The Club to my car, girding myself for a second go-around with Dale Pinkerton.

6

AS I HAVE SAID, there was a time during most of the period when Eugene and I were getting to know each other when he used me as a kind of safety check while he was away on referrals, locally and out of town. His calendar, his journal, his book of hours (there was no name for it; he called it his "safety valve") held all the pertinent information about where he was going, whom he would be seeing, when he expected to be back, and usually a number to call if he failed to return at the time indicated, since he was not unaware of the perils inherent in these blind encounters. On his return he painstakingly detailed the session in cramped, illiterate long hand. For a while I read this rambling account of abuse and low adventure until the repetitiveness of the situations described at painful length in crippled grammar and spastic syntax smothered my interest. He always got back on time, or called in if delayed for one reason or another.

This arrangement had stopped back in December when Symonds came into his life, and I had had no further interest in it until now. Eugene had begun his chronicle of bondage years before, and I was sure he had continued it after he moved back into the trailer. But I was also persuaded that no one else was holding it now for the simple reason that, until a few weeks prior to this, he had been consecrated to the sole use and enjoyment of Glenn, and there was no need for him to alert anyone as to the strangers into whose temporary keeping he might otherwise have assigned himself. It was also clear to me from the questions they asked, or failed to ask, that the National City police did not know of the existence of this little book, for by the time they came to see me the trailer must already have been combed through and sifted for all the damaging evidence it would yield, and if Gene's journal was not turned up among the clutter of papers and playthings to be found there, then he had it with him, wherever he was, or had stashed it away for safekeeping.

There was a third possibility, and that was that somebody else had gotten hold of it during the events of these past few hours and was keeping quiet about it.

Eugene his journal, Lawson his enigmatic visit to Ray's, the shadowy figure of the kid Cusped had referred to in calculatedly ambiguous terms, Faulkner . . . all were uncertain factors shifting around in my head as I headed south on India Street to the Hollywood Theater. Any one of these elements might furnish a key that would disclose what actually happened in that trailer and, more important, the reasons for it having happened. But the only one of them I had any means of approaching right now was Lawson, and Lawson, I realized, might have already slipped through my fingers.

The Hollywood Theater was a burlesque house on the north side of E Street. I had seen the show a couple of times because Pinkerton's sister performed there, and had been backstage once at her invitation. It was nothing special, no more than was anything else the town had to offer. Cruising around for a place to park, I pulled onto a side street and found one in front of an all-night hamburger joint. Hunger got the better of curiosity for the moment and, letting Pinkerton wait, I went in and ordered something to eat. There was a newspaper rack outside; the early Sunday edition was there with modest headlines and two columns of story. I read it at the counter while I waited for the eggs.

There was little in it that I did not already know. Symonds had been found clothed in leather garments (the expression was the reporter's), bound with "professional-type" straps and chains, gagged, blindfolded, and "unnaturally trussed." What they meant, but for some reason did not spell it out, was that he was strung up upside down. It was Eugene's favorite position, but Symonds, unaccustomed to being handled that way, and left to his own (under the circumstances inadequate) devices, appeared to have suffocated. Only the last was mentioned in the story. An autopsy was being performed to determine the exact cause of death. There at first appeared to be no signs of violence beyond the aforementioned trussing.

The peculiar mode of suspension in which they found him is a type of ligated crucifixion as old as society itself, and produces suffocation and unconsciousness, all things being equal, quicker than the more celebrated style of which it is a variant. Transfixion right side up is said to make it slower. The article mentioned evidence of a struggle (no surprise there), of sexual assualt (no more than to be expected under the circumstances, yet still a surprise to me), and (and here was the curious thing) mutilation, once the leather had been peeled off.

It would make interesting reading over morning coffee, all right. Sipping at the acidic brew the counterman set before me, I read on about how the owner of the trailer was being sought as a suspect by the police. Eugene F. (F. for Francis, F. for fuck-head; take your pick) Carmichael was 32, stood 6′2″, was light-complected, had brown hair and hazel eyes. They gave his build as husky and added that his manner was said to be mild but that he should be considered dangerous.

Eugene dangerous?

He had served a hitch as hospital corpsman in the Navy, then worked for a pest control outfit and a building contractor before getting a job at the shipyards as a welder, which position he had had for the past five years. He had no police record, and was last seen by the trailer park manager on Friday evening a little after six o'clock driving into the park entrance in his white Dodge sedan with someone in the front seat beside him. The car was gone the next morning. Several times during the night the manager and other trailer occupants reported hearing motorcycles in the area, although none had been seen or heard in the immediate vicinity of

Eugene's trailer. The Dodge's license plate number was listed and mention made of Eugene's bike, a black BMW which was found parked by the trailer.

Oh, and that BMW.

Symonds had originally bought him a Harley when their association began, but in less than a week Eugene rode up to L.A. and traded it in on a stingy-seated BMW—so quiet, who cared? He liked everything German, except the language, which he didn't understand and found too difficult to learn, and felt that the Hog was too assertive in sound and design for him to handle with any conviction or pleasure. Also, it was a test to see how Symonds would react. The reaction was immediate and predictable. Eugene showed me the stripes and bruises incurred as a result of his master's displeasure. "If you're happy," I said.

It was all very sick, and I was sick and tired of hearing about it and being shown the results of their little exercises. I had passed six years and one more skirting the borders of this largely harmless insanity, and had finally shucked Glenn Symonds and his hangups in order to get on with my own living. And although my mode of existence might have been judged by many to be warped as well, it was at least my own. And if it proved to be less stylized, modish, trendy, or what might be called old-fashioned, it suited me and was, for that reason (or so I told myself), more honest. Telling myself such a thing did not, to be sure, make it necessarily so. This I knew well enough. Particularly when I continued to be affected by these people. For a long while I had Eugene coming over to see me and running off at the mouth about what someone I knew like the back of my hand had done to him and expecting me to be breathless with excitement at his recounting of it. I could never detect a trace of malice or bragging in any of this, however; Eugene actually expected me to be proud of him for his catch. (After all, I didn't want Symonds again—did I?)

Of course, I could never be sure of how innocent Eugene really was when it came to his sexual vagaries—I mean, when it was a question of handing out violence as opposed to receiving it. Tottering on the rim of disaster acted as an analeptic to his otherwise work- and sleep-numbed faculties. He had been the fall guy for numberless con artists, including an older brother and his dear old mom, since he was old enough to say yes. (He never learned to say no . . . never. The closest he came to outright disclaimer or dissent was a typically ambivalent locution with a whining preface, "Umm . . . I don't know.") It might possibly be, as Lawson had suggested, that the worm in him had turned. Possible, but not very probable. Eugene was no versipel. Still . . .

Take the matter of the bike switch. His doing that was not any bid for autonomy, but rather a trial insubordination that left Symonds both irritated and pleased. It irritated him that Gene did not defer to his judgment that a Harley-Davidson 1200 was the only fitting vehicle to be seen astride on an evening's outing. On the other hand, it was somewhat pleasing to have his slave on a BMW because it further distinguished the one from the other whenever they went on formal runs, occasions rare enough in themselves to warrant special measures for maximum effect. Eugene had dared on that matter, but he stopped short of exchanging the Hog for a Japanese model, a breach of taste that might have ended the partnership then and there. As it was, he pushed his luck quite a bit just before Easter by grumbling about wanting to trade in the BMW on a Vespa. Symonds reportedly threw him in a closet and kept him there over the holiday weekend for that bit of sass.

People used to remark, after they got used to the arrangement, that those two were made for each other; something they never said or thought when Symonds had me, or when I was mixed up with Eugene. I reflect on this without bitterness.

The newspaper's background on Glenn was correct as far as it went. It described him as a successful and respected plastic surgeon from Beverly Hills, a graduate of Rutgers and USC medical school. He had come from Pennsylvania, and his parents were both dead. It was not known for the moment if there were any close living relatives. Besides his practice, he taught at UCLA, and was socially active in film and television circles besides having friends of importance in the world of business and finance. He was a member of civic and philanthropic organizations up there, and a collector of art and antiques. They tabbed his age at forty-six.

It was close enough. He had done his internship and general practice in San Francisco and his residency in surgery in Santa Monica. It was at the V.A. hospital in L.A. that he began work in reconstructive plastic surgery and, after receiving board certification, he set up private practice in Beverly Hills. And he had indeed moved in all those circles and enjoyed all those acquaintances and connections, mostly as spinoffs from his practice. The nuts he knew and moved among the paper did not mention, but would no doubt get around to as the case developed and word leaked out. The civic and philanthropic stuff was the usual window dressing expected of rich and prominent doctors, but the art and antiques were a fairly recent thing with him, an expression of the omniety of his interests at last being given elbow room after years spent in the confining company of his clientage and his colleagues.

For the rest, the sum of his years would come as a surprise to many who knew him. He looked much younger than forty-six, and those looks were not the result of the art he practiced on others. They came from the care he took of what good genes had bestowed on him in the beginning.

Yet all this summation was bare bones. Every statment in that brief account could have been expanded in interesting, amusing, sometimes fearful detail. Seeing it so, almost statistically tallied, gave an unfamiliar cast to it for anyone who had known him personally. And yet, in another way, the impression was just, as Glenn himself would have acknowledged had he been able to read it. He was the sum of what he was doing at the moment of his death. Glenn Symonds was a victim.

The only question now was, whose?

Having eaten for the first time in twelve hours, I paid the check, rolled up the newspaper for future reference, and walked around the corner to the Hollywood Theater. A line of men in Navy blues was out front stretching down the street from under the marquee waiting for tickets for the next show. Usually there was no wait, and the show had not changed since I had last dropped by two weeks earlier, but a missile cruiser and a carrier had put in for the weekend, and the movie was different: an old horror film. They almost always showed horror or cheapy science fiction grinds at the Hollywood, and there were some who came for the movie and never bothered to stay for the live entertainment.

Inside it was dusty and airless, and that musty reek peculiar to sleazy vaudeville theaters seeped out into the street when the foyer doors were open, as they were now. I went through the side alley to the stage door.

Pinkerton's sister, Bobbie-Jo Lane, was featured on the billboards. She was good enough to demand and get Saturday matinees off to spend with her seven-year-old daughter who did not know, or had sense enough not to ask, about her mother's means to a livelihood. Bobbie-Jo and her kid shared a bungalow out in La Mesa with Dale Pinkerton and his wife. God knows what the child thought about that living arrangement, whatever she may have guessed as to her mother's evenings away from the house. Bobbie-Jo and Pinkerton's wife Snuffy slept together while Dale was permitted the use of a spare room off the garage whenever he chose to join the family circle. His quarters were separate from the rest of the rooms and had their own entrance.

Snuffy was standing just inside the stage door that opened off the narrow side alley talking to the porter when I came up the steps. She looked at me and went on listening to some rigmarole the old man was spinning, motioning me inside with a peremptory nod. Backstage it was cramped and airless. I made my way cautiously around the cluttered dark to one of the flats from where I could watch the show's warmup attraction, a hypnotist who was yelling into the PA system from the apron. The stage was covered with volunteers from the audience (all male and most of them swabees) lying and sitting about in varying degrees of trance-induced stupor. He had a signalman out on the ramp doing mildly indecent things that aroused scattered laughter from the assembled voyeurs. The hall was about half-filled. It was early yet.

"Dale's expecting you," Mrs. Pinkerton grunted behind me. The old porter had resumed his seat at the battered desk, and she was coming on as if she wanted to pick a fight. It was the tone of voice she employed, and was just her way, although, as I had had occasion to observe, she was not averse to picking one if a ready adversary came to hand. She was not to be trifled with.

"Where is he?"

"Downstairs to the can, " she replied with an inimitable Pecos River twang and a jerk of her head which was covered with tightly curled auburn hair. Given a bolder tint of red in it she would have looked like Little Orphan Annie with pupils. A tiny thing, she could be mean as a scorpion when crossed. I was careful never to do so after seeing her in action once at the Calypso, a mixed bar in Pacific Beach, where she stormed into the men's room after a careless little girl who had gotten on her nerves over some snick or other and then sought haven from the ensuing storm in the only place she figured was safe, since the outside door was blocked by a crowd of standees and Snuffy was a fast sprinter. In this she further misjudged her adversary's determination and grit. Snuffy went in and dragged her out of that male sanctuary by the hair, kicking and screaming, as they say, and bent her over backwards on the edge of a pool table where she proceeded to go to work on the poor girl's face, alternating fists with nails (which were mercifully short from overbiting) while working a sharp knee into her groin. By the time the friends of both parties to this uneven contest, and who were enjoying the spectacle, waded in to break it up, the loser was in a state of blubbering collapse. Her face was a turnoff for weeks afterwards.

It should be noted here that at that time a "mixed" bar in San Diego meant that it catered to homosexuals of both genders, an arrangement that never worked out well, but was widespread for reasons no one seemed to understand, least of all the steady patrons who divided the space between them into two uneasy camps.

Snuffy was compact and wiry; a hot machine to the sisterhood. The encomia of those who discussed such things went so far as to proclaim her to be hung like a stud doughnut, an outworn quip that sounded fresh and apposite when applied to her. She did not any more than come up to Bobbie-Jo's navel, and to see the two of them dancing together—and they loved to dance in the ballroom style that was already out-of-date by then—was something to watch. Bobbie-Jo had all the slimmed-down, limber proportions of her raw-boned brother but none of his baleful looks. A male patron at another mixed bar had remarked once on seeing the two of them engaged in a tango that they looked like Chiquita Banana and a peanut. He was overheard by Snuffy and left in a hurry.

"He said tell you to wait."

Texas. The three of them were all from some place near San Angelo, and the peerless venerability of their shiftlessness—as common as pig tracks, to use their own vernacular—remained unalloyed and undiminished by the change of climate that found them here. They picked up money where they found it, and found things to do for it that required from them only what came naturally. Otherwise, as far as one could tell from talking to them, their goals were as middle-class as the next person's. They arrived in the area after Bobbie-Jo received a Mexican divorce five or six years before from a man they never mentioned by name thereafter. The marriage of Pinkerton to Snuffy was accomplished at about the same time, if not before, to lend a gloss of respectability to their *ménage.* La Mesa was full of people who liked things legal, and this was Bobbie-Jo's way of protecting the child from unsettling gossip and of covering her singleminded passion for Snuffy. Maybe it lent some appearance of respectability to Pinkerton too, for that matter; he had little enough of it as it was. But most of their neighbors seemed to be under the impression that Dale and his sister headed up a household united in holy (or at least common-law) bonds, and that the little girl *and* Snuffy were the blessings of their union. At least this is what the three of them believed. The truth of the matter was probably that their neighbors did not give much of a damn what the links and lineage of them might amount to so long as they kept the noise down.

"Is Bobbie-Jo around?" I asked out of politeness.

"She's still in her dressing room. The *Show Boat* number's first."

I remembered it: a travesty of Edna Ferber and Jerome Kern's classic that required the artful shucking of crinoline midway in the number leaving nothing but G-strings and pasties and the bare cage of hoops. It was an attempt at class which generally left the audience yawning.

"It's a terrible thing about that friend of yours," Snuffy burped. I agreed with her that it was. "A doctor." She shook her head. "That's a hell of a waste."

"Well," I said, resolved to match her philosophical mood, "these things happen."

She stared at me in the light from the stage as if she had caught me in a fib. Again, it was just her way. Her hair, frizzed out like a drawing of the section of a kidney, or the leaf lungs of a dogfish, appeared almost to swell and contract with the repressed irascibility roiling around behind those great round eyes. She had cast my remark upon the limited computer circuits of her brain and now cocked her head slightly while the mechanism whirred and the spindles turned. The altered look that registered on her face gave the impression that she thought I could be putting her on.

"They ought *not* to happen."

Like that, her pronouncements on local and national events tended to pithy, fault-finding utterances of dogmatic tenor. She expanded now upon this theme, her pupils widening in walleyed outrage while she spouted the regional blather about sedition and the dangers of immoral behavior undermining the foundations of public trust, to say nothing of private friendships. The discrepancy between her way of living and the cant she mouthed never bothered Snuffy in the least. I repressed an urge to engage her in orderly discussion of these issues; any attempt to clarify her thoughts would, I knew, result in further blurring of distinctions and an inflamed temper on her part. And besides the futility of arguing with Snuffy about anything, there was the disinclination I felt to render tiresome and dangerous that which was merely absurd and amusing. The entire clan was fairly successful at having it both ways. Six times a year the Pinkertons flew the colors from a pole socketed in their front stoop, and they professed to vote the Republican ticket under protest when they could not find some splinter candidate far enough to the right to suit their leanings. Law and Order; they were for it.

"There's too many nuts walkin' around on the streets in this town."

"Here and elsewhere." I thought it best to agree, with enough amendation to give the discussion some life in what was meant as a forum for whatever moral hickey was blemishing the simplistic smoothness of her long-ago-received opinions.

"Yeah," she conceded, "but this place is worse than anywheres—per capita."

I was genuinely surprised at this deprecation of the Promised Land. "I thought you liked it here."

"It's got nothin' to do with the climate," she muttered. "It's people bitin' off more'n they can chew."

It was my turn to stare. Was that meant to indicate some deep-lying eruption of malaise, or only another arbitrary turning her bridleless logic took when her sensibilities became frayed by some personal grievance too long unhealed? It was best not to probe. I did not press her for an explanation. Conversations with Snuffy Wiggins (she refused to go by her married name) were likely to take this tack, and I was relieved at that moment to see Pinkerton's rangy form rising like an amiable specter, circling wraithlike upwards in the gloom on the spiral stairs from below the stage.

"C'mon outside," he said, putting a long-fingered hand on my shoulder. "You can't hear nothin' in here with that maniac yellin' out there."

In the alleyway it was quieter. He leaned back against the brick wall opposite the stage door steps and drew up one long leg, hooking a cowboy-booted heel against the uneven surface behind him so that he looked like a crane in the shallows. After shaking out a cigarette from a pack in his shirt pocket, he offered me one, forgetting that I no longer smoked.

"I was told you had something terribly important to tell me," I began.

"Did Ray Reynolds say that?"

I told him it was something like that, and he shook his head before lighting up. The match flare highlighted long lines above his mouth and the thin beak of his nose. "Exaggerations," he said, expelling the smoke. "Isn't that always the way with sick people, though? They exaggerate and get ever'body upset."

"I'm not upset. And what you just said sounds like an exaggeration to me. And if you're trying to establish that Ray is sick, then save your breath. What's up?'

"No big thing, chum. I heard you'd been talkin' to Chuck Lawson, and I thought you might like to know he ain't reformed no more."

Trust this long-legged connoisseur of assholes to have checked him out. "I never believed he'd reformed in the first place," I said, "and I surmised he wasn't when I talked to him. Who cares?"

"The *po*-lice and the sheriff's office might care if they knew about it."

"Go on."

But he regarded me in silence through the smoke of the cigarette. He was stringing it our for a reason, and the reason was not to tease. There was not, as far as I had ever been able to determine, a gratuitously mean bone in Pinkerton's loose-jointed body. This is not to say that he would not sell his mother into bondage if the profit in such a doubtful transaction were enough to compensate him for the loss (and I mention the loss of a mother to illustrate a point rather than to establish a scale of values in his scheme of things. For that matter I don't remember him ever mentioning his mother). Scratch and the safety of his own hide were considerations next to which all else was secondary for Dale. He was a big, left-handed country boy who had made a good many scenes in his thirty-odd (very odd) years, experiences which had left him amused and no doubt saddened at times, but never so jaundiced as to want to take it out on anybody not noticeably ill-intentioned toward him. He had been beat up, cut up, rolled, run over, run out on, and stood up by friends and strangers past numbering, beginning with his family, or what for a while in his childhood and youth passed for such. Still, he managed to keep a relatively sane perspective on other people's point of view as it was handed to him—sometimes on the end of a stick. Besides being dishonorably discharged from the Navy, run in by the police of several incorporated communities, and forceably escorted to the city limits by the citizens of his native thorp, he was doubly indentured: to several dubious employers, and to the agency nominally established to put a stop to most of those very employments.

Since Pinkerton ran with the hare and hunted with the hounds, I have said that his work as an informant was only a sideline. But to be accurate, one would have to state that everything he did was more or less a sideline. He hardly attempted any coherent response to the question a stranger might put to him of, "What do you do?"—whether for love or money. People who knew him never asked. I learned about him only gradually, although he was one of the first persons I found myself in bed with when I arrived down there, and had sampled his gruntling good humor in a good many bouts since, for lack of anything better. By the time I met Eugene, the novelty and the satisfaction we gave each other had worn thin enough for us to forget the bodywork and lapse into an uncomplicated acquaintanceship. I had never had cause to avoid him, nor to increase the chance encounters that brought us together from time to time. Sexually Dale was a good performer (a professional, as I will presently explain), but beyond that there was nothing until now that gave him any but a minimal place in my attention. What detriment his connection with the law might, on occasion, have for those he knew never gave me pause to reflect, or any reason to worry ere this. But now I found myself weighing an entirely new set of conjectures while I waited for him to get on with what he had called me out for so I could stop guessing. When he finally spoke, I knew he had been stalling for effect.

With the sad, sly tone of one proposing that you join him in watching an indiscretion through a two-way mirror, he said, "Would you care to come and identify your former lover?"

7

THE ANSWER should perhaps have been no, but I did want to have a look, even though the prospect scared me. So, after attempting to find out why they had picked me to do the job, with no success beyond an ambiguous "You're the closest one that knows him," I said I guessed it was the least I could do.

We drove up in my car to Kearny Mesa where the morgue is, silent, both of us, most of the way. I was trying to steel myself for what I was about to see, and Dale appeared content to wait for what might turn out to be an interesting show.

The least I could do . . .

It was something anyone might have felt he owed an erstwhile object of affection, and I could argue with myself that it was right, the decent thing to do if for no other reason than that none but the dispassionate eyes of the medical examiner and various functionaries had been looking at him since his discovery that morning. But it was curiosity that led me up there more than did any moral, charitable, or sentimental considerations.

At the coroner's office I signed in, as did Dale, and then we followed a white-coated attendant down a corridor to a double-swinging steel door. Inside, he switched on the lights to reveal a cool, low-ceilinged chamber painted light green and trimmed in enameled metal and bright chrome. Through another doorway I caught a glimpse of what must have been the pathologist's room. It was all less grim, but no less disquieting, that I had anticipated. A faint odor of disinfectant permeated the place, and there was the constant low whir of a ventilating system.

The attendant led the way and Pinkerton followed me. Stopping halfway down a double row of what looked like oversized filing cabinets set into the wall, the white coat bent down to consult the card beneath the handle of one, then pulled out the long drawer which, emerging from the wall, reassembled the casket it temporarily was. Without ceremony, he flicked back the top of the sheet that covered the body, revealing the head.

"There he is," he said, and stood to one side while Dale and I peered down from the other.

There he was indeed, looking none the better for what he had been through. If Glenn used to abuse his own freedom in the contemplative game of shackling others as closely as he dared, he now presented an exemplary picture of that turnabout that is the fair game alloted us to play at whether we will or no. He lay in the shallow steel drawer, an allegory of defeat shrouded in shame.

One of Glenn's dicta had it that whenever you feel the temptation to sentimentalize about anyone, you have only to return to the physical. "The body," he said, "is sentient, not sentimental." He meant it as an antidote for passion, which, of course, it is not. But now, for whatever it was worth, Symonds' pronouncement was meaningless, for him at least. His body no longer felt anything, and the only sentiment that took hold of me, seeing him so, was something akin to horror mixed with pity and resentment that things should have come to such a pass. For, in the words of the Psalmist, Death was now his shepherd, and he would, like the sheep he had despised, lie in the grave and be consumed, to have no abiding, except as he would be remembered.

Someone else came in and joined us. I looked up and was not very surprised to find Don, the inspector who had questioned me earlier, facing us across the waist-high drawer with his silent side-kick, Pete, peering over his shoulder at me. Don nodded, saying my name.

"What did the autopsy turn up?" I said.

"He died from cyanide poisoning."

The surprise must have shown on my face. It may have been why they let me in, to see what my reaction would be. I looked down at Symonds again, understanding now the discolored features that contrasted so with the stramineous hair, short, but lying lank and flat against his head as if wet, rooted in the cyanotic skin. The face, now slate-colored, was furred with blond stubble.

"The broken ampule was still in his mouth," Don went on. "But that doesn't necessarily mean it was self-administered. There's something else . . ."

He took hold of the sheet which still covered the body to the neck and curled it back almost with a flourish, exposing the cinereous corpse to the mid-thighs. It was a body I had known almost as well as my own, but which was familiar now only in the sense that a luckless farmer's fields are familiar to him after an army has passed over them doing battle. The once superbly molded torso was trenched by a gigantic excavation made at the autopsy. The flesh was ripped apart in the shape of a Y: from shoulder to shoulder in a shallow V, with a midline cut running the ventral length from throat to cock where he had been laid open to explore. The prideful chest was collapsed like that of a lifelong consumptive from the snipped and everted ribs. With what efficiency had he been gutted! Heart, stomach, liver and lights, kidneys, bladder, testicles, and the stinking coils of intestines, all plucked up, probed, sectioned, and flung back into the carcass in fetid disorder as if to exhibit the vanity of all flesh, that is should come to this. I forced myself to look, and look well, at this mess of offal, all that remained of the person I had once loved.

For almost the last thing I let go of when I left him was the appreciation of the physical excellence of Glenn Symonds. That least durable, first noted, and most illusory of attributes in anyone we love remained to haunt me like a presence in the foreground of a fading scene, an all-too-clear memory that blurred even the changed outlines of his altered appearance when I had lately started seeing him again. Convinced now, as I realized I had not been before, that he was really dead, and viewing him, vengefully unsightly in his extinguished state, I put to rest at last the menacing aliveness of his memory, a pervasion of personality that used to saturate the space around me, keeping me always within his range. He was nothing but wreckage now.

The detective's finger pointed to the groin, all but touching the cock which, bluish like the rest of the skin, bore in still faintly discernible script the device tattooed in smalt along the dorsal length of it:

I leaven lees.

"What's that say?" said Pinkerton.

"Nothing," I said. "I don't know."

"That's not what I'm talking about." Don turned in disgust to the attendant. "Show him."

The morgue attendant pulled on a rubber glove and, reaching down, pushed aside the flaccid verge to expose a further piece of coltering.

"Did they have to cut his balls off too?" I said.

"His testicles were examined during the autopsy," Don said, "but somebody got there first."

We bent closer to look at the half-sutured wound around the sac.

"The medical examiner thinks he did that himself."

I looked at the attendant, and then at Don.

"You mean he tried to castrate himself?"

"Not necessarily." The three of them smiled briefly at this. "He tried to patch it up, it looks like. See those sutures?"

I had seen enough. I turned away, and Dale took hold of my arm.

"You all right, chum?"

I was all right. With my stomach under control, I forced myself to turn and look again, but the attendant had already pulled the sheet up over him. A bare left arm remained uncovered by a careless fold of the cloth. There, just below the swell of the shoulder I had so often cupped in my hand, I glimpsed for the last time his punctulate sign through the smoky chroma of once golden skin: his initial like a coiled whip laid across the raw red petals of a Tudor rose.

"What happens to his body now?" I asked.

"Somebody'll claim it," the detective said. "We're trying to contact some member of his family."

"There was only his father, I think," I said. "And he died early this year."

He nodded, probably recalling how I had denied much knowledge or interest in Symonds a few hours earlier. "Some other next of kin will have to claim it then."

"What if there aren't any?"

"Oh, with somebody like him there's always somebody that turns up." He almost smiled, looking at me. "We'll let you know—in case you'd like to claim it, Mr. Scott."

"Sure," I said, "let me know. I'll claim him."

If there had been doubts about Pinkerton's connection with the police, the foregoing dispelled them in my mind. He must have sensed this, but he offered nothing in the way of clarification on the way back into town. I parked in the same side street as before and we walked back to the Hollywood's side alley. The last show was almost over, but it was quiet outside. He took up his position near the steps and lit up. I asked him for a cigarette. My first in three years.

"That thing I said about Lawson don't have anything directly to do with this," he said, lighting mine.

"Then why did you mention him?"

"Well, the thing is, he was at a setup last night, and the police think Eugene might've been to it too."

"Have the police talked to Lawson about it?"

"They don't know Lawson was there."

If this were so, it meant that Pinkerton was holding out on his long-armed buddies in National City. I knew it would do no good to ask him why. Presumably he knew I was supposed to tell them anything I learned relating to Symonds' death and Eugene's disappearance. He also must have presumed, from the way he was handing this to me, that I would delay telling them what he had just told me. I knew my reasons for doing this. I wondered what his were.

"Where was this arrangement supposed to have taken place with Lawson," I asked, "and when?"

"Dunno." He picked at a piece of cigarette paper that had stuck to his full lower lip. "I heard it second-hand."

"From whom?"

He shook his head, he now of the tristful countenance.

"You wouldn't know the gentleman if I told you his name, and . . ." He grinned. ". . . professional ethics forbid me mentionin' it."

Maybe that was his reason for keeping quiet about it to the law. It still didn't explain why he was telling me.

"He wasn't in on it hisself anyway," he went on. "This person I'm talkin' about got it from a friend of his. They're both of 'em from outa town. All I know for sure is that one of the attractions called hisself Chuck."

"And the other one called himself Eugene?"

"Nope. At least not that I know of. But the description fits."

"And your friend had no idea of where his friend saw them? Didn't he even say what the surroundings looked like?"

"He ain't exactly a friend. Admirer is more like it. He's from back East too, and all *his* friend had to say about it was that it was somewheres between here and L.A."

"That's a big help."

"Yeah, ain't it though. An' for that matter, he may not even have got his directions straight. For all I know it mighta been down to Tijuana."

"Not likely," I commented, knowing Eugene's xenophobic avoidance of the border, even though (or maybe because) he had an older brother in some kind of murky business in that area. I decided to drop it in the face of Dale's affable impenetrability. The aftershock of seeing the wreck of Symonds still had me rattled, and I had to make an effort to hang onto the threads of this exchange. Bringing me to the morgue may have been the principal reason for Pinkerton's getting me down here, but there was something else on his mind as well. Ready to leave now, I asked him if that was all he had to tell me.

"That's about it, chum." He tossed his cigarette away and folded his arms across his chest. "How about answerin' a question I got?"

"Shoot."

"How far d'you figure on goin' in your investigations?"

"What're you talking about, Dale?" I was taken aback by his question. What I had been up to so far was nothing approaching an investigation, nor had I envisaged anything of so thorough a nature as to justify any such designation. "I'm just curious."

"What about?"

A sudden flush of anger took hold of me.

"Listen, you wasted fuck!" I said, letting my voice rise in the reverberating slot of the narrow alley. "I'm curious about everything, if it comes to that. Glenn's dead and Gene's missing, and everybody's asking me what I know about it. Because of your big mouth I get tagged by the law. Why shouldn't I be curious? Come to think of it, why the hell are *you* so curious? What're you getting out of this?"

He reached out and took me by the shoulder to calm me. I shook off that mancinist grip, but I had cooled down some.

"Well?" I said.

He reached in his shirt pocket for the cigarettes again. I refused this time, waiting. He lit one before he replied.

"You got ever' reason in the world to be curious, I guess."

I said nothing to this concession. He went on.

"But did you ever stop to think that what you're stickin' your nose into might bite?" He blew smoke my way, waiting again for some comment on my part before he continued. "Whoever done a thing like that on ol' Symonds ain't gonna look kindly on anybody that comes tryin' to sniff 'im out, d'you think?"

"You're giving me too much credit, Dale." My quick anger had spent itself, and I felt almost warm towards him. "I'm not primarily concerned with who killed Symonds, assuming it's murder and not suicide. The police will probably find that out before long without my help, I'm sure. I'm thinking about Gene."

"An' you just wanna find out where he's hidin'? Is that it?"

"Who says he's hiding?"

"Well now, he sure as hell ain't ignorant of what's been happenin' around here in the past twenty-four hours, d'you think?"

"You think he did it, don't you?"

"Don't you?"

"I know Eugene, and I'm pretty certain he couldn't have killed anybody, let alone Glenn Symonds."

"Shit!" He blew it out in a gust of smoke. "That kinda certainty ain't worth fish guts!"

He relaxed now into a long-jawed smile and chewed his lower lip for a moment, then stretched his tall frame lazily against the brick wall.

"Funny," he said, with a kind of sigh. "I never figured you and Eugene was that involved. Ever'body else did, but I never could buy it."

"What do you mean, *that* involved?"

"Oh, I thought you was just asshole buddies . . . roommates. . . . You know."

"Dale . . ." I tried to tone down the exasperation I felt rising inside me again, facing this dissembler. ". . . stop prying! I can tell you Eugene was a lost soul, and I liked him—let's say a little more than I ever liked you. But what he did and whom he did it with were of almost no interest to me. In fact, that part of him was one big bore. Period. But I am concerned about him. Whether I'm concerned enough to do any serious investigating, as you put it, is something else. I don't want him hurt, though, by the police or anyone, if I can do anything to prevent it."

"I ain't pryin'," he said. "But how about tellin' me where you was at last night?"

"What the hell for? I've already told that to those buddies of yours. They'll be checking it out without your help."

Pinkerton's smile had the effect of drawing the corners of his mouth down so that it was only the crinkling around the eyes that belied the resulting smirk. "I mean," he said quietly, "where was you *really* at, chum?"

I had always liked the easy touch and caress in his voice when Dale said "chum." But now there was something sleazy and calculating in the way he gave a sheen of menace to it, as if he were testing. I decided to test in return.

"For that matter, where were you?"

"Oh," his chest contracted in silent laughter. "I was all over the place, to tell you the truth. We had a little entertainment to put on."

"You and Bobbie-Jo?"

"Naw. It was a couple of the other gals an' me. It didn't amount to much."

As easy as that we had slipped out of the slippery discussion that was the object of my being there, and were into the prelude to my leaving. I asked him about Bobbie-Jo, as if an exchange of commonplace courtesies were called for before we parted.

"Why, she's just fine," he said, as if no kernel of discord had ever been ground between us. "She asked about you and said to say hello."

I told him to return the greeting, and refrained from inquiring about the "little entertainment" he had mentioned as the thing that occupied his time Friday night. Stag shows were still a thing of the present down there, and Pinkerton picked up some of his living wages as part of them. He MC'ed them, recruited live talent when that was called for, and occasionally shilled in them. This latter role was the one he liked best since it paid the most. Bobbie-Jo maintained it brought out his natural acting talents as well, and she ought to have known, since their highest bids used to come with the stipulation that they perform as a couple. Solo or half of a sister-brother team (although such a setup usually had them coming on as strangers), he did a beautiful job as a loner pushing his way out of the crowd to catalyze those around him. "I only get to do that now on out-o'-town setups," he once told me. "Arizona, Nevada . . . like that." Cities like L.A. offered competition better than he could match, and the smaller communities of southern California could seldom recruit enough paying celebrants at any one time to make it worth his while to put on a performance and pay the protection necessary to keep the local vigilantes quiet, even though they were usually part of the audience at these affairs. Then too, topless dancing had recently reached the boondocks, and this was letting off enough steam to further restrict his act's already limited circuit. And, of course, they knew him too well by now in San Diego County for him to come on like a stranger in a gathering of drunken ranchers, business, and professional men, acting the part of an eager stud yokel.

Stud he was though. He had enough cock to feed the poor. His soft-spoken boast that he could lay ten quarters on it hard was a standing bet he would rise to whenever challenged, using coins out of a broken roll the skeptical had to supply, and the remainder of which he kept with the fleshed-out ten as forfeit of the wager won. He was seldom averse to proving it to whomever wanted to watch, and who had the means to pay for what he saw. It was an easy five dollars to come by.

And Pinkerton only looked the part of a yokel, if yokel be taken to mean a victim of rustic naïveté in the face of urban sham. No citified bumpkin he: Pinkerton was every bit as sharp as the uncountrified clientele he served, and occasionally duped—although on this last I had only heard talk, and no solid evidence, unless his connection with the police was what was intended.

His sexual proclivities (aside from what he did for pelf) were nothing unusual, except for the acquired taste that Eugene claimed to have given him in the art of shoving his fist up someone's ass. He had a long arm and never tired of regaling the gullible listener with descriptions of impossible feats of penetration. "I've been there, Dale," I told him one time in an attempt to cut him off at the pass, "so never mind. I know all about your operations, with your hand and arm slathered clear up to the elbow with Vaseline or surgical jelly or Crisco or butter, if it comes to making do in an impromptu situation."

"Not surgical jelly, for crissake!" He shook his head in something like a professional's disgust at the mention of shoddy methods of procedure. "The pores absorb it and too much friction develops."

"And I guess it must be plain hell then, mustn't it?"

"Well, it always pays to use the proper thing, that's all."

"Ever think of using your dick?" I said.

But that was just to rag him. I didn't feel much like joking with him now, or waiting to see if he had any more surprises to spring on me. I thanked him for the tip about Chuck Lawson.

"Any time I can be of help . . ." He lifted his arms above his head in a lazy stretch. ". . . just lemme know, chum. And," he added, coming out of the stretch with a long, heavy forefinger bearing on my clavicle, "think about what I said, you hear?"

"I'll keep it in mind, Dale."

He nodded, lowering his eyes and his eyelids, smiling like a demure and lecherous turtle caught without his carapace. I found myself wondering, as I turned to go, if Pinkerton had ever been in love with anyone, ever in his whole life.

8

AT THE SAME TIME I wondered, after seeing the devastation worked on Symonds, if I could ever be again. The physical basis of love is never more evident than during the strain of its act or when the object of the action lies stilled in death. Pity as well as fear, I well knew, can serve as a portal of lust, but having beheld Glenn dead and ruined in a coroner's drawer provoked no compensatory desire in me. The strongest urge I felt at that moment was to turn aside and give it up. I mean, give it *all* up. But the sight of him had reanimated the guilt I felt at having run out on him the night before. It was a self-accusation I was in no shape to face or reason with; it was the thing that kept me out and running almost directionless, pointed only to where next seemed indicated by the immediate press of events as they happened. Anywhere, I said, but home, and with anyone so long as I was not alone.

I sat in my old convertible outside the all-night cafe and tried to sift through the accumulated data and the impressions the data were making on me. I had had four or five hours of sleep in the last forty-two, and the dexadrine I took at the lab was about used up. I popped another one from the small supply I kept on me in case of need and forced myself to consider the implications of the colloquy I had just had with that mackerel-gutted canary. Because, in spite of the small regard in which I held Dale Pinkerton in all his guises, he had shaken me with the unexpected comments and questions following our return from the morgue. He knew something. And what he knew seemed to point to what Lawson came out to Encanto to talk to me about but had thought better of after spotting the law parked outside my house. Could either of them have been in contact with Eugene since Symonds was last alive? It was something to consider. Ordinarily Eugene would not be so incautious as to leave a message for me through anyone else, least of all by way of either of those two. But this was no ordinary situation.

Then too, it was not so much Dale's words themselves as the omissions in what he said that left me puzzled. The thinly veiled warning about sticking my nose into this might be no more than a fatuous display of self-importance, and his pointed question as to my whereabouts the night before could be dismissed as a mere fishing expedition. With his kind one had no way of knowing when he spoke as the oracle of his keepers, the *po*-lice, or when it was some personal motive that stirred him to utterance. Curiously, I found myself relying on those detectives to sustain me in this matter. My stay at Crystal Cove covered the time span the law was interested in, and that was the main thing as I saw it.

Yet what caused the greatest strain in my dealings with these developments was the varying and often contradictory reactions of those who knew that it was a former lover and apparent friend of mine who had been killed. To be sure, I was largely to blame for the confusion in their minds. I had for too long played down my association with Symonds by trivializing it, or in one way or another squelching the subject whenever it came up. As a result, few were prone to bring it up now with an eye to ferreting out the deeper influences that may have been lingering under the surface indifference I displayed in the way of self-defense. They were either overly solicitous, like Chris, or blithely indifferent, like Cusped. None were really callous, merely careful, like Lawson. Added to this was the fact that when it comes to a death, people now find it difficult to respond with any outward grace to a phenomenon which fundamentally offends or threatens them. Old-fashioned manners dealt with this kind of thing in accepted, clearly patterned ways; the new fashion is to deal with it as little as possible. So, as if to make up for others' apparent lack of curiosity, I found myself, while I sat in the car waiting for the dexadrine to take hold, paraphrasing the questions earlier put to me by the police. To wit, what was Glenn Symonds to me, or I to him, at the hour of his death?

Well, I hardly knew what to predicate about Glenn that was not personal history. The effort expended to get clear of him had involved so much concentration and control on my part that by the time I saw him again, eighteen months after our separation, my emotional circuitry was all but shorted out and my perceptual capacities dulled by a determined refusal to be impressed by him any more. After seven years of total immersion in the volatile force field of his dominant and dominating personality, my ability to respond very warmly to others took a while to recover from the hibernal insensitivity dissociation from him produced. Eugene Carmichael coming on the scene was the harbinger of better weather. And taking him on, even as a limited, one might almost say shallow, commitment, seemed to signal an end to the long captivity.

It was for this reason that the reappearance of Symonds, just when I had reached the point where I could forget him, was so shattering a blow to my restored self-confidence. Retrained, as it were, and recovered, as I supposed, I almost suffered a relapse then and there when Eugene walked into the house back in December, reeking of him. This retrogression into old, bad habits of the mind had shoved me for a while back into a laconic and introspective mode from which I emerged only enough to ward off the inquiring intrusions of those few who knew me well enough to think they discerned the cause of so great a change in one whose repertoire of social response, for all they knew about it, had never included either passion *or* the sulks.

It takes doing to get things straight, and that was our problem from the beginning: Symonds and I never really got things straight between us. Not, say, in the easy way that Gene and I worked them out from the beginning of our short affiliation. Eugene was never any problem, not even when Symonds reentered the picture—which, I see now, he was bound to have done. Symonds was always the problem; Eugene never. But then, Eugene was never more than a counter in the game.

The silent distemper I contracted this second time around, however, contained elements that, if not foreign to, were at least different in intensity from, the first and longer complaint. Glenn's plucking up of Eugene—hapless, credulous Eugene—as a way of getting to me forced me to a not-quite-amused reassessment of his constancy and consistency. Glenn Symonds had more guts than a dead bear and more insight into my weaknesses than I ever gave him credit for while he was alive. But it was this very insight into other people's vulnerability and a willed assertiveness to act upon it that had, in great measure, served to build him a lucrative practice and a solid reputation in two very different fields of accomplishment. Amazed and beguiled, I gradually gave up fighting him off, now that he was showing up down there, and got to know him again, on different and better terms—or so I thought.

In the beginning, getting to know Symonds was a process of forcing my way through the screen of his attributes, searching for what lay behind them. I soon discovered that he was in substance the sum of those attributions, that he had become in reality, long before I met him, the various masks he presented to the world. To in any way strip him of any of those roles he played would, to the extent of the dispossession, deprive him of reason, and with it his reason for living. This I never wished to do, even in our bad season, and, sensing this, he came to trust me as no other in that he could feel at ease with me, knowing I knew how little of substance there was to him and that, knowing it, I did not care. The bond we formed had about it a quality of the surreal, as one of the practitioners of that school described the genre: the fortuitous encounter upon a nonsuitable plane of two mutually distant realities. He was speaking to a theory of art, but it defines exactly the nature of our relationship to each other.

When I first met Symonds, I had no idea of what I was getting hold of. Over two decades ago the name, and the man behind it, already possessed an aura. There was, of course, his professional reputation, established and expanding then, but of this I was ignorant, having no knowledge or need of plastic surgery—although, after knowing him scarcely three months, I humored him by letting him perform a theliectomy on me to remove the barely visible blemish I bore over the region of the right twelfth rib. In different circles his name was uttered with respect or mentioned with misgivings by people who knew of him only by word of mouth. In other, allied quarters, it was mouthed with feigned contempt, even with indignation. These latter did not count, being themselves objects of amusement or scorn. But I, living in a different part of the state and caught up in the self-centered isolation of the undergraduate rat race as it was run in the Eisenhower era, had never heard of him.

It was during an Easter recess. I had come down from Stanford with some dorm friends who lived in Glendale. They were not into the same things I was, and before the long week was half over I was ready to cut and head back north from boredom and frustration. Came Good Friday and I managed to shake them in a Westwood coffee house and take off by myself for some heavy cruising. In a bar on Santa Monica Boulevard I met an older man who had wheels. He took me on a tour, showing me some places in and around Hollywood until I lost him in one that I think was called The Madness, but that may have been before its time. At any rate, it was a very leather and bikey scene, and it interested me.

Being an undergraduate posed no problem where bars were concerned; I was a good four years older than those whose house guest I was. This was the result of indecision after high school, followed by interference from the world in the form of the Korean War, in which I volunteered to escape the draft. Just the same, my ID was checked at the door. The bar was in one of those county pockets unabsorbed by Los Angeles city, unincorporated plots existing like stable cysts wherein was drained the accumulated suppurations of repressed naughtiness enflamed elsewhere in that sprawling urban blister.

The first thing that caught my eye after the initial confusion wore off and the cultivated darkness of the place lightened to the adjusting eye was Glenn Symonds standing in the restless crowd like a stag at gaze, his head mounted on that tall columnal neck and capped in the amber sheen of short-cropped hair. His eyes were narrow-set and so deeply recessed under a creaseless brow that I could not tell their color. He was, no doubt about it, the best looking thing in a bar that was filled with what, to me, then, were exceptionally good looking men. Hollywood was like that. One expected it.

Symonds told me later that he had noticed me immediately even though I was out of uniform, so to speak, and not the kind of thing someone there would ordinarily pay much attention to. I believe I was wearing Levis and a T-shirt under a jacket of brown suede. I would not be surprised but that I had on loafers and white socks as well. Not many college guys came in there, but those who did did not come in like that. I was very green, but the reason I sought out places like this instead of other, different places was that I liked the more physically emphasized presence found there. The term *Leather,* and the visible witness of it is nothing but hypostatized maleness: supple and toughened skin—balls. It was a way I had, to seek epitomes.

Emerging slowly through the mash of bodies, Symonds gradually revealed himself from neck to heels in close-fitting, intricately accoutered leather, gleaming black. I did not move from my post at the bar, but studied him frankly as he came on, pausing en route across the room to acknowledge a word or a nod from people he evidently knew, or who knew him. Once he stopped and took his time to look the crowd over, ignoring me as if he had thought of something else more interesting to do. But he resumed his progress, while I watched him coming.

He was still in his thirties then, at his peak. Seeing him, almost immediately aware that he had noticed me and was slowly clearing a path to me, I shifted and shuffled, and looked around a lot, embarrassed and pleased. And worried. It was his air of underplayed intrepidity combined with all that complicated leather armor that got to me. I thought of angels; more precisely, of some kind of destroying angel zeroing in on me with ambiguous intent. I was young, and still very much of a romantic behind the cultivated cynicism I used to hide it, even from myself, and the latent narcissism left over from adolescence, the self-love on which the romantic impulse battens for its drive, came to the fore in a flood of excitement I cannot remember enjoying before or since. Finally he was standing in front of me, asking me questions in a voice without emphasis or shading of any sort, as if it were coming from a programmed machine—which, in a way, it was.

I told him where I was from and how I came to be there, and he rapped out a few more queries that required only yes or no for answers, then told me, having decided I would do, to come along.

Expecting nothing less than a ride on a bike as the opening garnishment to this encounter, I was somewhat put off to find instead that he was traveling in an Oldsmobile. It was quite a grand Olds, to be sure, one of those baroque, heavily chromed Belchfires of the '50s. A black convertible with its black top up, it looked like a state barge. He moved the thing out into the sparse late traffic with the smoothness and aplomb of a space voyager getting under way on an intergalactic mission.

These were the kinds of thoughts that passed through my mind as I settled back into the leather-cushioned comfort of the seat beside him, stealing glances at the profile, thinking he had to be an actor with a jaw and eyes and hairline like that. The eyes, I saw, on leaving the bar, were a murky blue.

I asked him if he had a motorcycle, and he told me, "You don't have to talk," as if that were an answer. This initial putdown did not turn me off—nor on, for that matter. I kept quiet out of fear that he might change his mind and drop me on the curb if I didn't accede to his implied wishes. And anyway, I reasoned, lots of people don't care for small talk before the action, nor after, for that matter. Idle chatter and the delinquent urge to rehearse personal history and confess were not my style and I could respect a desire for restraint on the part of others. Still, his total silence made me nervous. As if sensing this, and to discourage further utterance, he switched on the radio. For the remainder of the long ride out to Topanga Canyon I listened to some orchestral crashings from a "good music" station and kept to my side of the seat.

He did not have to ask me to keep my hands off him as well as my mouth shut; his attitude and set proclaimed that injunction. It was that that did not sit so well with me. Before we arrived at our destination I was starting to have second thoughts about the whole thing. Kooks I had had in the past, and enjoyed having on occasion, but I had and have a respect for my own person and capabilities. If this was to be a hands-off arrangement or some kind of watch-and-jerk session, then I was ready to tell even so splendid an object of desire as sat beside me to go stuff it. I had seen too many other good tricks in that bar to waste my last night in L.A. on somebody else's ego trip.

Still . . .

I looked at him from time to time (this apparently was permitted) in the light from passing traffic. There was all that shiny black second skin, proportioned as it should be, with here a glitter, there a sparkle: chrome and steel accenting whatever messages this sort of sartorial code spelled out. And from above the rally collar rose that head like something out of a pioneer movie with a cap of hair the color of honey clipped as short as a Nazi tank commander's and receding from a high, sloping forehead that joined the nose by a prominent glabella; the this, and that; the promise stenciled on the whole damned package. . . . The romantic gurgle inside me cautioned patience to sullen pride. And pride listened, chained to lust.

Where he lived turned out to be a single-storied ranch-style house set back from a winding road behind some old sycamores. It was dark, but a dim light shone in the hall when he opened the front door and motioned me in. The light came from a small spot in the ceiling that accented a pair of tall, immaculately polished black riding boots standing at attention against the far wall beneath a beveled mirror which mimicked our entrance in an incongruous rococo frame. He went ahead past an open archway through which I caught a shadowed glimpse of the living room, to a door at the end of the hall near the boots. Opening this, he preceded me down a flight of steps, already lit, which ended at a landing and another door. Somebody had told me once that houses in L.A. and San Diego did not have basements. I have seen several since then.

The room we entered was exceptional, as basements go, but its impact was minimal that first time. The lighting was dim, but not shadowy, the floor padded. There was enough equipment around to indicate that it was basically a gym—or had been. I did not pay much attention to the chains, pulleys, hooks, odds and ends of rope and saddlery and other atypical items attached to walls and ceiling along with the more familiar equipment. Not that I was totally ignorant of what these, coupled with the way my taker dressed and acted, might represent; I was neither worried by, nor particularly interested in, any of it, having at that time no appreciation of the refined tastes and special learning necessary to the proper management of such attachments by their devotees. All I cared about was him. His liking for boots, uncommon apparel, and the unconventional decor that seemed to go with these was something that had nothing to do with me. That he probably liked contests involving eccentric and strenuous versions of riddles and forfeits left me in no way abashed. With the crassness of the middling young and the innocent singlemindedness of one in hungry rut zeroing in on a choice cut, I brushed aside these implications. He could play those other games with other people.

"What do you do?" I asked for openers, as crassly as might be expected. Now that we had come to roost, so to speak, I thought perhaps the ban on conversation might be lifted. He looked at me for a moment as if I were an intruder who had followed him in without his knowing it. Then he asked me what I was talking about.

"I mean," I faltered, once more almost put off by his look in which there was a frown of annoyed surprise—having recognized me, was he now wondering why he had dragged me back with him? "I mean, what do you do for a living?"

"Oh." He turned and strode across the gym pads to another door. With his hand on the knob, he looked back at me. "I'm a doctor." And, like a doctor, added, "Take your clothes off," then disappeared, closing the door behind him.

This, I thought, was more like it. After getting out of the few things I had on, I leaned against a padded buck and waited, ready for him. He took a little longer to divest himself of all the crap he was wearing, but when the door opened again and he reappeared, my willful obtuseness in refusing to see what might lie ahead gave way a little before the picture he now presented. Stripped down, he redeemed the physical promise his close arrayal had earlier made. It was plain that he actually used, or could with ease make use of, the athletic furnishings of this private gym. Yet, sauntering toward me, he revealed himself still clad in the implements and doodads of his peculiar hangup, and, although there were fewer of them, it was a case of less being more. He was shod in a different pair of boots, sole to swelling calf; a leather band covered with diamond-shaped studs encased his corded wrists; a high, wide-open, silver-conchoed leather vest inadequately covered his rolling shoulders and spread open over the deeply defined contours of his chest. It was then I noticed for the first time the red and blue hierogram tattooed on his left arm just where the swell of the shoulder ended. A heavy belt stamped with a webbed pattern cinctured the lean tuck of his waist, and on one side, like a scabbard, was snapped a curled and knotted crop while, tucked between the leather and the lean flesh of the other, a pair of wadded gloves hung down like the tassle of a sash. The only other thing he wore was the barest string of a pouch, likewise of leather, hostering the tool I had expected to see full out. Well, it was his house and his move. I let him make it.

There was nothing rough about him at first. He carefully took the gloves from his belt and sheathed his hands in them, taking his time, pulling the fine kid down to each finger's crotch while I stood watching him. Then he cased my body as if I were not in it, the way a prospective buyer might test the strength and build of a slave on the block. His hands were firm and yet exquisitely gentle in their appreciation of what he felt. I reached up to touch him in return, but he parried the move, almost without seeming to do so.

"Lie down," he said. "You need loosening up."

All right. I did as I was told, looking up at him standing over me.

"Uh-uh." He shook his head, smiling faintly. "On your stomach."

So over I turned. He arranged me to suit himself and knelt over me, straddling my legs. The touch of his hands again, authoritative, warm even through those flesh-like gloves, hypnotic, dragged a groan of pleasure from me as he worked over my shoulders, back, legs, with every once in a while the gentlest pat on my ass.

Well, I thought, a mesmerizing haze of good feelings investing me, so what? Guided by his hands into an erotic doze, I vaguely registered the increasing frequency and sting of the arrythmic smacks laid on my buttocks. Then I felt the cold, hard-edged slide of leather drawn across my thighs; a pause; and even as I was half-bracing to receive it, got it—*crack!*—right across the butt.

I twisted around and jerked myself up on one elbow to see this now unbelted angel in a sweat above me.

"Skip it!" I said, the hardness gone out of my crotch and into my voice.

I waited for him to make his next move in response to this while I got my legs in a tangle trying to get clear of him. He did nothing more than stare at me from those deep, shaded sockets. I reached for my clothes nearby while he watched, estimating. I slid out from under the arch of his thighs and started pulling on my socks, avoiding his gaze. Then, when I at last did look up, I saw that all his tenseness appeared to have deserted him. He tossed the belt away and stood up.

"All right," he said, as if disinterested in any further play.

"Yeah, all right," I muttered, indignant, sorry for myself with my wasted evening. "You can do this sort of thing with somebody else, *doctor.* Why didn't you pick one of those weirdos out of that place you found me in?"

He gave the impression of being amused at this.

"Why did you come here with me?" he said.

"Because I liked your looks," I mumbled, fumbling with my T-shirt, "but your style is something else."

Still avoiding his eyes, I reached for my Levis. I was embarrassed and becoming angry because of my own confusion in the face of his calm. I pulled the legs outside out again, and looked at the faded denim for a moment. Something like a sigh breathed above me, and I heard him unsnap the wristbands and the pouch, and toss them away. I heard him slip out of the vest and drop it on the pad near me. He pulled off the gloves, and they brushed my thigh in their fall. Still standing, he removed first one boot and then the other, and then I looked up and saw him revealed the way I had wanted him in the first place: just like anybody else, only more beautiful. He dropped down beside me on the mat, and we made it with no explanations, inhibitions, or complications—again, just like anybody else, only more beautiful.

I fixed his wagon good that night, but through all of it he never uttered a word. Only the enraged and captive quality in his stare told me that he felt anything. But he was the accomplished sexual athlete I expected he would be, able to take the best that I could give. And when it was over, and we had rested a while, the only reference he made to what we had done was to ask me, "Do you always keep your socks on when you do that?"

It was June, two months later, before I saw Glenn Symonds again. In the interim I found out a lot about him, and what we had done was not what he was known for doing. It bothered me that he would lay for what he could not master. And it worried me too that that might have been a way of purging his interest in me—

assuming he had any. I was too busy to do anything about it during the rest of that spring, but by June I had my degree in hand and was in L.A. again, this time on my own. His telephone was unlisted, and I did not want to call him at his office, so I went by his house. On the third try I found him at home, busy with somebody else down in his work room. He came to the door in Levis and a shirt partially unbuttoned. For a moment he looked as if he did not recognize me. But he did. He asked me what I wanted.

"I want to fuck you again," I told him. "It's been on my mind for the last few months."

"I'm busy right now." There was no hesitation or change in his voice when he added, "Do you want to wait?"

"Sure," I said, "I'll wait."

He motioned me in and turned on the lights in the living room. "It'll be a little while. Fix yourself a drink." He pointed to a sideboard. "And you can turn on the television if you want to, only keep it down."

I fell asleep waiting for him, and when he woke me up, it was early in the morning. He was fresh and showered and smiling, and he took me into a bedroom this time. I spent the rest of that summer with him, and I was his lover for seven years after that.

9

SHORTLY BEFORE one o'clock I pushed my way into The Club again. Chris was bubbling away behind the horseshoe, playing host and helping his help in the frantic last hours before it would all come to an end and the crowd would be turned into toads and shoveled out into the street. A lissome covey of fluffy sweaters was paying court near the cash register where, leaning on his knuckles, he babbled away, wagging his head like one of those dolls some people stick in the rear window of their automobiles. I penetrated the woolly heap and confronted him with as friendly an aspect as I could summon up, smiling, as I remember. The conversation, if it could be called that, ceased as he caught sight of me and fell silent.

He made a face that indicated he was ashamed of me. Not totally disheartened, I asked him for a beer and then inquired if the fuzz had given him any trouble over the incident earlier.

"Baby!" he caroled for all to hear, "they *never* give me any trouble. I run a clean place." And followed it with the vaudeville laugh while he reached under the counter for my brand. After having 86'd me, he was willing to be friends again, it seemed. I put money on the bar and motioned him closer.

"Got any idea where they went when they left here?"

"Who?" He straightened up, indignant, resolved, I suppose, that everybody in the place was going to be privy to what I preferred to keep a private exchange. "Those bums? I couldn't care less! I hope I never see them again—any of them! Starting a fight in here . . . in *my place!*"

"What was it about, anyway?" I kept my voice low.

"I don't care what it was about!" He jerked his head back and to one side in disdain. The others continued to stare at me. "And I've got no patience with you people and your vicious carryings-on."

More heads turned.

"I'm not vicious, Chris. And I don't carry on."

"You used to." He adopted a pout, as if I had ruined his evening by reminding him of the incident.

"Not really," I said, still keeping my voice down.

"*Not really!* Do you expect anybody to believe that, knowing that facelifter or Eugene Carmichael, and knowing you lived with them?"

"I don't give a shit what you believe. And I didn't come here to discuss what I do in bed."

"In bed? Oh, my!" He touched palm to forehead, as if about to swoon. "If you did it in bed it might have some redeeming qualities to it. But a bed's not good enough for you people. You've got to have a bare floor, or a sawhorse, or a table, or a rack!"

"Hey!" I said, "I think you've got some great ideas there!"

"Is that so? Well, you tell me. Isn't it the truth that Eugene Carmichael gets his jollies hanging by his heels from the ceiling—just like he did it to our late friend?"

"He also gets them lying on his back in a bed."

Chris lifted his gaze heavenward. "Will wonders never cease!"

"Chris," I said, trying to bring my voice back down to where I had maintained it before he got my goat. "Chris, will you just tell me what happened and, if you know, where they took off to?"

"I don't know anything about it." He turned to the wide-eyed group in front of him and went on about what a nice thing he did maintaining a decent place where these boys could get together without the city breathing down their necks, and how his good works were being jeopardized by irresponsible hoodlums. Looking at me, he snapped, "With a clientele like you, who needs bankruptcy?"

"You aren't closed yet, Chris. And you're not about to be."

"No thanks to you and your friends!"

"Chuck Lawson's one of your boys," I reminded him.

"Used to be, dear. *Used to be.*" He sighed deeply, and appealed with great, resigned eyes to the cluster of deferential admirers who formed a circle around me. Then, shaking his head, he intoned, "Lie down with dogs and get up with fleas."

"I accept your apology," I said, and left him still fulminating to an audience made up of the sort that regard the favor of public-house proprietors as a fount of social prestige.

Fading back into the crowd with the bottle of beer, I suddenly found myself staring at the evening's second smiling caricature of the Old—or was it the New?—West.

"This is the *New* West, pardner," it said in a stage drawl that was not so much thick as viscid, and extended a small, hard hand to shake. It was Cactus Jack Straub, decked out in a carefully beat-up flat Cheyenne, split cowhide short jacket that must have been making him sweat over the checked flannel shirt and purple string tie in the press of bodies around us. A pair of tan kangaroo-hide rundown dogger-heel boots with tooled avocado-green appliqué and white stitching set off his short underpinnings whose every contour was revealed by white drill dress pants so tight they looked as if they would unman him if he sat down in them. He was begging for that long, lean look, but he would never make it, not even by tottering about on those three-inch heels. Not that it mattered—except to him, of course.

I liked Jack. He went his own way, doing what he pleased and changing to something else when it got to be a drag. Right now he was in a crossover stage between Leather and a cowboy fantasy trip. The Western thing was a long-standing fetish of his; the Leather bit a recent excursion into the dark.

Jack owned a radio station in the San Fernando Valley that offered twenty-four hours of country-and-western interspersed with enough newsbreaks to fulfill its public service obligations under the terms of the FCC franchise. There was a lot of Nashville yearning in L.A. County, and he was making a bundle answering it with a simple format and small overhead. He drove a pale lavender Cadillac convertible with matching leather upholstery tailor-made in Tijuana. Purple seemed to be his color now. The year I met him it had been chrome yellow.

"What brings you down here?" I said.

"The news, what else?" He grinned like a frog, as if hooks were inserted in the corners of his mouth, so that a smile split his face in two along the line of a square jaw, a line accentuated by the straight-razored trim of heavy red sideburns. Jack picked up walk-on parts in TV westerns for fun and further profit, but mostly for the myth-generating satisfaction it gave his unabashable ego. Bearing this patina of professional status with the Western and Leather crowd was the kind of standing he nurtured. Cactus Jack was authentic, at least according to the canons of the New West.

We had known each other for several years. I met him through Symonds who had erased some unsightly marks left on his face and shoulders from a crash he had survived during a race at Laguna Seca in his salad days, which might have been called his James Dean period. As was the case with a good many of my acquaintances north of San Diego, the standing I enjoyed with him accrued to me as a side benefit of my connection with Symonds. Jack was just another luminary of about the sixth or seventh magnitude in Symonds' star chart, but Jack's admiration for Glenn was boundless. The news of his death must have had enough effect on him to drag him down the coast out of curiosity to know more. I asked him if it had been on the wires.

"I tore it off the sheet myself this afternoon," he said. "Called you, but you weren't home, or not answering. Then I found out everybody was down here." He said something else, but his voice was lowered and the music from the jukebox drowned it out.

"Speak up!" I said, bending low.

"I said when did Symonds take up with the Mafia?"

"What're you talking about?"

"That's what I got. The authorities think it was a mob killing."

"Good God!" I looked up at the ceiling and laughed. "I wonder what authorities they had in mind. The ones I've been talking to haven't said anything about that. What in hell do they think the Mafia's got to do with the sort of thing Glenn was into?"

"Drugs," he said simply. "A guy I called at the *L.A. Times* said they've got narcs down here all over the county now."

"Shit!"

"What's the real scoop on it then?"

"I wish I knew," I said, rather lamely.

"Well, it doesn't sound like the kind of thing Symonds would ordinarily walk into, does it?"

"It isn't the kind of thing that anybody would walk into ordinarily. And it seems to have gotten an extraordinary response from out of town."

"You talking about me?"

"You and some others. Glenn would've been touched."

"How're you taking it?"

"How should I be taking it?"

"I don't know. Thought I'd ask, that's all."

He went on mumbling, but I interrupted him.

"Does the name Chuck Lawson mean anything to you?"

"Rides a white Harley?"

I nodded.

"Sandy hair and an ugly face?"

I nodded again.

"I've run into him before."

"You sound unimpressed."

"The only thing that impressed me about him was his big mouth. He hasn't got manners enough to feed slop to hogs."

Jack's image was turned up full and coming across fine tonight. I had to laugh again. "Sounds like he got the better of you one time."

"At no time!" he snorted. "Hell! he couldn't get the better of a one-legged man ridin' a rollerskate. He's got a big mouth, that's all.

The next time I see him I'm gonna stick a fish hook up his ass."

"Maybe you can get your chance tonight if I can find him. He went out of here more or less under duress a couple of hours ago with Cusped. You know the mad harpist?" He nodded. "The two of them and another type I never saw, but I understand's from L.A., and who looked like he was in charge of things, got into a scuffle here and roared off somewhere."

"You'd think they'd hold it down a little after what happened to Symonds, wouldn't you?"

I told him I had said as much to Cusped. "But it seems as if everybody I've mentioned, and a few I haven't, were either down here, or had contracted to be down here, for some kind of arrangement before Symonds got it. In fact, Symonds was suppose to be part of the arrangement. In fact, *he* maybe arranged it."

"I don't believe it."

I shrugged. "It's what I heard."

"Where's Eugene Carmichael?"

"I thought you'd never ask. Didn't the newswires mention him?"

He shook his head and waited, his upturned face blank with unknowing.

"Missing," I said. "Presumed guilty."

Jack whistled soundlessly. "That doesn't figure."

"What doesn't? That he's missing, or that he's guilty?"

"I mean it doesn't sound right." He looked at me hard, tilting his hat and then his head back so that his thick neck swelled as if he had inflated it in display. "You know Gene."

"Everybody keeps reminding me of that."

"Well, what the hell kind of a reason would he have for doing Symonds in? And even if he did have a reason, he's not the type. D'you think?"

"Anything's possible. Or so I'm beginning to believe. The police want him, and I think they think I know where he is."

"Do you?"

"In a word, no. Right now I'm mainly interested in knowing where Chuck Lawson is—or was. And another thing. . ."

He waited. I did too for a moment, turning over in my mind

whether or not I should mention this to him. Finding no reason not to, I asked him, "Do you know anyone named Owen?"

"Owen Keller?"

"I don't know his last name. I think he knew Symonds."

"I've met him. He's gotten to know a lot of people lately. I haven't heard anything on him recently though. What about him?"

"I've heard the name mentioned, that's all. Where is he from, anyway?"

"Up in L.A. somewhere. I only met him a couple of times, but I don't think he'd be a bottom man in anything like what Cusped was talking about."

"Cusped doesn't always know what he's talking about; that's why I asked you. Anyway, Cusped wasn't naming any roles."

Jack eyed me quizzically. "Are you working with the cops in this?"

"No." I had to smile. "Frankly, I'm not sure what I'm doing. I'm not responsible for Eugene and what he gets himself into, but I'm fond of him. And I'm concerned that they're going to try to pin this on him from the way it's shaking out."

"Where does Keller fit into the picture?"

I shook my head. "It's just a name. If I'm the only one around here who doesn't think Eugene did it, I've got to be interested in any names I don't know. You know?"

"You don't think that asshole Lawson did it, do you?"

"By no means. I want to talk to him, that's all. He came by to see me tonight, and then thought better of it. I want to find out what was on his mind, and to do that I've got to find him."

"He hasn't got a mind. And I thought you said he went out of here with Cusped."

"Among others. But I don't know *where* they went."

"Want me to inquire around?"

"It might help."

"I'll see what I can turn up."

He moved away to be swallowed up in the heaving boister of standees. I turned back to the bar for another beer, heading for the far end, away from Chris who was still jawing with his clique at the cash register. Thompson, a barman Chris had lately stolen from competition on the hill, handed me a full bottle. He possessed the bland good looks often seen in liquor and cigarette ads, and he disbursed a spiritless amiability to all comers. It was his job. He had a lover who was the High Church rector of a Low Church parish, but they lived separately so as to keep things seemly. Aside from earnest philosophizing on the techniques and benefits of weightlifting with familiar patrons at a local gym he frequented and the superficial sociability he engaged in during working hours, Thompson's only intercourse was with his closet cleric. This, coupled with good looks and an easy manner, made him something of a challenge to the more sporting element found in The Club and an object of inarticulate longing on the part of other, less aggressive viewers-from-afar. He was good for business, and, while never a participant, he acted as recipient and transmitter of information on impending action that passed back and forth across the bar on a weekend evening. On the off-chance that he might know something, I asked him about after-hours scenes shaping up.

"There's a party at Chris'," he said in a remarkably lifeless voice, his smile a caul of courtesy covering the stillborn conviviality preserved at room temperature behind it.

"That's not quite what I had in mind. Besides, I'm on his shit list."

There was a mechanical quality to his smile; like an indulgent but bored playground supervisor, he eyed his watch and said, "He wasn't serious. He's forgotten all about it by now."

"You've made my evening. Thanks."

Thompson went on up the line dispensing drinks with temperate charm, and I pushed my way back to the corner by the cigarette machine to wait for Jack and settle into my beer, watching the crowd without paying much attention to it.

But what a strange assemblage it would be if reassembled now. It was a smooth-cheeked age ago, for neither beards nor mustaches had we then. All fresh and clean we looked, and even the Levis and chinos that were standard, informal dress among the slacks and sport shirts and sweaters gave no impression of willed poverty or plastic hip. Yet what was doing then was the foreshadowing of what was about to happen in stages throughout the country, the seedbed of an explosive influorescence that would obliterate the once clear demarcation indicating for most people what is and is not a natural thing to do.

Watching the almost imperceptible hunch and thrust of a kid by the jukebox, responding in solitary pleasure to the repetitive beat of a Beatles record, I realized that even down there the change was happening almost without our being aware of it. A sense of freedom about to open out, a conscious daring on the part of a few, watched by others, betokened by such things as the way this one was flexing in onanistic imitation of coitus, was a posture almost too new to speculate upon: exhibitionism more than latent, but not yet bold enough to provoke a serious reaction from those quarters it ultimately threatened.

These musings, by the strange pathways of the mind, brought me back to my reason for being here. The hints and tag ends attaching Lawson to Eugene lent anger to the urgency I felt, and it was centered on Chuck Lawson. I meant to make him cough up what he knew (for I was convinced he knew *something*), and what I learned from his disclosures I might retail to Don and the National City constabulary—or I might not. If their minion Pinkerton had not seen fit to alert them to this lead, then why should I come forward with it? Dale no doubt had his reasons for withholding; so also did I. Anyway, I had already skirted the bounds of felony so closely in the past twenty-four hours that a little more poking around the edges was not going to make much of a difference.

Still, Pinkerton's smarmy warning about sticking my nose in things gnawned at me. Not that I was afraid—for myself, that is. Rather, I had the impression that critical information was being intentionally withheld. Take Jack's news of a drug connection. If the media knew something like this, then the police did too, and probably Pinkerton as well. Why the silence?

In selecting from among alternative theories, I had come, by training and habit, to choose the one least likely to require unreasonable assumptions. And it was unreasonable, as I knew them, to consider that Eugene would commit murder or that Symonds would allow someone like Gene to get the drop on him, or that either of them were mixed up in drugs. So what were the other alternatives? There were none, since I was unfamiliar with who else was involved.

While absorbed in these thoughts a sudden urge came over me to quit The Club and go back home. There was a chance that Eugene might have tried to telephone me, and would try again. There was no one else I could think of whom he might try to get in touch with except his mother. But then, like a cautioning voice, it occurred to me that Pinkerton may have been encouraging me obliquely to return to Encato and stay there. Because the roads up to Radio Heights were being watched?

With that thought the urge to leave subsided. The wind-curved telephone lines out on those hills might be tapped and waiting for me and for whomever might call me. I was not going to go back and sit there waiting for any call, not with what I had zinging in my blood right now. I would find Eugene my own way, or the police could find him in theirs.

For what I had in mind, though, I was moving too slowly and drinking too fast. The beer Thompson had given me was already gone. The feeling that I was standing still while events were moving by me at an accelerating speed had been growing on me ever since Symonds reentered the picture as Eugene's teacher. Things were now almost a blur. Symonds, Eugene; Eugene, Symonds. . . What the hell *had* been going on all this time anyway? Probably I had been unconsciously asking that question for months, and the concern in Eugene's voice over the telephone Friday infected me enough to drive me down here to stick my nose in it after a long season of letting well enough alone. The proof of this, for me, was in the reaction I sustained on seeing Glenn in that trailer, subordinated to a hovering figure who, whoever he was, rendered them both strangers to me. Embarrassed, fastidious, angry, I left the scene of something I knew was not right, one in which I ought not to have hesitated to make my presence known. But would anyone questioning me believe my confession of shyness and scruples?

So, discounting, as I did, Eugene's responsibility in the deed (although he had had enough abuse worked on him over a span of thirty years to warrant his putting the shoe on the other foot for a change), there remained the possibility that some kind of a mistake had been made even though the use of cyanide would seem to preclude it. Yet another coil of thought brought back the question of Owen Keller. Assuming he was the one Symonds had so smugly alluded to while talking to me Tuesday night, and granted that Eugene was sorely put down from having been cast aside by Glenn, would any of this point to a provocation sufficient enough to trigger so fundamental a turnabout as would have resulted in the final tableau revealed in his trailer? Just such a case, I felt, was being constructed by the law. But although the engineering of it bore his neat and thorough mark, the surgery was a messy attempt indicating haste, or panic, or dementia, none of which bespoke Eugene.

Still, killing has a way of getting out of hand and knocking hell out of a person's usual way of doing things.

I flagged Wendy for another beer and let my body respond quietly to the music from the box, "You Can't Hurry Love," reflecting that the refrain carried enough truth in it to serve as the leitmotif for all the principal ingredients in this melodrama that had been shaping up over the last few days, months, years, lifetimes . . . generations. You can't hurry love, but there's no stopping hate. That would be the classic liberal view an attorney for the defense might argue with a plea of diminished capacity should anybody be hauled before the courts when the dust of this hunt finally settled: You can't hurry love, but hate has a way of suddenly overtaking and dismantling settled natures and of wiping out indifference by a heedless stroke of unequivocal and devastating involvement. Had I hated Symonds?

I kicked it around like this while waiting for my beer, unhurried for the moment by the day-long itch to keep on the move. For I had not stopped, except for that solitary hour passed at the lab where I had been attempting to bury the roil of apprehensions besetting me by tackling some descriptive taxonomy on a batch of specimens collected weeks before. (In the face of uncertainty, describe. A good rule of thumb which, even if it yields no essential answers, tends to calm the nerves and set things in some kind of order.)

Wendy appeared with my beer.

"You want another one before last call?" she asked, staring at my change on the tray.

"I've had too much already." I drew her back as she started to move away. "Do you know anything about a heavy session tonight, Wendy?"

"Sure," she replied, a plain woman a little makeup and a smile would have done wonders for, "two or three are firming up. What d'you have in mind?"

"Something Chuck Lawson might be mixed up in."

"Try Chris'."

"Thanks," I said, "but no thanks."

She shrugged in utter disinterest and moved off.

Cactus Jack took her place.

"It's at Chris'," he announced.

"It? It what?"

"The arrangement with Lawson in it."

"I don't believe it. Chris wouldn't let something like that go on over at his place. Particularly after what happened here tonight. Who told you?"

"It's the only action I've heard about where it could happen. I put one and one together and came up with a pair, that's all. What d'you think?"

"You've come up with nothing. What makes you imagine Chris won't know?"

"I didn't say he wouldn't know. But it might not be in his apartment. That's a big building, isn't it?"

"He'd find out quick enough when he gets there."

Jack pondered this for a moment, then he asked, "Who did you say Lawson went out with besides Cusped?"

"Somebody from L.A."

"Wasn't he talking to Dale Pinkerton?"

"Earlier, yes."

"Maybe they went to Pinkerton's."

"Pinkerton left here by himself before them. And anyway Pinkerton's wife doesn't appreciate him bringing people home. And he doesn't go in for what I think they've got in mind for Lawson."

"You sure?"

"I'm at least sure about his wife." Dale's *pro forma* marriage contained a clause in the unwritten contract permitting him bed and room by himself and board at the family table if he so desired on condition that he keep it that way and not bring any of his filthy male friends into the house of an evening. It was an arrangement he respected scrupulously, with the result that he spent about half his nights away from the little bungalow in La Mesa. Any arrangement put together in his place would come to an abrupt halt before it even got started, with the lot of them being thrown out in the street by an enraged and very badmouthing Snuffy Wiggins Pinkerton.

The beer must have been reaching me, because I was smiling, imagining the tawdry scene in all its yawping disruption of the peace.

"All right." Jack interrupted my bemused imaginings. "Where would Lawson ordinarily be staying down here?"

I sighed, accepting the obvious. Everything was pointing to the place I considered the least likely, for I had forgotten that basement room. "At Chris', I guess."

"Well, what're we waiting for?"

I set my beer on the cigarette machine and took his arm. "We'll check it out, partner," I said. "But I can't guarantee they'll let either of us in."

"You mean at Chris'?"

"No, I mean down in the basement."

"What basement?"

"You'll see—if you're any good at this sort of thing."

10

OUTSIDE we crossed the street to an all-night bottle shop, then, with Jack tailing me in his lavender Cad, I drove west toward Washington Avenue, and past it to Pringle. The final section of freeway still being worked on opposite the airport had left a disaster area of the residential rise it cut through north of where the canyon debauched into it. Chris' colony of upgraded shacks crested the south slope of this rundown section, and his big apartment house was above and about half a block beyond them on the crown of the hill. When the freeway would be completed, he was sure this section of town would come back fast. Canny Chris had bought cheap when the oldtimers fled the change, the dirt, and the confusion, and at a time when few others were willing to take a chance.

His termite-ridden, vine-covered cottages might just some day fulfill his predictions. They possessed a certain ramshackle charm qualitatively different from the usual gone-to-seed dwelling in San Diego. They could be reached on foot by climbing a steep, stroke-inducing flight of wooden steps anchored onto the south slope of dun-colored cliffs crusted with ice plant that came to life in day-glo colors in late winter. A less exacting, but more circuitous route was indicated for wheeled traffic.

Jack followed close and almost rearended me when I put on the brakes at the entrance to the paved parking lot bordering the apartment house. It was filled up and an unoccupied VW blocked the driveway. I cruised further on down the quiet street and we curbed ourselves half a block away, then walked back.

Inside the lot, shadowed by eucalyptus and pepper trees, were a half dozen bikes, among them the BMW I had seen earlier, Cusped's Indian, and three Harleys; Lawson's white trash wagon, a chopper with fancy sissie bars and twin exhausts sticking up in the air like a chrome bustle, and a saddle-bagged job I recognized immediately. I went over to make sure, however, and confirmed that the classic blue was Faulkner's all right.

They had a guard out, a big-bellied guy who looked to be from out of town too. He was wearing bike chaps over Levis, and had an AMC patch on his jacket shoulder; the overlay was no club I knew. He ignored Jack and came up to me, piping an uncertain greeting.

"Hey man, you a bikey?"

"No," I said, "I'm a trucky."

He spat, just missing Jack's stovepipes. "What's a trucky?"

I straightened up from examining Faulkner's machine and rested a hand on the ape bars of the fancy chopper next to the blue. "What's a bikey?"

"Well . . ." He looked uncertainly from me to Jack, and back to me again. "I mean, you draggin' leather like that. . . ."

And so I was: an old but well-cared-for, black calf hide jacket, a one-of-a-kind designed by Symonds for me and made to measure by

Lenny up at the Cloak and Dagger in San Francisco three years before. I kept it in the trunk of my convertible for chilly nights, and had donned it as an afterthought when I came out of The Club with Jack. When in Rome, or headed there . . .

"Is Nelson Faulkner inside?" I indicated the apartment building with a jerk of my head.

"Does he know you're comin'?" he said, relaxing somewhat.

"He won't be surprised to see me."

He laughed with the sound of a leaky pressure tank; someone's low man put out on this chore while his big friend, who might be smaller than he was, was having all the fun inside.

"Tell 'em to send me out a beer, will you?"

"You bet," I assured him, and we left him there, doing his duty and glad, underneath all his fretfulness, to have been chosen.

There were four floors and fifteen units to the building. Two rear stairwells, open to the weather, connected all the kitchen landings to the basement service area where there was a coin-operated laundry room, storage cubicles, and an almost airless utility room that Lawson had the key to.

"We'll go upstairs first," I said, leading the way around front to the foyer.

"Why?" Jack was trotting along beside me. "I thought you said they'd be down in the basement."

"I'm not ready for them yet. I need a drink."

We went inside to the elevator. Another couple had just entered it and held the door, then the four of us rode it to the top in silence. The shamefaced quiet of our chance companions had a squelched quality to it that was a reaction to Jack's attire, at which they cast covert glances while trying to avoid our unblinking stares. What would now be called the homophile community in San Diego tried, by and large, to look as much like the rest of the ruck as possible at that time. Consciousness-raising among the real and imagined oppressed was a thing of the future—near, but still some time away—and their manner of dress was chosen on the basis of protective coloration rather than distinctiveness of style. The blatant individuality of someone like Jack constituted an anxiety-provoking affront to most of those he came in contact with outside L.A. In bars or at private parties they would loosen up and allow themselves the thrill of imagining what it might be like to take someone threaded out like that to bed, but few of what Eugene used to refer to as the clutch-and-hug set ever let go to the extent of actually making it with the likes of Jack since it would entail being seen in public with him, even if only for the distance and time it took to walk to his car—itself no mean attention-getter. The same prejudice was directed against the leather set, or what was to be found of it down there. Besides Eugene, who used to underplay this aspect of his personality until he took up with Symonds, and Chuck Lawson, who never underplayed anything in his life, there was only a handful or irregulars who proclaimed their bent by the clothes they wore: weekend hellers, bike-and-bed buddies, and some straight cyclists who cultivated a boonie's image of the Hell's Angels and got themselves and their women into enough trouble to make all the fuss and expense worth their while. Cactus Jack probably fitted into no known pattern of ostentatious obliquity they had yet encountered, but they would relax once they got into the swing of things upstairs.

On the top floor, we stepped out into a narrow hall and proceeded down it to Chris' door where we pushed the button.

Chimes.

It was opened almost immediately, letting out a wash of babble and music and smoke. Duke, a strange youth who did handyman work for Chris and attended classes in a desultory fashion at the local junior college, stood before us, blocking entry. He looked at Jack, nodded sleepily to me, and ignored for the moment the pair behind us. He was wearing a half-assed version of cowboy vesture himself. Boots and a broken-down moonshiner's hat was all he could afford, I guess, but he wore them well. I said hello, and he voiced a quiet good evening to us from his six-and-a-half-foot height, motioning us on past. I pulled him down slightly by the shoulder to ask him if he had seen Cusped or Faulkner or Lawson here yet. The two who came up with us swept on by under his scrutiny, then he gave me that curious smile that suffused his bland, little-boy's face, wrinkling up the freckled bridge of his nose and exposing the gasps between ill-cared-for teeth. Earnest, and with a will to please, he played the part of greeter and bouncer with equal poise. "I guess everybody's here," he said, as if that answered my question, and directed us hospitably down the hall of the apartment. We left him smiling some more through the door. The way he was churning them in, everybody probably *was* here, I thought, or would be shortly.

Jack kept turning around to get a look at Duke's trim backsides. His pygotropic tastes tended to tall ones.

"Go back and get it if you like it," I challenged him.

"Is it available?"

"If you've got a tattoo."

"Dammit!" he said, darting one last glimpse at the impossible dream, "I knew there was something I forgot to do."

The hallway was not long, and it made a dog's leg midway where the kitchen and dining room took over the rear of the apartment, which had three bedrooms, a living room, and two baths besides. All six rooms were filling up with everything that knew Chris, and quite a few who did not. Singletons floated free, or had attached themselves to the walls, waiting. Twosomes, trios, and clumps of drinkers stood around talking, laughing, their eyes never still. From the living room came the sound of the Supremes belting out "I Hear a Symphony." A few people were dancing. Jack and I received the momentary attention paid fresh boarders on a bus, then we blended.

There was no bar, but glasses and setups were spread out on the kitchen counter, and the setups were already empty. Nothing else; not even ice. The refrigerator had a lock on it. We fixed ourselves drinks from the tap and the bottle of Seagrams Jack had bought when we left The Club. Somebody came in and said Chris was sending up some ice and more mixes from the bar as soon as he was done there.

Jack settled uncomfortably in a cross-legged lean against the counter that just about caught him in the kidneys, and squinted at a darkened double doorway that opened into what was ordinarily a dining room. A barely discernible crowd was working out in the shadows inside. The table and chairs had been pushed against the walls, probably by Duke, who set these things up, and, out of respect for the convives who were busy in there, only the kitchen range light was left on; the other bulbs had been removed. The faint acridity of grass floated out into the kitchen from the cherry-red glows that flared up periodically in the dark. And every so often there was a soft "pop!"

"That's not it," Jack muttered.

"No, that's not it." I motioned to someone being done standing up against the refrigerator near us. "Nor that either."

"They're downstairs, huh?"

I nodded.

"Want me to go check?"

"Go ahead," I said. "You're good at checking."

He set his glass down on the counter. "Watch it for me."

"I'll watch it. Give me a cigarette before you go."

It was a measure of how I was getting along that I had been craving smoke ever since the law came to see me. Alone, I lit the thing, surprised at how good it made me feel after a three-year lapse.

While I smoked, people came and went through the kitchen, passing into and out of the darkened room in front of me. From the vaguely moving welter inside it there came a sound like a prolonged, whispered sigh, as if a vast reach of water or plain opened out in the blackness beyond. It was the breathing of many bodies, the rustle of cloth, and the shuffle of feet, repositioning the occupants; the surrealistic sound of some formless monster or primal matter stirring to delineate itself into a shape. Every so often a figure became detached from these corporate heavings, took on the lineaments of a recognizable body, and materialized out of the gloom to come into the kitchen, there to stand for a moment pulling its clothes together, then pass on out into the hallway, in search, I suppose, of the bathroom.

People kept on arriving; the front door was chiming as regularly as a department store floor signal and the kitchen was filling up fast. Every time somebody flared a match to light a cigarette the shifting mass in the dining room was highlighted for an instant like a page from Doré. A few who came in that I knew I nodded to or exchanged hellos with, keeping always to my corner, casting a spell of reticence through which they did not bother to penetrate beyond the initial greeting.

This was, after all, no social evening for me, and there were only two persons I was interested in talking to. What I was doing was drinking to get my courage up. Not that I feared Lawson, or Faulkner either, if it came to that. Rather, I was uncertain as to how to get to the first, now that he was in the hands of those others, or *the* other, since Nelson was, I assumed, in charge of things. I was glad enough to have Jack down there scouting out the territory for me. I faced with scant enthusiasm the possibility of having to force a response out of that feckless West Virginia clown. Nothing that could be going on down in that basement room appealed to me, but, having come this far, I was resolved not to leave until I got the information I was seeking. At least it would keep me busy until I was tired enough to sleep from the total exhaustion I was courting. As it was, the specter of Symonds laid out in that metal drawer threatened to make an insomniac out of me for the rest of my life. I was almost in a mood to grab someone, anyone, to go to bed with just to keep from having to sleep with myself and all the ghostly company that would join me after the silence settled in.

I looked over and saw that the mouth job by the Coldspot was accomplished. The donor disengaged himself and put his detumescing tool, unwiped, away, then moved out into the hallway while the squatter rose and engaged in platitudinous conversation with a couple of the onlookers who had gathered round to witness the climax and stayed to interview the co-star. Finding no takers among them, he left, clinging for a moment to my belt with one hand in passing, intruding on my spell-wrapped solitude, but letting go to be swallowed up in the frantic, welcoming darkness of the room beyond. I found I wished he had stayed a while. Turning, I bumped into Cactus Jack who had come up behind me. He was beginning to get a belly on him, from the feel of it.

"Did you find them?"

"Not yet," he said, then leaned closer, lowering his already lowered voice to a whisper. "There's a dame in here."

I looked around. "Where?"

"In one of the bedrooms. I had to take a leak, and I opened a door out in the hall, looking for the bathroom, and there's this broad."

"Maybe it's somebody in drag."

"No it isn't. She didn't have any clothes on. It's a dame."

"What's her name?" I asked, thinking Ray might have shown up with his buddy, Linda. They often crashed these things together.

"How in hell should I know her name? I didn't ask her, for chrissake! What's a dame doing here anyway?"

"What did it look like she was doing?"

He nursed his retrieved drink for a while in silence, but the scene he had witnessed rankled his sense of propriety.

"They ought not let dames in," he grumbled, setting his glass on the counter again and fishing for a pack of cigarettes. He tilted his hat back and ran his other hand across the short, close-clinging mat of auburn ringlets receding from the temples of his bullet shaped head. After lighting the cigarette and blowing the smoke over his warm drink in an audible "Whew!" of disgust, he went on griping.

"There was a guy in there wearing leather too. Can you imagine that? He was with her."

"You should've asked him where the rest of them are."

"I wasn't asking him anything. He was plowing her. Jesus!"

"Well," I said, "why didn't you wait 'til he got off, and then ask him?"

He refused to grace that with an answer. Instead, he shook his head as if the evening were ruined for him. "I didn't know there was going to be any dames here."

"Neither did I," I tried to console him. "Maybe Chris'll throw her out when he gets here."

Jack continued mumping about this contamination of the festivities.

"On the other hand," I said, "she may be part of the entertainment."

"Part of it!"

"Take it easy, Jack," I crooned. "This is San Diego; you've got to allow for the perverted vagaries of normality down here. We don't get a pure essence the way it comes to you in L.A., boy. Now quit your belching and get on down to the basement and find out what's really going on."

He grunted a sour assent to this and started to move off when a willowy shadow from the other room, one who had evidently overheard part of our exchange and taken a notion to possibly look into something different from what he had just been sampling, asked us what was going on down there.

"Nothing you'd be interested in!" Jack snapped at him, pleased to find an outlet for his pent-up outrage.

"Oh?" The wraith, with wide-eyed impudence, was determined to provoke him further. "Well, what is it you can't find here, doll?"

Jack glared at him as if he were responsible for the abomination he had witnessed in the bedroom. "Leather-lovin', boot-lickin' sex!" he barked with all the gutty punch he had learned how to give it from paying attention to a score of John Wayne movies and the graffiti in a thousand barroom heads.

The other drew back in slack-lipped revulsion.

"You must be sick!" he hissed.

"Friend," Jack growled, leaning far back on three-inch undercut heels and tucking his chin into the open collar of his elaborately silver-embroidered shirt, "If you only knew!"

They both left, in opposite directions, and I consulted my watch. Chris ought to be showing any minute, and I was bemused for a moment by the prospect of witnessing his discovery of a woman on the premises. Chris had an almost superstitious horror of females, and considered it immoral for queers to have anything sexual to do with them. This did not prohibit him from renting to them, however. He rented to anybody who could pay, and used the proceeds to support the tenants he liked who could not. He said this made him a kind of modern Robin Hood. Besides Wendy, who did not count, since she worked for him and was a lesbian, Ray's companion, Linda Morrison, was about the only woman he could stand around him without succumbing to a case of the jitters. But if she was here, using one of his bedrooms to screw in, it might stretch the fragile cordiality he extended her to the breaking point. I started to move out in search of Ray among the crowd, since Linda never went out at night without him.

One of Chris' tenants, whom I recognized, wandered into the kitchen, which was getting hard to move around in, and attempted to strike up a conversation. It went on for a while with him giving a running commentary of his recent movements, most of which I had heard before, and none of which required anything from a listener in the way of response aside from an occasional yes or no. A nod sufficed for most of it. The one who had gone down over against the refrigerator reappeared from the dining-room grapple and stood beside us, listening to the monologue while he groped me. The liquor on top of the beer was reaching me; I was at that stage where, were I wise, I would ease up and maintain the alert buzz investing me now. I was also feeling the work the cocksucker was doing on me, and although it was not what I felt like having just yet, it was an anodyne for all the problems and portents I had been pushing around over the past few hours. Sex is like liquor and dope; it can take your mind off more important things. The trouble with all three is that they can become a habit, and they never make the important problems go away. Pushed far enough, however, they can absorb them so that, by an elemental and disastrous simplification, they are supplanted as in a blanket loan: one big bill in exchange for all the small ones. I was almost yearning for what I imagined that kind of simplification might offer in my present state, but the bottle Jack and I had brought in with us was no longer on the counter, which showed how far my attention had strayed. It did not matter; there was probably something to drink downstairs whenever I chose to go that route.

And now suddenly here was Cusped, in a change of costume. Something thrown on in haste, it would appear: an old pair of Army coveralls half zipped up in the front. His spurs still peeked out from under the fatigue green of the pant legs, and through the slit down the front the telltale gray of chest hair appeared more pronounced in the dim light. He should have dyed that instead of the slicked-down Valentino pompadour on his head. If one had taken the liberty of searching through that head of hair under better light, the tiny, thread-like scars might have been discerned around the ears and above the hairline where Symonds had pulled his face together three years before this. On stage, at a distance, or up close in light like this he looked all right if you were in search of a special kind of excitement. It was obvious *he* had been excited recently. Sweat beaded his chalky face and his deep-set eyes were sadder-looking than ever. He shook out a cigarette from a battered pack, and I asked him for one. While he lit it, I got rid of the leech gently, but firmly, annoyed that Cusped had regarded the sight with a questioning frown.

"Your cigarettes stink," I said.

"You plan on coming down?" he said, "or does this scene interest you more?"

"Just warming up." I decided to reciprocate his condescension. "What's the matter, you winded already?"

He disregarded that, took out a Kleenex, and blew his nose while staring at the dark writhings in the dining room.

"That San Fernando cowboy you sent down disgraced himself," he said thoughtfully, the faucal plosives thicker than usual in the afterworkings of his nasal passages following this oblique rebuke.

"Cactus Jack?" I smiled. "In what way?"

"No control."

"That's too bad. Maybe he'll do better next time."

"Maybe. He said you wanted to talk to me."

"I didn't want to talk to you," I said. "I've already talked to you. I wanted him to find out for me if Lawson was down there."

"You knew that."

"No, I didn't *know* it. If I'd have known it I wouldn't have sent somebody down to inquire. I wanted to be sure."

"You shouldn't send amateurs like him."

"Who would you have me send, one of that bunch in there?" I indicated the dining room.

"How come you're interested in Lawson all of the sudden?"

"He provoked me all of the sudden."

"I guess that's a good enough reason all right. Only, you ought to come down yourself."

"I mean to. Who else is there?"

"You know everybody except a couple of guys from L.A. They're okay."

"Who's in charge?"

"Who d'you think?"

I knew. "I'd like to talk to him too."

"I don't think he's in a talking mood."

"He'll talk to me."

"You're probably right. C'mon."

We left the kitchen and made our way through the hallway, now so packed with people that it was one long grope to the door.

"What's going to happen when Chris comes home?" I said, once we were past Duke, still nodding pleasantly to fresh arrivals, and out in the corridor.

"Why should anything happen?"

"For one thing, there's a woman in there. And for another, Chris doesn't like the sort of thing that's going on in the basement."

"How do you know?"

"I heard it from the horse's mouth."

"Chris is a horse's ass. And he knows what's going on."

I turned that over in my mind while we waited for the elevator to disgorge some more people, among whom was yet another woman. When we were alone inside, Cusped punched the button marked B. As we descended, he leaned back against the side of the cage and regarded me with a look of conjecture.

"What did Chuck Lawson do to you?"

"He didn't do anything. I want to ask him a few questions, that's all."

"Does it have to do with Eugene?"

"It has to do with him—Lawson."

"So, you couldn't have come at a better time. He's in no position right now to refuse you anything."

Followed by that insane cackle.

11

HALF THE BASEMENT area was a garage. The elevator doors opened next to the door leading to this. Along two sides of the remaining area were padlocked, partitioned, latticed bins, storage facilities for the apartment renters. There was a door at the rear that opened directly to the outside and the steps that connected all the back service porches of the six floors. In addition there was a laundry area with a coin-operated washer and dryer. Lawson's room was squeezed in between the rear and garage walls and the elevator shaft. Cusped knocked on the door. A voice above the music coming from a radio inside said, "Yeah?" Cusped identified himself, and someone I had never seen before opened the door a crack to look out, then motioned us in.

I was nearly bowled over by the stench: stale sweat, fresh poppers, and air blue with tobacco smoke. It was also blue from the illumination a tiny seven-and-one-half watt colored bulb screwed into a ceiling socket shed on the scene. A kind of shadowless glow, like moonlight, bathed the dozen square yards of floor space and the objects in it like a tableau in a daguerreotype.

Nothing was happening at that moment. The first thing I focused on was Lawson, become the optionee for the evening, anchored in the middle of the room, the center of attraction, just as he always wanted to be. Some people look their best strung up—Eugene, for example. Lawson, on the other hand, looked best strapped down. They had him spread across a modified, leather-padded sawhorse that gave an upward tilt to his furry butt. But no, no longer furry! Stepping over for a closer look, I saw that he, or somebody, had shaved it. A pretty sight, really! A quite pretty sight, dotted with eschars from the glowing butts of previous bouts, and mottled like a matron's thighs from the concussions of the night's preliminary attentions. And there! As if someone had been determined to drive a stake through his heart the long way, his flue was disrupted by what looked to be the business end of a baseball bat. I placed the flat of my hand around one firm, yet pliant bun and felt with satisfaction the shudder this caress produced. I leaned over to see his face, and he, restrained as he was, tried with a strabismal rearing of the head to get a glimpse of who I might be, without success.

"Not yet, you sonofabitch!" I whispered. "Not yet."

They were all watching me, as if it were the quiet hour and they at rest from a complicated piece of work that had them stumped. I stepped aside as an abdominous member of the visiting team—the one who had admitted us—stepped ponderously over and, without ceasing to drink from the bottle of beer he held tilted back over his head, began in a leisurely manner to hose down Lawson's locks as if it were the most natural way in the world to relieve himself. Chuck squirmed under the rinse, whether in protest or from enjoyment it was hard to tell—probably both.

Stepping out of the way of the splash, I surveyed the rear end more closely and saw that they had attached a testicular cangue secured by stout cord and a bosun's knot to the crossbar of the sawhorse. Every time he squirmed it pulled tighter with a cruel bite into his manhood. And dangling in semi-tumescence through a hole in the board, hung a prick intorted like a pig's. "It's a good thing you don't like to fuck, Lawson," I told him once. "You'd tear somebody apart with a thing like that."

Propped up near the door with a glass in his hand and nothing on but his boots was Cactus-ineffective-Jack, averting his eyes beneath a fulvous mat of hair. Duke had found his way down here too; he nodded to me with a complicitous smile. Then, nearby, what did I recognize but the angular figure of Pinkerton, drawn up like a grasshopper in what might have been sleep, fully dressed on a palletless steel cot in the far corner of the room near a wall sink.

Pinkerton included in this? There was something wrong. As I looked back at him, Jack took this moment to slide with perfect aplomb down the wall to the floor where he sat regarding me in a baleful stupor, now thoroughly drunk from drowning his shame at what no one else cared about: his shot in the dark.

Against the wall opposite the door, under a small painted-over window set high up against the ceiling, was a table at which sat the L.A. type I had seen at The Club. That would be Rublatz then. He still wore the leather I had seen him in earlier. He was bent over the table, staring at a handful of cards. He appeared to have taken no part in the action just concluded on Lawson. Facing him across the shallow boards of the table sat Faulkner, likewise contemplating a fan of cards. It was a wonder they could make out the suits and numbers in the cloudy light.

To see Nelson Faulkner in this clownish arrangement bothered me even more than the presence of Pinkerton. I had last run into him on a street in San Francisco about six months after I came down to work at the museum. I was up there for an interview at the Academy of Sciences; he lived up there. We had talked briefly, about nothing in particular. I did not know if he knew of my leaving Glenn; I told him I was living and working in San Diego, and trying to get out. He wished me luck and made no mention of Symonds, so neither did I. I don't know what they were playing, but Faulkner gathered in the cards and began to shuffle them. That done, he held the pack in one hand and reached for his cigarette that burned on the table edge. Only then did he look up, regarding me quietly, if that inflamed, accusatory glare of his could ever be said to look quiet. He had on bike boots, leather codpieced breeches, policeman's-issue webbed belt and chromed buckle, and a plain black leather vest. On the table near his glass lay a pair of wadded, black kid gloves—the same kind Glenn had favored. He nursed the cigarette and picked at the full, recurved portion of his lower lip with the fingers of the same hand. There was a terse and fluid grace to these movements which augmented the impression one got of controlled ferment within. I nodded in greeting and received only his continued stare in return.

Nervous, limited, concentrated, Faulkner was one of the few in this line of play that I respected, for no other reason than that he was dedicated to it wholly. As someone put it, his was a life laid waste by a single idea. He made this scene the center of his world, and with the grace of a panther and the lonely prowl of a wolf he stalked and found the only kind of prey he fed on. If Eugene was the complete bottom man, then Faulkner was the absolute top. Eugene his trailer in National City and Faulkner his workshop in San Francisco constituted the negative and positive poles about which clustered, like iron filings around the horns of a magnet, the patterned fragments of this highly structured world. Faulkner was a motorcycle mechanic by trade, but he possessed some small talent as a draftsman as well. He was paid marginal hire for the humorless pornography he labored over on his evenings after work, depicting in preposterous detail the extravagant dreamworld of this action he lived to breathe life into and to perfect. These shadowy ancillaries to the real thing earned him room, board, and fuel for his wheels, so that now he worked on bikes only when he felt like it. His own, specially reconstructed Harley, and all the clothes he wore on his back, were gifts from admirers and the victims of his art.

The impression one got with Faulkner was that the usual riddles posed by the fall of events in everyone else's life seemed to have been waived in his. It was idle to ask for an accounting of what he had done, was doing, or expected to do with his lot in this world; he was, and always had been, doing exactly as he pleased. And if his deeds were too circumscribed and questionably refined for the admiration of the world at large, they were nevertheless performed with a distinction none could challenge, and he experienced no problems in the pursuit of his calling save those of supply and demand. He had, in the course of a dozen years' devotion to his vocation, done a variety of other things, skirting, but always keeping within the limits of, the law (the guardians of which he was known to have an inordinate respect for, as witnessed in the contrived and repetitious plots—if they could called that—of his cartooned fantasies). It was about the only attribute he shared with Pinkerton; with this difference, that in Faulkner it was a matter of choice, whereas Dale had little to say in the matter. Nelson, in his thirty-nine years, had worked as a night watchman and a process server; he had driven an ambulance, dug graves, tended bar; was a licensed pest control inspector, and skilled in the maintainance and repair of two-wheeled vehicles in the way some general practitioners are said once to have been in regard to their patients.

I first ran into him when I brought Symonds' bike into the Harley-Davidson place on Page Street in San Francisco to have some work done on it when he, Symonds, was back on the East Coast attending some meeting or other and I had the use of his machine while he was gone. Faulkner and I hit it off well from the beginning. It was no affair in the ordinary meaning of the word, nor was I the plaything of his usual wont. I became a partner of infrequent moments of relaxed and surprisingly tender sexual release.

He knew I did not talk to others about what I did, or with whom, but early on I got the impression that he would not be put off were I to mention to Symonds that I had made it with him. I never did, but Symonds found out—I don't know how—and simmered over it for months before delivering himself of an abusive harangue against a man I did not even know he knew, and the gravamen of which was that Faulkner was not to be trusted. Which may have been so. For all I knew, on reconsidering it afterwards, Nelson may have taken me to bed for no other reason than to provoke a man with whom he felt he might compare himself, but I did not believe this to be the case.

As it turned out, my connection with Symonds was not by any means the least of the factors contributing to Nelson's interest in me. It was an interest which waned, however, after I mentioned Symonds' eruption to him. I had at that stage no desire to cause Glenn any hurt, and I believe he imagined that Faulkner was getting away with something on my body that he, Symonds, had never accomplished and was then in no position ever to do. Only once had I visited Faulkner after that (I was still doing graduate work in Berkeley at the time), and it was then that he told me of the last encounter he had had with Symonds five years before this. Glenn, it seems, had made the tactical error of attempting, offhandedly and in the presence of others, to admonish Nelson for tampering with a slave he was mastering at the time. Faulkner, if his own version of the scene is to be believed (and I never asked for, nor heard, anything on the subject from Glenn), put him down publicly. "I told him," he said, "I'm gonna take him every chance I get. I'll fuck his ears off. And if you don't like it, then keep him home."

It sounded like Faulkner all right. But I felt a twinge of disloyalty at having listened to it. Hearing the man I loved, and who trusted me, denigrated by another who took for granted my complicity in his disparagings gave me no particular joy. Embarrassment and concern for Glenn in his discredit, plus a further unnerving revelation that I cared about what this peculiar circle of acquaintances said about him, left me silent. There was this too in Faulkner's account of throwing down the gage before Symonds in a public exchange: it was an indication of some aberrancy in behavior I would not have believed possible on Glenn's part had he not already reacted to my knowing Faulkner with an intensity and vehemence out of all proportion to the provocation as I reckoned it. After all, we neither of us were shackled to each other in the bonds of absolute fidelity. Glenn had his route and I had mine. Allowing himself to be placed in a compromising light by assuming a competitive stance was a trap Glenn Symonds was careful never to put his foot in. I wondered who that slave could have been, but, feeling as I did, I never asked.

Looking at him now, I was struck again by what a strange and fastidious actor Faulkner was. People like Cactus Jack or Cusped, or even Symonds, as close as he had been to being superb in this region of the mind, were none of them quite real regarded in the same distorting light that Faulkner walked in. They were Sunday painters, trifling or expert in a talent requiring for its full fruition the abandonment of all other devotions before it. They had other, ultimately more important, things to do with their lives. Cusped had his music, Jack his money, and Symonds, whatever else he did, was absorbed in his professional career and the virtues stemming from it. Lawson hardly counted at all, but Eugene came close, in the total attention he gave to it, to matching Faulkner's excellence. I had often wondered why the two of them never got together, but I knew now why, really. The issue of so equal a bout could only have led, step by logical step, to the end that had occurred last night in Eugene's trailer—only not by accident, but by design, and with Eugene on view at the morgue—although I doubted that Faulkner would be hunted for it. Eugene, be it noted, had tried, but Faulkner never responded to his overtures. Maybe I thought, he was saving him for his old age, the *summa summarum* of his life's work.

But these others hardly counted at all in the visionary schemes and feats of Nelson Faulkner's universe. He tolerated their company as an actor does the setting. They were part of the equipment, or the action, or sometimes the subject of his intentions, the necessary leaven to the other, more substantial ingredients of knowledge, skill, and the determination to put them to use that constituted the essence of his charge. If all poetry, as has been said, is based upon the antagonism that exists between reality and the dream, then Faulkner was a poet. He was also, like Symonds, not quite right in the head.

"Hello, Nelson," I said, looking down at him seated at the table like some maleficent and erotic claustral prior to those proceedings, aware of all that was going on, but thus far taking no part in them. There was the suggestion of a smile on his uncharitable lips,

lips that were vermilion hued in daylight, but made black by the peculiar illumination of the room. The look he returned to me held no surprise in it, as if he already knew what business I had come here to transact, and was expecting me. But . . .

"What brings you here?" he asked, like an alienist confronting a new analysand. The voice was a rasp, as if his vocal chords were scarred.

"I've got an interest in that," I said, motioning to the helpless Lawson behind me. "You through with him yet?"

Faulkner's look changed by the barest crinkling of flesh about the eyes. "I haven't touched him."

I glanced back at the procumbant form. The head still dripped. It waggled as I looked at it as if to prove that it was still alive. Everybody else in the room was watching us, with the exception of Faulkner's L.A. friend who had taken the pack of cards from him and was dealing himself a poker hand. Pinkerton had come awake (if he had ever been asleep) and gave me a greeting with a finger raised to his brow, smiling. I doubt if any of them could hear what we said above the music coming out of the little radio that was on the table along with a bottle of rye.

"Well," I said, "it sure looks like somebody's been working on him."

"That was Cusped." He leaned forward on his elbows, both hands cupped around the cigarette. "And that short shooter you sent down."

"Sorry about that."

He waited, watching his partner snap the cards down on the table top.

"There's something I want to ask him—Lawson, that is."

"What?"

"Just something private. Nothing very important."

"What's stopping you?"

"He won't talk, I don't think."

"Is it about Symonds?"

He meant it as the question it was, but also as some kind of expression or acknowledgment that he had taken into account the effect Symonds' death might be having on me.

"No," I said, "it isn't."

He nodded at Lawson. "Go on and make him talk. Nobody's going to stop you."

"I think I could use your help."

"Don't want to get your hands dirty, is that it?"

"More or less. How about helping me?"

"Why should I?"

I shrugged. "For whatever's in it for you." I wondered if I could lure him. "How's this for starters? He might not talk even if *you* worked on him."

He smiled at that, and picked up a white Benzadrex inhaler his friend had lying next to his glass, fingering it while he turned the idea over in his mind. I looked at the little plastic container too, reviewing while I waited, another of this kind of life's unexpected ironies. Errhine vasodilators should have been Symonds' most enduring claim to fame, since his remarkable cosmetic achievements were, by their nature, ephemeral, and, as a result of their success, largely unheralded. But this sidelight to his career will probably remain forever unacknowledged because he never published his revolutionary little discovery, the times being what they were when it came about. It was around the period when I left him that he recommended the therapeutic use of amyl nitrite to a colleague, a psychiatrist who had the wife of a major studio mogul on his hands crying with chagrin over an orgasmless affair with a young man who was otherwise making life again worth living for her. The inhalation of one ampule induced multiple charges the intensity of which made her a new woman in no time. The strange thing was that after that very successful initial run, the use of poppers never caught on with the straight set for nearly ten years, while the more adventuresome or jaded segments of the homosexual community who needed something to get them through a night of sheer hell or flaccid boredom immediately appropriated this dubious genie, carrying it in a variety of canisters like phylacteries around the neck. The surge of sales of the drug in just a few years provoked the FDA and comparable state agencies to place what had largely been an across-the-counter item, sold for the most part to little old ladies with thrombotic tendencies who carried them in their handbags against a time of need along with aromatic spirits of ammonia and nitroglycerin tablets, on the list of prescriptible drugs. Those canary yellow tins of Burroughs' Vaporoles™ became tickets to instant and controllable highs at about the time when other, less manageable, more expensive, and sensationally publicized philters were exercising the attention of the guardians of public order. Nothing succeeds like success, and with the withdrawal from the open market of Old Reliable's five-minim hit, a river of the volatile hydrocarbon, manufactured at some peril to limb and property in kitchens and bathrooms and basements throughout the gay ghettos, flooded the market at inflated prices with little more than its stink to recommend it. But the reputation remained when the true essence was nowhere to be found unless you knew a doctor who would prescribe or a pharmacist who could filch.

But this was before the days of general, one might almost say vulgar, knowledge, and everything in that basement room partook of what was then the arcane and the seldom encountered. No item in the severe sexual furniture there but was made to order with the pride of craftsmanship and ingenuity that one-track minds can bring to bear on a proposal hardly admitted in open congress. Thus, no specially manufactured cylinders of aluminum or silver or gold (so chic!) existed to contain the pungent vapor. Inhalers from the drugstore served. The tampon soaked with nasal decongestant was discarded and the chamber cleansed by washing and aeration, after which the ampule could be crushed (the "pop!" was half the charge) and inserted, and the cap screwed on against the moment of whatever lift one was looking for. I often wondered about those ladies and the contents of their handbags.

Faulkner interrupted my ruminations, having put down the inhaler which he, in any case, never used. "What d'you have in mind?" he said.

I had never considered it, any more than I had ever seen this paradigmatic figure in his proper setting. Here he was not the man I had known under other circumstances.

"Well," I said, "anything you have to do to get him to tell me what I want to know."

He smiled again, faintly, and said nothing to that.

"Just give me his head," I went on, warming to the subject, "and you take the rest. I'll let you know when he's cooperating. I think if you take it easy at first and work up to something gradually, it won't be necessary to . . ."

"You can have his head, friend," he interrupted me again. "But don't tell me what to do, or how to do it."

"Sorry," I said.

He nodded once more, and stood up, sleek and availing, dropped his cigarette to the concrete and ground it out under the sole of his boot while he pulled on the pair of fine kid gloves, never taking his eyes off me. He fingered a fresh smoke out of his vest pocket, lit it with a wooden match he took from another, and then said, "How do you feel about Symonds?"

"I don't feel anything right now. How about you?"

"I haven't felt anything in years—about anyone."

There was whiskey on the table, and I asked to sample it. He motioned to it in assent, and I helped myself from the mouth of the bottle. Setting it back down, I found the L.A. type, Rublatz, staring at me with distaste. He pushed a glass in my direction, but I had had enough, and was ready to go to work.

The others were alert now. Seeing Faulkner move, they almost visibly tensed. The evening was going to pick up at last.

"One thing more," I said as we stepped over to Lawson. "Could you keep the rest of them away from me? This is . . ." I searched among my suddenly limited store of words and settled for, ". . . something private."

"You've got the head all to yourself, friend. You nod when he won't cooperate. If he doesn't satisfy your every need, I'll take his balls off. How does that grab you?"

I told him it sounded like it might work, but I would as soon he did not go that far; I had seen enough of that sort of thing for one night.

As I squatted down and leaned close to Lawson's ear, the others, as if at a signal, rose and grouped themselves into a semicircle around Faulkner, that jack-of-all-trades and master at one. Well schooled, they kept their hands to themselves; cautious, they waited to see what would unfold. With one hand I took Chuck Lawson's chin and gentled his head around so that he could see my face. His bloodshot eyes blinked as I delivered to him a prepared speech, spoken in a whisper only the two of us could hear.

"Lawson," I said, "you came out here smelling of kerosene, cold biscuits, and cooking fat to get in honest people's way and make trouble for them with your devious manners. Now I mean to bring some order into your life."

I don't think he had seen it was I that came in the room. There was surprise in his look, and a note of panic in his grunted "Huh?"

"Shut up!" I told him, and cradled his head in the crook of my arm to make it easier for me to address him and hear whatever he had to say in reply. In any situation but this, it might have been taken for a loving embrace. He had expressed a wish to make it with me in the past, but I had never given him even the hint of a nod. Now he would remember me as the front part of an engine; the part that had nothing but gentle, yet implacable consideration for what he had to say.

"Lawson," I went on, "you don't, never did, and never will do a thing for me. It makes me sick to look at you, and I resent having to demean myself to come down here and be a party to teaching you a lesson just to find out something you should've told me in the first place six hours ago."

He scowled and, predictably, he answered, "I don't know what you're talking about."

So I gave the nod to Faulkner.

12

OLD WINE in new bottles. No one came to interrupt our cellared struggles with old Iron Ass. Neither Chris, nor any of the celebrants under his roof, nor the supposed sleepers on the floors between, descended to inquire at the bolted door as to the cause and meaning of so much squalling. But it must have carried at times even above the volume of the radio turned up not so much to dampen as to supplement with apposite rhythms, screaming ads, and heedless, mindless lyrics, the damage being executed there.

Clamor and fetor filled the confines of that basement chamber where I opened Lawson's mouth—with a little help from my friends. I never would have thought him capable of holding out as long as he did. Under the working over it took to get from him the name of the whereabouts I sought, he was marked in ways he might keep for the rest of his days. It was, in short, a night to remember. Symonds would have been proud of me, I supposed. Or would he have? In the course of our exertions I had time (because neither Lawson nor the others were in any way disposed to hasten the issue as I had set it up) to reflect upon this question.

For what we were doing was not to my liking.

I have perhaps protested too much to this effect already, and it may be asked—and was soon to be asked by those whose public mandate gave them the right to an answer—why, if I did not like the sport, did I frequent the field? And all that I could, and can now, say in reply was that I did not find them incompatible, these players, in anything other than their ludicrous pastime. It had occurred to me that this was being less than truthful with myself, but I believed I might keep clean of what I regarded as the blemishing moral stench of them by staying away from the instructional cockpits of their exercises (bars excepted; bars are no more than the holding pens of the arena). But now . . . this.

Consider, as I considered on the middle burners of my brain while they were at work back there, that the hold I had had on Symonds (although neither of us had ever acknowledged it to the other, or to anyone else) consisted for the most part in the resistance I presented to his early attempts to entice me into the circle of lunacy that was his other life and, after that, to my continued tolerance of it as an aspect of him that had nothing to do with me. For that especial otherness in him, that separate existence of his, did not in any way diminish the power of his attraction for me. I constituted a kind of third presence for him, whereas he threatened all the time I knew him, clear up to this day of his death, to usurp the entire field of reality as I viewed it. I loved him in spite of and, I came to realize, because of this by-play. With increasing misgivings, I came to recognize this antagonistic affinity as the strong point of mutual resistance by which our fragile union was supported and our separate identities maintained in what otherwise had been a passionate muddling of personalities. He his work, I my studies—and this. And the longer we continued together, the more any weakening of either of our established postures threatened to collapse the entire structure that had grown up about us like a protective and sustaining shell secreted by the symbiotic reaction of seven years' struggle to get and to maintain the upper hand.

In Symonds I found someone I could count on, but could never understand and be at peace with—and peace, an end to strife, was what I wanted. In me he faced an adversary he could not subdue, but who never challenged him on ground of his own choosing. Instead, I knew from the start the way to finger out the secret, mollescent susceptibility that was his deepest concern to hide and his greatest pleasure to have quickened. He was a man of willful oppositions and terrible ambition who, in reflective moments saw in me one who defamed, however privately, his vast, mistaken self-esteem. Once, in those last days together, in an unguarded attempt to have the last word in an argument, he affirmed that he had taken me on after our first meeting only to insure my keeping silent about what we did. Which seemed to him to indicate that my bondage was somehow greater than his own. He may have been right, for I don't believe he ever comprehended how much I admired and loved him, and it may be that I gave too little credence to the exalted emptiness inside him that denied the enormous tributes paid into it in the specie of success and reputation, hate, esteem, and love and fear by all who knew him, and knew of him. My attachment to him gradually assumed the characteristics of an addiction to something that demeaned even as it stimulated and gratified. His to me had about it the fascination for an object he could not master, but who, on the other hand, could provoke a desire and indulge him in a fantasy he could not respect in himself. I never wanted it that way.

By some curious osmotic exchange in our personalities, Symonds and I absorbed the potent qualities in each which held so much a hold upon the other. His cultivation of fear and doubt, and the excellence of discipline over desire, of will over caprice, coalesced in what, for my part, had been a flowering of affection, openness, and the easy-going license of fortuitous pleasure. It was a quiet softening in him, I thought, a gentling unnoticed by any but myself.

That link which bound us to each other was composed of much more than sex. Glenn was a demanding, but also a generous, lover, and much of what I became that was good I owed to his encouragement. His finest gift was the awakening he stirred in me to the hidden faculties I had never bothered to search out and test. Good doctor that he was, he nursed me out of a sodden sleep of two decades' duration, and I have never gotten any rest from thinking from that time to this.

For, before we came together, I had been comfortably drifting toward the great wasteland of wage-earning corporality. There was nothing I sought for love nor money that was really out of my reach. All I wanted seemed to come easily to call, or was already at hand. Anything beyond the available held little interest for me. I was not competitive. In this last I was like Symonds, but for entirely different reasons. His refusal to compete was a response both to self-esteem and to fear; mine stemmed solely from laziness. At the time though—the time before I knew Symonds—I never gave any thought to such considerations. College (at least the undergraduate part of it) was a walkthrough; comfort and security of the sort the '50s offered was the goal. I believe I was almost happy before I met Glenn Symonds. I know I have never been so since.

And here, now, Lawson pleading ignorance. He knows that that, above all factors human, is no excuse. Perverse recalcitrancy is a pungent savor to the appetites of those who want to master and have their mastery ultimately acknowledged. I remember learning, over a period of years from lectures by Symonds in his gratuitously instructive moods, that you can beat someone attached to you as long as you caress them afterwards. You can beat them as often as you wish and caress them very little, as long as, when you do, you caress them well. In this, most people are like dogs (it is Symonds speaking), which may be why dogs come to resemble their masters, and why visitors so often try to change or spoil the habits of other people's dogs, being unable to make any impression on the people themselves.

I heard and heeded all this sort of thing from Glenn without, however, embracing the philosophy behind it. And anyway, in the matter of caresses, the dagger of the flesh will get to the heart of things better and more fundamentally than a hand, for where we strike for instruction and fuck for love is the seat of the soul, if we are to believe the Greeks, and the Greeks, as is well known, were masters of this sort of thing (Symonds still raving), and that loin-bedded acorn-headed shaft incorporates in its venereal reciprocity the ambiguous guise of both a beating and a caress.

Poor Lawson. I fed this sort of thing to him between the questions and replies, such as they were. I'm sure he thought that I'd gone mad, which was as good an impression as any for accomplishing what I had set out to do.

I left him for a moment to go piss and knock back another draught from the bottle on the table, unguarded now since Faulkner's friend had joined the others. The action stopped and everybody waited, eyeing me. Was I, as might be thought by any unschooled in this sort of thing, the master of these games? So thinking, they would have been wrong. I was as much the object of it as was Lawson who had never learned to keep his nose clean or his mouth shut so as to attain a vantage point of safety in a world of snipers bent on picking off the strays. I was about the only one in that room who did not entirely believe that much is forgiven the strong, the alert, and the confident, but nothing is overlooked in the weak, the forgetful, or the self-condemned. No returns, no repeats, no renewals on the options are allowed. Not even a little bit. I put the bottle down and went back to work with Symonds ranting in my head.

"It's a pattern, Lawson," I murmured the way a nurse might croon to a feverish child. "It's a pattern cut to size from the beginning . . ." And like the sweating, cretinous, burly infant that he was, he refused to accept the point of my chafing lecture, shaking his head instead in mock, complicitous confusion. ". . . and never changed."

I gave the nod again to those back there that they might resume work on this stubborn, harnassed bull. I marveled at their unanimity, the smoothness of their functioning, as one unskilled at tinkering might regard the gleaming movement of an expensive timepiece. One winds, admires, and uses it for what it's for without tampering with the mechanism, appreciating the intricate escapement that makes it tick, but content to let it go on telling time untouched. I had, in a sense, wound them all up, and Faulkner functioned as a perfect lever and pallet in his timing; Faulkner, whose sadism resulted, I think, from a frustrated quest for perfection.

Well, a sadist, as a sign on the wall of a smart-assed bar in L.A. once proclaimed, is someone who takes great pains and gives them to other people. So Symonds and Faulkner, and others like them. For sadists, true sadists, are perfectionists *manqués*, as are all absolutists of whatever stripe, aware that nothing in the life of man, or any man, is perfect or perfectible. They are a sad race, sadists, shamed by their sadness and full of anger from their shame. And Symonds, since he only rarely perceived the hint of pains I suffered suffering him (one learns to play the game if the stakes are worth it—and Glenn, I was convinced through most of that long sabbatical, was worth it), came, by degrees, to falter in the pursuit of discharging his wrath; for while I found myself playing a role that was compatible with my abilities and inclinations, he more and more was at a loss to say just what his role with me amounted to, other than the shame it brought him in his own eyes. He wanted it, and had it, yet hated me when he got it.

Love is enough, if one will let it be. But the weakness love uncovered in him drove him further to delineate more clearly, if this were possible, those qualities he cherished in the role he played with others. My coming into his life had not so much created a need in him as it had given substance to what was already waiting there in form, a form he feared. But if this self-determined failing (as I saw it) was not my fault, I was still to blame in his eyes for awakening him to the liking of a dominance he expected with the most poignant longing, succumbed to with almost ritual resistance, and reacted to with loathing and horror after the pleasure of what I gave him had subsided and he was himself as he liked to think of himself once more. Few, I know, would have put up with the trouble it took to lay him, but if all this nonsense was not very healthy, it was gratifying in the extreme. Symonds was not unlike a cat in heat.

His professional calling, as I have mentioned, was the single strong pinion shoring up what I began to see was the collapsing framework of his personality. In a field filled with quacks and gulls, he maintained an integrity and honesty in the face of esurient disfigurement that has seduced lesser practitioners to quicker wealth, but not to greater reputation. For at least a dozen years the undermining elements of his hidden weaknesses of character (pride and despair) had been eating away at the imposing structure of his public manner like termites infesting a mansion built on sand too shifting to support the weight of it. By the time I got wrapped up in him the process was already well advanced, and I suppose it might be said that I hastened the rot, although that was hardly my intention, nor was it my view of what was going on at the time.

In all of this there was no calculated enterprise in either of our actions, neither in his proposal which was never spoken, nor my acceptance of it which was simply taken for granted. Again, if sadism be a perversion in which gratification is got by torturing the person loved, then what happens if the torture is not recognized as such by the one on whom it's worked? This is the way it shook out in the beginning, and by the time the picture cleared, the course of our efforts on each other was so deeply channeled as to preclude any measures we might have undertaken to change, because it worked, and worked so well.

To what end we did not know, although we thought about it in ineffectual ways when some contretemps or other put us in an obdurate mood in which we chewed, each in his own corner, on the question: Who, after all, is the bottom man? Requited hate, kept in the dark, can be a breeding ground for passion, and confusion of roles is the mother of violence and decay. In situations such as we had worked ourselves into, the fomentation of cruelty in each other in all its subtle guises became, in time, the only recognitive signal of sexual desire—of love, if you will—to which we were attuned to respond. In an attempt to cushion the disastrous effects the taking of him was having on his cultivated sense of immunity to passion, I yielded to Symonds on almost every other point, a concession which took no great effort on my part. Symonds usurped the center of attention wherever and with whomever he was. Countless times I listened to his recitations of theory and practice, and the dreary details of feats he had accomplished on the bodies of others in one or the other of his fields, which he differentiated as his sacred and profane pursuits. In company and in private he either regaled or inveighed, and often what he said made sense. But sense or nonsense, he was a consummate actor.

I preferred the company of his professional colleagues to the gatherings of those I found around me now. Rhinoplasty, hair plugs, breast implants, the pinning back of ears, mouth peels, and honed down jaws . . . these did not interest me particularly after a first hearing, no more than did surgical shop talk and what I called the cosmetic gossip they engaged in over dinner and drinks, but what they performed and talked about at least made sense to me, and Glenn's worth in their eyes was a coin I could honor in place of the other medium of exchange he seemed to prize as much, if not more.

In that closed span of half a dozen years while I was advancing in my own studies, I came to see Symonds in a different and brighter light than the equivocal gleam which emanated from his dark side. The old aura that used to surround him by virtue of his being a surgeon never completely dimmed for me, but at the same time there grew up a deeper understanding and respect between us, a cohesion in itself not without stress. Glenn was content with what I had chosen to do with my life and was amused as well. He had the practical man's distrust of the theoretician, while I was often filled with a benignant pride at being so close to one who accomplished so much. I never told him this, of course. There are limits to which interest and sentiment can be carried where somebody else's work is concerned.

It may be that he had no inkling of how I looked upon his work—his sacred field. A doctor brings a particular aura of magic to his practice by the very reason of being a doctor; it is a vocation like that of priest, or artist, or advocate. The scientist's magic, on the other hand, is less personal and more uncertain in that it is loaned him in a general way and depends upon the prestige of his area of work and the fashions of the age. A doctor is a kind of thaumaturge, and Glenn, for me, carried with him that special presence which prevailed no matter what discords rent us at any particular moment, although, as I say, he was never aware of this.

He tried, in the beginning, to encourage me to go into medicine, and I toyed with the idea for a while, but gave it up when the limitations of my own personality and character became clearer to me. The desire to heal, or even an interest in changing people, has never been a prominent ingredient in my makeup or my ambition. Nor has the idea of service, the accumulation of wealth, or the hunger for recognition out of the ordinary driven me into what I do. Being neither clever, nor conniving, nor well-connected, I have, for as long as I can remember, relied on luck and a certain bone-bred intelligence to see me through. It has worked, barely. But then, luck frequently involves a close finish, and it was luck that ultimately counted in what I had accomplished thus far—luck and the native wit to profit from the chances as they came. Luck often fills in for the other qualities of success (or failure) if they are in short order. I was lucky most of the time, and saw myself surrounded by others who were not, and who served in their lucklessness as exemplary lessons to me, because the luckless are legion in any war, and in the lifelong struggle that is living they get in the way, blocking the field of fire or exposing you to that of others along the line. I could no more than concur in Symonds' belief that they must be pushed aside, these luckless, if one is not to contract the energy-consuming malignancy of spirit which is their directionlessness. As corollary to this point of view, I had avoided the sort of encounter now confronting me, even as Glenn, for the same reasons, cultivated it. He drew strength from a formal putting-down of those who would be stepped on, while I established limits as to how much I would listen to, and how far I would go, in dealing with evident bottom men.

Now, however, I was listening to a new voice determined on an answer, and in doing so I was going very nearly the whole way in this den of fanatics, acting here as querist to a pig-faced nut who would not tell me what I wanted to know.

And the way they watched me instead of Lawson, on whom they were nominally at work! For as we tunneled our way into the yielding honeycomb of his will to resist, I glanced at them back there from time to time and saw them all engrossed in me, as if I truly knew what I was about in this work and had been hiding my science from them all along. Well might they have imagined it! The curious thing to me was the ease with which I brought my energies and craft-learned-second-hand to bear in something I had heretofore shunned as foolish and unworthy of a serious man's endeavor. No wonder they were fascinated! Only the concentration of their attention on the details of their own work prevented them from being amused by what was going on. For here was I, who had been the anomaly in Symonds' life, the joker in the deck, the one no one could fit in with his scheme of things, revealed now as one no better than he should be, and after all no different from themselves save maybe in a finickiness of touch, a certain delicacy regarding the forceful application of the principles being acted on, a deputizer unprone, as Faulkner put it, to dirtying his hands. And with Lawson—a pretty sight!

I needed another drink, and a cigarette as well.

Faulkner came up behind me, followed by Pinkerton who, as far as I could see, had acted only as an observer to what was done, since no one had wanted his long hand in the thing. Nelson glanced over his shoulder at him, a motion that served to stop him midway between the two of us and the others who stood where they had been left, waiting.

"Are you all right?" he said quietly.

"I'm all right." I took a drink, again from the bottle. "I could use a cigarette."

He gave me one and lit it for me. Then I asked him if this couldn't be hurried up a little. He looked at me with half a smile. "You're doing all right," he said. "We'll get you there."

It was as if he were acknowledging the successful prospect of a long-postponed initiation, one which, by the unprecedented delay in its consummation, commanded exceptional regard and special handling. I was, it appeared, at last being accepted. Too tired and too disgusted to take issue with unspoken assumptions, however false I thought them, I went back to the work I wished now I had never gotten myself into.

And no, Symonds would definitely not have been proud of me. He would not even have been faintly amused, but would have regarded this as one more attack on his integrity. In his eyes I was not for this sort of thing at all, and to have me mixed up in it with somebody else's hand directing the proceedings, even if he knew he would have been dead at the time, would have constituted an affront to his judgment and his capacities. Faulkner's presence at any time would have incurred an automatic condemnation for treachery. Even my taking up with Eugene, when it got back to him, as it did, must have been a source of annoyance to him. But as to that, it was not until Symonds reappeared on the scene that I sensed the further damage I had done his image by taking on the leading local ballbuster in the masochist's trade. And it did not matter what I might have said to others on the subject; the evidence was there. But I, in my folly, was fool enough to imagine it was no one's business but my own—and possibly Eugene's.

And . . . taking on.

Certainly I had taken him in. He had just returned from a bad season in Phoenix where, as the latest notch in his spiked and holstered weapon, he had carved a record of gullibility unmatched in the earlier annals of his nitwit career. Swindled by an aged cozener who passed himself off as some kind of sourdough sitting on a fortune in copper ore, Eugene had abandoned his job, sold his trailer, cashed in his savings, and handed over the entire proceeds of this disvestiture to the old man, a quiet dodger half his size, twice his age, and possessed of five times his awareness of how the world was set up to deal with ninnies. He regaled Carmichael's credulity with tales of the Old West where men were men, women knew their place, and treasure lay all around for the enterprising to grab hold of. For several months before he finally made the move and cashed in his assets, Eugene raced 250 miles across mountain and desert, and back over desert and mountain, laden with grub, liquor, money, and whatever else the whining old larcenist had said he needed the week before. A rattle-boarded old shack sitting on fifty desolate acres of rock, saguaro, and cholla constitued cover for the wealth of Araby supposedly deposited by the beneficent hand of God beneath an otherwise God-forsaken expanse of waterless, sun-scorched wasteland. Money for mining engineers and equipment was all that was needed to make the place blossom with profit for anyone with enterprise enough to scour it out. Money and patience.

Eugene supplied both, and by the time the trap was sprung, the old man had taken on the lineaments of a foster father to him; a parchment-skinned dream parent, mean, foul-mouthed, demanding, thankless, and hung like the stallions he raved about in his drunken mouthings on those lonely weekend nights they spent together planning, planning, each one something different. When it was all over, Eugene could come up with no other comment than the rueful, almost satisfied admission that his taker had been one mean sonofabitch, which seemed to somehow make it all all right.

By the time I came across him in a downtown bar that catered to sailors, truck drivers, and drifters who weren't sure what they wanted but were open to suggestions and liable to battery once the play was carried through, he had been taken as planned, and was setting himself up for the more ritualistic punishment this expensive and time-consuming scheme had probably been meant to justify in advance. Not that Eugene ever engaged in anything so cerebral as personal soul-searching, or as philosophically complicated as justification by bad works; he was just looking to get the hell beat out of himself.

And, like Lawson here, it was no easy thing for him to come by what he ultimately sought. Well over six feet tall, weighing something above a hundred and ninety pounds with no appreciable fat on him, and possessed of features as craggy and shifty-eyed as a made-to-order villain out of that Old West he had so lately courted, he was not a specimen anyone took on lightly. If you hit him and did not get out of the way, he could cripple you by taking you with him under his fall. Only a sleazy indefiniteness in his conversation and the feckless whine in his voice would alert the connoisseur to what lay behind the rugged veneer and underneath all that bigness.

I knew on sight. And I took him home with me knowing it. But I did with him the kind of things *I* liked to do, which did nothing at all for him—or at most very little. Still, Eugene was not one to give up on a trick he thought he could educate. I must have given hints of portals closed, but behind which lurked the undeveloped delights he had in mind, because he undertook to school me in what he wanted, needed, had to have. And it amused me to go along with him up to a point. I was lonely, and I did not dislike or feel contempt for Eugene as I found him. It was an attitude he was unused to in the people he made it with.

So, an oppugnant stand on violence in bed was not necessarily a contraindication where the treatment of Eugene was concerned. As with most, and however little he may have been aware of it, pleasure and care was his aim, whatever else he sought in seeking to be known by another body. The way to his heart was through his ass, and that—need it be said?—was where I came in. But, as I would not be his master, so he never was my slave. I let him hang around, that's all. He came, came often, lingered long, moved in on a temporary basis, stayed, and we adapted to each other as the time passed. I unbent so far as to allow him the minimum of fantasies necessary to his getting off, and he, before enjoying the procto-manipulative intercourse that was his heart's delight, accomodated my more usual form of gratification—and that with a right good will. He taught me how to string him up with a padded, fur-lined bilbo, a hangup, as I finally pieced it together, stemming from a time in adolescence when he used to depend feet first at the end of a long rope from a cliff a hundred feet above the foaming Pacicic in Del Norte County where he was born, and masturbate into the roaring brine below, dreaming of capture and ravishment the while.

It was not imagination, but singlemindedness, that got him onto this. That, and the fact that he was strong as an ox. He liked to play dead, passed out in bovine ecstasy, during the hoisting operations, although a word of technical correction might issue from the slack, pleasure-swooned lips if a contretemps over the complicated straps and snaps and buckles and cuffs by a clumsy neophyte or a too-zealous veteran threatened the fantasy. It was no work for weaklings, and this travesty had resulted in an inguinal hernia for at least one incompetant who sought to master him on his own terms. I took to block-and-tackle when the mood to use him to his liking hit me.

But the rest of his game I would not play. He had to go to others for the rougher stuff. And go to them he did. He subscribed to a Canadian publication, a venerable precursor of the *Barb*, the *Free Press*, and the *Advocate*, but differing from them largely in the apolitical editorial policy in its pages and the lack of humor in its classified ads. With singleminded sobriety, these announcements catered to the fantastic and illegal conceits of a continent of pen pals dedicated to realizing their demented notions and willing to travel far and spend much in these pursuits. Eugene's out-of-town excursions sometimes resulted in a month of weekends away from Encanto, an absence I welcomed for the most part. In fact, I minded nothing that he did so long as he left me at peace when around the house and did not bring his pickups and correspondents home.

My restrictions on his bringing anybody back to the house finally induced him to move out, since the loss of tricks with no place to go proved too much of a waste for his conservative and conserving nature to bear. After the Arizona debacle, he got a job with an outfit he had worked for before he became a welder, pounding nails for about six months until there was an opening again at the shipyard. Then, with a few paychecks and a bank loan, he bought back his trailer from the dissatisfied owner who found its modified interior did not conform to his ideas of comfort, and the money necessary to remodel it could be better spent in purchasing a new one.

After Eugene moved out, I saw him fairly regularly down on National Boulevard; if anything, we became closer now that we lived apart and the tensions attendant on close proximity and unavoidable contact were eased. I felt in a way protective toward him, and he looked upon me as the only person he had ever known that he could completely trust—this despite his resistance even to the desire to trust anyone, since his whole bent cast him in the direction of self-betrayal where human relationships were involved. I became, in a sense, his cautionary.

There was another factor in our relations with each other that was almost never mentioned, but which hung like a shadow, darkening the field between us. And that was Symonds. Early on, Eugene had found out about my severed connection with him, but he respected my reluctance to talk about it. Looking back, I believe Eugene must have found it at once inconceivable that anyone could bring himself to part with such a prize as Glenn Symonds, and at the same time he admired what he regarded as the tower of strength I must have been not to have collapsed from the emptiness resulting from so grand an abandonment. Yet if, as was my way of coping with that very real emptiness, I feigned indifference, he, to my intense annoyance, accepted as substance this appearance I projected, and used every opportunity to capitalize on his association with me so that, by second hand and without my knowledge, he brought himself into a position to be noticed by the one he considered truly worthy of trying the mettle of his resistance and of tapping the neglected richness of his victimhood. Panting to be undone in ways he could hardly imagine, he was like some Turandot of the leather world alerted to the advent of the Unknown Prince. There were times when I had to tell him plainly to shut up and get out, for, in a kind of blasphemous parody, I was finding myself cast in the role of the Baptist preparing the Way of The Master.

The first mistake I made was in doubting that their conjunction would ever come about. The second was in not preparing myself for the appalling hurt I felt when it did.

It was Lawson who first alerted me to the rumor that Symonds was down there sniffing around, having been seen here and there in full regalia—now they saw him, now they didn't. And, it was also Lawson, a few hours before this, who had let drop in his shifty way that Symonds had taken up with something new, a rumor he had copped from Gene—maybe. Now I wanted to know *where* he had copped it.

I twisted his head around to look at him. A sorry sight. And at that instant whatever they were doing to him dragged forth a bellow of pain from his gaping mouth, a howl so elemental that I thought they had finally gone too far. I knew I had passed my limits some time ago

"Lawson," I pleaded, "for God's sake quit this nonsense and talk!"

He murmured, whimpering something that approached assent by way of faltering protest. Was it yes, no, or maybe he would play along some more?

"Listen to me," I whispered. "If you don't tell me now, I'm going to leave you completely to them. I'm all that's holding Faulkner back as it is. Now talk, you sonofabitch!"

He grunted and slobbered; the drool from his mouth danced in fragile beads, stringing down my chest and belly as he shook his head.

"I can't."

"Can't, shit!" I held his filthy head in hands made filthy handling him, and whispered into his wax-begrimed ear, "Just tell me where you were. I'll keep it to myself. There's nobody here who's interested in it except me, you asshole!"

Looking up over the sweaty, welted ridge of his back, I saw Faulkner's face watching me, and behind him, watching also, Cusped and Pinkerton. The others, I sensed, because the bulk of hidebound Lawson hid them from me, were observing everything from the line of wall, cot, and table.

"Tell me," I whispered, feeling suddenly desperate. "Now! Tell me, or I'll leave, so help me God! And if I go, you'll be lucky if you're able to crawl out of here tomorrow."

"Tell 'em . . ."

"Tell *me*!"

"Tell 'em . . ."

He was impossible. "Tell them what?"

"Tell 'em . . ." He was blubbering now, terrified, but I could not bring myself to believe that he was not acting until, in panic-stricken clarity, he uttered the word, ". . . *stop*!"

His swollen fingers twitched against the chains holding the restrainers taut around his wrists, chains attached to rings socketed in the concrete floor. I shook him.

"Tell me, goddamit!"

His voice rose again, siren-like, a yell that mounted to a scream. What had they done that my jolting him could produce a reaction like that? I found that I did not care; I was too worn out to think. I was fed up with him.

"This is the last time I'm asking you," I hissed in his ear. "Where were you last night?" And with an invigorating surge of pure rancor, I shook him again, with the same result. Someone turned up the radio. Tom Jones; "What's New, Pussycat?"

"Drink!" he said with a gasp. I thought it was the beginning of an answer, but his mouth was dry, that was all.

Well, why not? I needed one too. I let go of his head, and it dropped, bringing forth a yelp of lesser magnitude than the last. Stiff from crouching, I hobbled over to the table where Rublatz was again seated. He eyed me impassively as I took the bottle, but reached out a staying hand.

"Here," he said, in a voice so quiet I almost missed his meaning. He pushed a cloudy tumbler over to the table edge.

"Use this."

I poured out three fingers from the bottle into the dirty glass and went back to the bounden hulk. Rolling a tongue wash of the cheap rye around my mouth, I surveyed him with disgust and anger. Except for the speculum in his ass and the inflamed tracery across his back and bottocks, he appeared to me no worse for wear. Respiration, perspiration, temperature, trembling, complaints, puke, stink, and drainage from all nine body openings, and from others that were not normally such . . . everything was as was to be expected.

Crouching again beside him, I felt like Sisyphus as I once more took up the burden of his head by the greasy locks and placed the tumbler to his lips. He slobbered at the booze, and coughed, almost breaking the glass against his teeth. I helped myself from it again and set it aside on the floor, empty.

"Now," I said, "Shall we talk, or shall I tell Faulkner to pull your guts out? He's getting impatient back there."

"Get me outta here, Scotty. . . . Please!"

"Sure." I wrapped my arm around his head. "Where were you last night?"

"First get me loose."

"Uh-uh. Tell me where you were; then I'll get you out."

"You *gotta* get me out."

"If I say I will, I will. And I will." I twisted his neck and looked back at Faulkner, nodding. "Now . . ."

"Scott! Scott!" he pleaded. "I can't take no more!"

"I can see that. Talk."

And he did talk.

"Stallion Oaks."

So softly did he whisper, I had to make him repeat it. His breath foully bathed the side of my face turned to hear him say it, "Stallion Oaks."

"I'll accept that," I said, patting him on the shoulders and then let his head down easy. It was a measure of my tiredness and the disorientation of my seeking that the obvious had not occurred to me before this. Aside from Eugene's mother, and his brother somewhere across the border where he would not venture, the only likely place nearby where he might go to hide was the place Chuck finally revealed. It might lead nowhere, but I was at the end of my resources and ready to go to bed. I motioned to Faulkner, and all of them in the room moved toward me.

"You can let him go now," I said, aware, even as I spoke, of the breech of decorum I had committed.

"Let him *go?*" Cusped protested. Pinkerton, Jack, and Rublatz said nothing. The big-bellied sixth member uttered an expletive. Faulkner turned and told Cusped to shut his mouth. Then he looked at me as at a stranger.

"I gave him my word," I said, too tired to argue, knowing I was also too tired to stand up for Lawson if it came to that.

"That's interesting." Faulkner looked at me with eyes that must have matched the emptiness I knew must be in my own. "I don't remember giving him mine."

"Can I talk to you a minute, Nelson?" I dropped into one of the two chairs, the one he had been sitting in when I came in. "Alone?"

He stood over me, waiting, expressionless. The others, except for Pinkerton who had stretched out on the iron cot again and was stroking his hollow cheeks as if to squeeze out the yawn that was hollowing them even more, stood apart, exchanging glances and waiting to see what the outcome of this palaver would be. The ending of a session, as Symonds once put it, is harder to define than its beginning.

"Aren't you about as tired as I am?" I said to Faulkner. "Can't we just call it a night?"

"*We?*" He half-smiled, mocking me with fury-tinged eyes. "You want to leave?"

"I want a drink."

I poured out another one in the glass I had carried back with me to the table. "And yes," I said, after setting it down. "I want to leave, and he does too."

"No one's keeping you," he said, with a slight accent on the 'you.'

"Look, Nelson . . ." I stood up, meaning to implement my declaration, but a sudden rush of lightheadedness and proto-nausea made me stagger. I reached out for him, and he for me. The others swam in my vision, blurring, as he helped me down to the concrete, still regarding me as he might a stranger stricken on the street, and gentled my head at last beneath the staring moon of Lawson's undoubtedly still conscious head which I had thought to see no more of that night. It occurred to me in that brief moment that Faulkner might not have done all that he did merely to accommodate me, as I had supposed when I first proposed the evening's action to him. Self-esteem soars boundless when nothing acts to rein it in, and self-flattery is the greatest dodge of all.

By what would be daylight in the world outside, it was now all over for me. I slid into a deep, dreamless trough—drugged, as it turned out.

Sunday

13

I AWOKE sluggish and aware of enveloping stinks. What little light there had been had been turned off, and in the seamless dark, my eyes throbbed with the grandfather of headaches. Someone had been thoughtful enough to throw my jacket over me, but it did little in the way of warmth; there was nothing between my bare back and Levis but the cold concrete. A pleuritic ache was settled in one side of my chest. I moved my head, and the effort brought my stomach lurching up my throat. Keeping that down by lying quiet for a moment, I next lifted an arm to see what time it was, but the luminous dial of the watch refused to glow. It took a few seconds for an unadrenalized brain to register the simple information that I was not wearing my watch. No watch, no shirt, and no firm grasp on where I had left off before the blackout dropped me. Only the final impression of slipping to the floor under Faulkner's guidance, and his hands cold on my belly—a man of poor circulation, I thought; my last thought. I brought my arm back to rest over useless eyes, letting the salients of pain define themselves in Very lights, etching a skull-and-crossbones pattern on the quivering retina. Something like a whisper of traffic on the canyon drive below seeped in from far away like breathing. Otherwise silence, and acute discomfort. Aching skull, neck, thorax, a rising stomach, and now the bladder, swollen with the night's inswillings. Distended, it complained of neglect and crowding, a fractious lodger of the lower gut accusing the kidneys with vicious jabs and demanding relief from me, the sluggard guardian of the sluice.

Slowly, like a casualty crawling from a foxhole, I derricked myself up to an unsteady sit, wondering in which direction the lavatory bowl might lie, all dark as it was. The stink of excreta and something else—unvolatilized lees of amyl—momentarily appalled my overreceptive and squeamish faculties. Standing at last, I staggered as the bilious dark clawed at my guts, and caught myself on a hard beard-stubbled flank. Lawson, by the feel of it, about as I last remembered too. From the lay of him, I got my bearings and headed, unbuttoning on the way, for the corner basin, there to retch and spew, clutching the vitreous enamel for support until the heaving subsided. After bladder and stomach were emptied, I rested from these labors, supporting myself on the sink rim, catching my breath and wondering why no one else had stirred. Moving crabwise along the wall, I came to the door and the light switch. The discreet blue bulb nearly blinded me, but through the ache and sickness I made out Lawson in the center of an all-but-empty room, for only the sparse furniture kept us company there.

This egestive pilgrimage had brought out the sweats, and my stomach howled for the horizontal. I let myself down again, more or less where I had awakened, near Lawson's lolling head, and, with eyes closed, allowed the visceral churnings to subside. In a little while I felt stable enough to open my eyes once more. There above me hung the puffed and pockmarked face of last night's star—assuming, with uncalculated modesty, that I myself had not been the main attraction. It was partly his stink, heavier than the close air of the room, that bathed me, and the sighs I had mistaken for street traffic were his wheezing exhalations: bad breath from doctored booze. At least he was alive.

Letting the headache have its way, I surveyed those parts of him that came unsearched for into view. Saliva still drooled in a thin string out of the curl of lip and had attached itself, like the anchor thread of a spider's web, to a forming puddle on the floor beside me. With difficulty I moved my head, observing it, and where beside it lay several crushed ampules, little saffron-colored silk purses mashed and sullied into sows' ears and smelling like old tennis shoes. I gagged, but nothing came up. Whatever had been dropped in my drink—chloral hydrate was my guess—had carryover effects perhaps unanticipated by the donor, or donors.

Heart and breathing had quieted enough for me to sit again without too much fear of being bowled over by the effort. But with the advent of physical equilibrium came the dangers of lucidity and a more disquieting prospect. For the muddle posed by my even being there was enough to floor me again with a craving for that oblivion from which I had hardly awakened. The thought, like a carrion flower, bloomed in my ravaged head: Could all this have been planned for me? The foulness of its implications almost made me sick again. I tried with the breath of reason to dispel them. How could any of those people have been sure I would turn up in that basement? Or, having appeared, how could I be counted on to do what I did?

Unable to sustain any cogent line of reasoning, I turned to more solid matters, things touchable and, for that reason, possibly responsive to whatever came to mind concerning them. Lawson, for instance. I looked at his hands, nearly black from the pressure of the restrainers that pulled his arms down from shoulders hunched over the padded sawhorse. No one had bothered to loosen anything before they split. "Get me outta here!" he had pleaded. But his present situation was mute evidence of the influence I enjoyed with the crowd that had gotten him down. It was also a tribute to his toughness and stamina that he could still breathe.

I yanked the panic snaps that attached the wrist restrainers to the rings in the floor, releasing his arms, then rested a moment before undoing the leather cuffs. Such oversight—if it were oversight—was not what I would expect from Faulkner. This discrepant element was added to the others as I watched the big, ham-like hands ball up slowly into fists, and then more slowly relax again as the blood came back into them, changing their color from the blackness the room's illumination lent them to something approaching gray, and which, in normal light, would glow in swollen erubescence. The release of tension had taken some of the pressure off his diaphragm too; he began to snore.

Near the sawhorse was my shirt, lying where I had doffed it during the heat of the action ages ago. On it, a half-pack of cigarettes—Faulkner's. I fished one out and lit it. The shirt was clammy and smelled of the sweat it had soaked up before I took it off. The first drag of smoke nearly made me sick again, but at the same time it started me going. The cigarettes I had caged the night before were a trial run, but this one constituted the resumption of a habit I have never dropped since. My breath evened out and the heartbeat became steady again, but my head felt as if it had a crack in it, a fissure extending from behind one eye diagonally to the back of the skull. Leaving the shirt unbuttoned to warm and dry on my back, I sat there on the floor with my arms clasped around my knees, sucking smoke and trying to remember what had happened.

Something. Someone. Or more than one. . . . A name.

I was not worried, merely puzzled that this confusion of impressions would not gel into anything coherant. Conflicting data accumulated over the past week pointed nowhere, or rather, it pointed wildly in all directions at once. A sorting out was due, but now was not the time. Once the body was in shape again, the mind might function in its usual orderly channels. But . . . something. Someone. A name. . . .

Crushing out the smoke on the floor, I got to my feet, using Old Beat-Up as a purchase. An overall view revealed him now in all his squalid splendor, the very picture of impudicity, with upturned, precatory rump and obligingly furcated thighs. A studded bike girdle still protected his kidneys against the beatings he had taken, but the hide above and the gibbous buns below it, as well as the hunch of muscular legs and shoulders, bore the cross-hatched evidence of the long night's work. I ran a hand down the length of him. Most of the heat had gone out of those welts, and he certainly felt nothing at that moment, but there would be pain and discomfort to last him for days when he came to his senses. His ass alone must have felt as if a Roto-Rooter had been at work up there; a slight hemorrhea trickled like a discolored spring rill over the impasted surplus of lubricant cratering the inflamed pucker of his flue. It looked as if he had been ganched and then reprieved when the executioner was already halfway through the job. If his padre could see him now there would be hell to pay.

His feet were the next piece of unshackling to do. No sense in letting gangrene set in; one botched job was enough for the newspapers to chew on over a single weekend. When I unsnapped the ankle cuffs, his legs sprang free and rose a little, flexing as the tendons tightened. That must have hurt, because he moaned. But ah! The boys had really been sloppy: one turn in his sleep, had he been able to turn, would have torn his balls off. Releasing the thongs that held his manhood in thrall was easy, but undoing the cangue presented difficulties. The slide bar was almost buried in the folds of swollen sack. Bloody too, it was, but nothing serious; no more than was the discharge still seeping from his abused rectum. It was lucky for him he was still out; what I did to free him must have caused him agony had he been awake to appreciate it.

His scrotum felt like a hairy bag of ice. Within the hour it would swell to epic proportions and hurt as if it were in a tightening vise. "Par," he would later say with a braggart's modesty, "Par for the course."

These corporal works of mercy done, I looked around the room, taking stock. My watch was on the table with keys and billfold, money and cards intact. Perhaps I had taken them out and off before I went to work on Lawson last night. I could remember nothing.

Someone, something. . . .

The bottle and the glass were nowhere to be seen. Whoever had done the drugging had taken the possible evidence of it with him. I slumped down in the chair Faulkner had occupied, pocketed my stuff, and, staring at Lawson's sleeping form, sorted out the limited possibilities that presented themselves as agents of the past night's happenings.

There was not much to go on. Two of us drugged; everyone else gone. Rublatz had shoved the glass at me, but who was Rublatz? He was Faulkner's friend—or, at least, acquaintance; Faulkner could not really be said to have friends. For this reason I discounted Rublatz, because nothing that went on in that room was done without Nelson's knowing it. Of this I was convinced on no more evidence than what might be called faith; the belief in Faulkner and his capabilities. And since, as corollary to this, none of the others would have done it to me without his knowledge and approval, it followed that Faulkner had to be the one. Had another of them acted independently, I was sure they would already have accounted for it to him, and Nelson would be here in the room with me now, because he never did anything to anybody who did not want it done to him.

But there . . . if this were so, why would he do it to me who neither knew about nor wanted it? And leaving Lawson shackled like that?

I got up and performed a last kindness before leaving, Grasping the cycle girdle, I tugged, and Lawson slid off the sawhorse onto the concrete with a plop that knocked a grunt out of him. A final check of the jewel pouch revealed something like burgeoning elephantiasis, but at least the coloration had livened from its former ashen hue to what in normal light could probably be likened to hot strawberry, for it was now anything but cold. By the time he got around to waking up, he would stagger out of there like an old man carrying his balls in parentheses. It irritated me that all this solicitude on my part was wasted on him. He would look back on this past evening as a storehouse of exciting memories, and would remember me in the approving light of an immitigable doublecrosser. And were he to learn that I was the one who cut him loose, he would take it as an offer of protection to which he would respond with fawning promises of obedience.

He risked catching pneumonia lying there on the floor with nothing between it and his sore flesh, but there was nothing in the room to cover him with. I shook him, then slapped him hard a couple of times across the sleep-puffed face. Finally, in exasperation, I kicked him in the butt. To no avail. He didn't even groan. His breathing was as regular as a healthy child's; his sleep as profound. The cold concrete was a kinder bed than the bare steel laces of the cot, even assuming that in my present condition I could have hauled him over to it. There was nothing more to be done with him, so I left.

Upstairs, I pressed the button outside Chris' apartment and waited, expecting his complaining face to respond to this interruption of his Sabbath rest. Instead, after a second mashing of the button, Ray Reynolds' girlfriend Linda opened for me. She stood blocking the doorway with nothing on but a man's red flannel shirt that reached down to mid-thigh. She had not bothered to button it—if indeed she could have. She made no move to invite me in, but seemed to be coming to grips with the tangle of sleep and surprise that were fighting it out under an expression of puffy annoyance at having been summoned to the door from someone's bed.

"What're you doing back here?" she yawned, her good humor triumphing with a smile over the bad as she came more awake.

"Who says I was ever here to come back to?"

"Oh, I heard." She stood aside to let me in and closed the door behind us. "What a night!"

"Where's Chris?"

"Asleep, I guess." She turned and preceeded me down the cluttered hallway, dirty glasses, napkin wads, ashtrays, plates and the spillings from these all under foot, the stale after-litter of the party. The kitchen was a health hazard.

"I want to talk to him."

"At this hour?" She pawed among the debris on the counter, retrieving a cup here, a spoon there, a percolator from a cabinet above the stove. "Coffee?"

"What time did the last one leave?" I said, checking my watch against a wall clock housed in a Delft windmill over the refrigerator. It was somewhere around 8:15.

"The last of them are still here, but most of the rest were out by dawn, I guess. At least that's when the noise stopped. I need a cigarette, and you don't smoke."

"Surprise!" I pulled out the pack Faulkner had left me and offered her one. The last match in the book flared out in lighting, so she bent down over the stove, shifting the coffee pot with one hand and holding her hair back with the other while she lit it from the front burner. I waited in silence while she got the coffee started, then she turned around, leaned her meaty hams against the counter, and breathed a thoughtful billow of smoke my way. The used flavor of it acted as an analeptic to the ache in my head. I fished one out of the pack for myself.

"I'm sorry to hear about your friend," she said, handing me her cigarette for a light and eyeing me closely for whatever reaction her remark might provoke.

I nodded. "It was a bad scene."

"Do they have any idea who did it?"

"For all I know it could've been an accident."

"Nobody gets hung by accident."

"Who told you he was hanged? " I felt myself falling into Saturday's scenario all over again. "I understand he was stabbed to death."

"Ray said you told him something else."

"When did you see Ray?"

"Here."

"Is he here now?"

"In the middle bedroom." She ran a hand through her black hair,

hair so mussed it looked as if several bed partners had come in it during the night—a not unlikely conjecture. She seemed to read my thoughts. "You look terrible," she said. "What went on down in the basement last night?"

"Nothing much. We were playing tic-tac-toe, push-pull, who's-got-the-button. Good clean fun."

She smiled, lizard-eyed, letting the smoke curl out of her nostrils with a suggestive little grunt. I had never cared much for Linda. She was a pushy broad who had a peculiar hold on Ray and resented his taking up so closely, as she imagined he did, with me. I did not mind that, but her nosiness was sometimes hard to put up with, and she subjected Ray to grillings whenever she suspecting him holding out on her. Ray, for his part, could deny her nothing. Besides her teaching job, she was working on a master's degree, or a secondary credential, or both, out at the state college, and these chores, together with a roaring case of nymphomania, left her little time for anything else, least of all for her ineffectual husband and infant daughter, both of whom she believed, and said, were the cause of her internal unrest. I could not see in her the splendid companion that Ray did. Her appearance, her heedless, wagging tongue, and the action they pursued together—action which cemented their curious relationship—was not, to my mind at least, good for Ray, whatever benefits it held for Linda. A word about this minor variant in the behavioral patterns of the human fauna I was dealing with down there may put things in a better perspective.

I have mentioned somewhere above that Ray got his kicks second hand. I should have added, "for the most part." During a period before I arrived, at the time when Dump, Linda's husband, was receiving the benefits of reality therapy up north, Ray and she had taken to going out cruising together. These forays developed into a standard format that covered a search for stud singletons who might be turned on by the kinky overtones of a weak and possibly impotent husband (Ray) who was forced to endure the degradation of seeing a better man than himself plow his wife (Linda) into gasping satisfaction before his very eyes.

Ray played the part well. When they spotted their quarry, they would begin the act by staging a connubial quarrel which climaxed with Ray's retreating into the shadows to brood while Linda nursed her drink in sultry solitude at the bar. After which the fingered stud usually moved in on the bait. When he had nursed her, as he supposed, into a good humor again, Ray would saunter back into view, and the trap was smoothly sprung.

They had performed this peep show so often that a goodly number of the straight drinking establishments in the county took them for man and wife. They claimed to score every time they went out, with Ray getting his rocks off at the side of the jiggling bed with here-a-feel-there-a-feel—liberties permitted him by the huffing mark as no more than fair, and anyway further proof of spousal inadequacy and an appreciation of superior merit in one who knew how.

For Linda, fleshed out on a rented bed like some kind of geographic feature sprawled across the landscape, it approached the ultimate high she dreamed of: to be fucked until her mind caved in before a crowd of witnesses rapt in silence, the better to hear her getting off. On a few occasions she had almost found herself playing the secondary role of observer, but Ray was careful not to overstep by too far the bounds of friendship and ruin a good thing.

Although neither of them realized it, they had a kind of crazy love affair going between them. Linda's husband, who took care of their baby daughter and kept himself company after she was put to bed by leafing through girlie magazines with one hand, and Ray's putative lover who came down from Fresno to see him once a month, were no more than token holds maintained on the outer world as stable reference points from which they could map out these other paths of relatively harmless, intermittant lunacy.

Linda's was the managing voice in their association, but this was not enough for her. She was forever trying to extend her dominance to include and encompass others they knew. Distance saved Ray's Fresno friend, and the pleasures of Onan sealed Dump off from any interest in what she thought or did. Her opinionated bossiness made a little of her go a long way with most people; I was one of the few who saw them on anything like a regular basis, and I put up with her only because I found Ray amusing enough to tolerate her sticky advances. I was also one of the few who had played their little game and seen them again after it was over with.

For that was how I met Ray, not long after I came down to San Diego. Loneliness beyond my resources to combat got the better of me one night, and I sought out a go-go joint east of town where I discovered them yammering at each other from an adjoining bar stool. I had sort of felt sorry for a fat lady like that, and the rest of the farce was played as cast. Their pageant took my mind off myself for an evening, and Ray, more perceptive than she, picked up my number before we were fairly started in the thing. It was Ray who, a few weeks after this, found my house for me just across the hill from him.

After that initial run they got the message that it was also the final one. Or, at least, Ray did. Linda was a little jealous of our friendship, although she had no reason to be, and her reaction to it was to push for another chance. Her cutting sass was, I realized, a form of sexual provocation opposite in manner from Lawson's backward ways, but no less certain in its aim. It was obvious now that she and Ray had been playing around with the sweepings from the party the night before. They did not usually work this side of the street for the reason that Linda might very well find herself watching while Ray exercised the catch. I asked her about it.

"Ask him if you're so interested," she said.

"Is he alone?"

"The last time I saw him he wasn't, and that was when you rang the bell."

To be sure, he was not alone, but the figure on the far side of the bed was asleep with the sheet all but covering his head. Ray beamed a smile at me from the hither edge, wide awake. He drank little and slept deeply as a rule, and could consequently afford such a bright, good-morning aspect.

"Did you find out?" His conspiratorial whisper was loud enough to waken his companion who, however, did not stir. I nodded, approaching the bed, where I sat so that we could talk quietly.

"Did you see Cusped or Pinkerton or Faulkner, or anybody from downstairs come back up here within the last few hours?" I said.

"I've been in bed for the last few hours."

Ray could be exasperatingly exact at the wrong time.

"All right," I said, "at *any* time while you were here?"

"No." He twisted his round face into an oval of concern. "Should I have?"

"You're no help." I looked over at the figure sleeping under the blanket next to him. "Somebody doctored my drink, and Lawson's too from the look of him. Then they all took off—unless that's one of them here with you."

"Not likely," he smirked. "Want to see?"

I was of half a mind to pull the covers back and check but I could be sure it was no one who had been down in that basement. I said his trick could remain faceless and nameless for all I cared.

"Why would they put something in your drink, Scott?"

"Ask me something simple."

"Where's Eugene?"

"I said something simple, by which I mean easy to answer. Have you and Linda been gleaning up here?"

"He's a very nice young man," he said by way of answering me, lowering his voice and turning his head toward the sleeping mound. "You mean you don't know anything more about where he is than you did last night?"

"I know where he might have been."

"Where?"

I thought about it for a minute before I answered. Ray would probably spill it to Linda, but then, everybody who counted knew it already. "Stallion Oaks," I said.

"Oh, that bad place! Are you going out there?"

"I might. Where was Chris all the time we were downstairs?"

"He came in about three, but that's all anybody knows. What were you doing down in the basement?"

"Linda says nobody's seen him. And I was minding my own business."

"She means *she* hasn't seen him. Do you want me to stop asking questions?"

"Yes." I stood up. "When you get time you might check on Chuck Lawson. He's still down there."

"Where are you going?"

"Home to bed."

It was something I had been saying to people for a dozen hours, and I did not mean it now any more than I had before, although I needed sleep badly. There were a few calls I wanted to make, but I did not want Ray or his girl friend knowing about them.

"If you see Chris, tell him I said he runs a cruddy house."

"Can I come by and see you later?"

"I'll call you when I wake up. I feel like shit right now."

"Umm . . ." He snuggled down in the covers next to his companion. "You look divine. And after all that nasty stuff you said you'd never do . . . I never realized you were so bad!"

"Bad?"

"Bad. Those terrible things you people do to each other."

"You ought to try it, Ray. It relaxes the nerves and heightens the senses."

He squeezed his eyes shut and opened his mouth in a caricature of horrified disgust.

On leaving the bedroom, I was granted a vision of Linda approaching along the hall like a temple whore, a mug of coffee incense-steaming in each hand. The flannel shirt blew unflatteringly open in the breeze of her progress, revealing widespread tits and bush; the tits walleyed, the bush overhung by the full tuck of her belly.

"Coffee-e-e-e-eee?" she trilled, managing it with the last of the cigarette I had given her stuck to her lips.

I took a scalding mug and sipped, then gave it back.

"You're not going, are you?" she said.

"Sleep," I yawned. "I need sleep."

"Could I have another ciggy before you go?"

"Here, take the pack." I put it in her pocket and buttoned the shirt closed near the top. "You don't want that nice young man in there to see you like that."

"I'd like to've seen what you looked like last night," she smirked.

"Linda," I said, bussing her on the cheek, "you'd never have believed it."

She cocked her head in amused disdain. "You sure are branching out, for a prude."

"Survival training," I said, and opened the door for her. She went in, giving me the heavy-lidded smile again over her shoulder before kicking the door shut with her foot.

14

OUTSIDE, the neighborhood lay quiet and deserted; the warm, nacreous morning was clear under low cloud cover—what in San Francisco would be called high fog. Through a break in the trees I could see the Coronados standing out in tones of blue and gray on the horizon. Closer in, a corner of Mission Bay showed two more cruisers anchored, newly arrived in the night. The Hollywood Theater would have lines around the block for the eight o'clock show, and it would be a good time for tourists to drive down to Tijuana to watch the sailors eat.

My ten-year-old heap of an Oldsmobile sat all by itself where I had left it. Cactus Jack's purple shaft was gone, and there was no sign of bikes in the parking lot. Save Pinkerton, they might all have been halfway up to L.A., or simply having breakfast somewhere, together or by themselves. My head felt as if there was a brick lodged in it, and the persistent depression of the night before still hooded my thoughts, but I did not mind being alone so much now that it was daylight. There were two more calls and two visits I wanted to make; by the time they were done I meant to be ready to drop so that the coming of night could take care of itself without my worrying how to get through it.

It was too early to telephone anybody, so I decided to do my visiting first. I needed food, but the stomach still felt too queasy to hold anything, and when I put the key into the ignition, my hand shook. Wrestling the car away from the curb and turning it around to head back out the way I had come left me breathless and near to being sick again. It was that bad. I wove my way out of the bungalowed streets and spun the dust of Old Town from my wheels, turning into the sparse traffic on Interstate 8, heading for Chula Vista and beyond, where lay the phony ranch Lawson had come up with before I passed out, wondering all the while what I was going to do when I got there.

Stallion Oaks was a stud spa that had been opened a year before this by two aging Hollywood body queens who had some film history and money behind them. The history was just that, but the money was present. They had bought a parcel of land, about 150 acres, with house, barn, and some dependent buildings on it, and turned the place into a kind of amusement park, a secluded retreat for discriminating hedonists. This abridged spread was all that was left of what had once been a legitimate livestock operation which had sold out, bit by bit, to tract speculators who, in turn, went broke when the anticipated surburban growth of San Diego County did not make it that far. These two, Ted and Walter by name, remodeled the house and barn, built a glorified bunkhouse, restocked with a few gentle horses for show, and did a general refurbishment on the entire piece, achieving in the process the kind of atmosphere they had in mind to make it attractive to out-of-town and out-of-state clientele with money enough to support out-of-the-ordinary tastes. From what I had heard, it managed to pay for itself—barely. Tax writeoffs must have been as much an object as profits.

The ranch sat on a gentle rise of land a dozen or so miles inland from the coast in uplands dotted with slonks and gulleys where I sometimes came to replenish the frog population of my garden pond. The lower edge of Cleveland National Forest surrounded part of it and the border town of Tecate was not more than twenty miles to the south. Cottonwoods, scrub and mesa oaks, some stunted growth of laurel in the mostly dry stream beds, and the omnipresent eucalyptus groves hid it from unpaved road that ran along its southern fence line. Plateaued hills rose on every side, giving the land an aspect of Mexico, and bald outcroppings of gray rock lent an air of outlaw country to the section. Low mountains rose in the shades of blue and gray further to the east where the county ends, dropping off into the great Imperial Valley. A hint of that dry desert climate reaches down into these ranchlands where the temperature rises with every mile of distance from the ocean until it changes abruptly into the blistering heat of the Anza-Borrega Desert beyond the crest at the county line. The morning had already lost its freshness when I turned off the two-lane blacktop and onto the gravel road that followed a dry creek bed screened in dusty cedar breaks. It was about a mile to the high, chain-link fence.

From Eugene's journal I had learned of Stallion Oaks, and he had filled me in on how it operated. Whenever a big night came around, a call went out for helpers. In return for "help," the respondents received food, booze, a place to sleep it off, and from five to twenty-five bucks, depending on their looks and/or the contributions they might make to the night's entertainment. Tips from gratified guests could round off their earnings to a sizeable take if they were particularly accommodating and the visitors more than usually grateful. Closeted business and professional men from the East and a few from in-state funded these get-togethers in which was tolerated a gambit of pleasures that ran from quiet hand-holding to the more exaggerated realizations a patron's pent-up imaginings might, under the indulgent but prudent eyes of T. and W., lead him. The line was drawn at actually killing and most battery of a serious sort, although simulations of both were regular items on the agenda of a really big weekend. Helpers and patrons might, for fun or the gratuities it earned them, find themselves chained naked to trees, mildly charred with cattle brands, playing scary hide-and-seek, suffering capture by posse in Western garb and trial by kangaroo courts, after which they could expect to be frightfully used, condemned, and sent to the gallows. There were staged fights at the poolside bar and in the bunkhouse. Gang-bangs were standard fare, or sometimes just happened (most entertaining), and the barn and hayloft were equipped to handle just about anything else that might cross the jaded mind of a visitor out for a night to remember and with the money it took to make it happen. The animals were known to get into the act too. One of the four horses was an amiable Palomino whose piebald penis was a source of giddy fun to those impressed with size rather than performance. For performance there was a male Boxer who had been trained from puppyhood to mount on command, although it took a bit of contortion on the part of a human being to effect entry with any hopes of successful completion of the act. A few of the help managed it though—for a fee. It beat hell out of Knott's Berry Farm.

The only full-time employee on the place was a kid named Chumly whom Ted and Walter had picked up with the horses. He took care of the stable, cleaned up after parties, and was a live-in "helper." I had met him when I drove Eugene out there once just to get a look at the place. I met Ted and Walter then too, and did not like them on sight; they, in turn let me know they did not care to have non-paying visitors dropping in on them. Chumly was interested, however, and took down my phone number from Eugene. We had made it on an average of once a month since then, which was shortly before Eugene and Symonds started being seen together. Usually I picked Chumly up out on the road near the ranch and brought him to Radio Heights. He had to do the calling, since Ted or Walter would block any attempt to reach him on their telephone.

When I drove up to the ranch gate that Sunday morning with the dust roiling behind me, I found Chumly leaning on the long swing bar above the high-hanging sign board that said, in smaller lettering below the deeply branded name of the place, NO TRESPASSING. I turned short of the gate, having no intention of driving in, and came to a stop under the shade of a low-branched oak. Chumly slipped the latch on the gate and came over to the car.

"Hi, Chumly!" I called.

"Hi, Scotty!" He came on at a lope, a lithe, but homely kid who had trouble with his adenoids, but otherwise appeared to have not a worry in the world. He was a rare variant, a bodybuilder whose narcissism began and ended there. He kept his flesh in shape for show and use, it hardly mattered to whom. Pick up any jackoff body brochure of that era and you will more than likely find him inside or on the cover, flexing. He had more pen pals then Eugene, but did less with them. A relatively simple soul, he was incurious, without ambition, and incapable, as far as I knew, of speaking ill of anyone. He liked the chores required of him at the ranch because they kept him out of doors and meshed with the image he had of himself. The plain, open, smiling look of good faith, good health, and amiability mirrored the basic honesty and undiscriminating accessibility that lay behind it. He had a way with animals and posed no problems for the human element's appetites when it came to their using him. He played no games, but was a damned good lay—concave or convex.

"Here it is Sunday, and everybody's ringin' at the gate," he said, leaning in at the window and putting a friendly, calloused hand on my shoulder. "What're you out here for?"

"Would it make your day if I said you?"

His smile of pleasure was genuine, eager.

"Who's 'everybody'?" I asked.

"You mean here? Well, there was Nelson Faulkner. An' just before him there was that tall, skinny guy . . . the one's married to the stripper."

"Dale Pinkerton. He's not married to her; she's his sister."

"Yeah, somethin'. He just left, and now the police are here."

"What police? The sheriff?"

He shook his head. "It ain't the sheriff's car, and they ain't in uniform. There's two of 'em up at the house now, talkin' to Ted and Walter. Everybody's stayin' outta sight. Scared."

"I can imagine. Maybe they're inquiring about the party you people had this weekend."

"Could be, but I dunno why. There wasn't anything happened." He waited, ingenuous, obliging, composed.

"Hop in, Chumly," I said. "I want to talk to you, but not out here. I'd as soon the fuzz didn't see me right now if they finish their business up at the house and come back out the gate."

"Sure, Scotty." He trotted around the front of the car and climbed in beside me in a puff of body odor compounded of the stable and everything else he had been around for the past few days. I no doubt matched him in this. "Where're we going?" He looked at me. "Walter's gonna be lookin' for me soon as he's free."

"I won't keep you long." I started the car. "I just don't want to be seen."

The oak scrub offered concealment in depth a few hundred feet off the road in the stony rise. We rocked and rolled slowly over the uneven shale until I judged nobody could see the red paint job on the car. The heat under the black canvas was bringing out the sweat in both of us; I started the engine again and undid the windshield latches, then pulled the knob under the dash that activated the top. With a gratifying whine, if lifted up and back, letting in a flood of light and air, hot but fresh.

"Now," I said, reaching over the top of the seat and enfolding his shoulder in my hand. "Tell me about Friday night."

"Sure, Scotty. What d'you wanna know?"

"Was Chuck Lawson out here? You know Lawson, don't you?"

"Yeah, I know him. He was here."

"What about Eugene Carmichael?"

"He come in late, and then he left again."

I squeezed his shoulder and ran my hand down the firm, corded muscles of his arm. "Are you certain?"

"I know 'cause I saw Walter chewin' him out for bein' late."

"What time was that?"

"'Round midnight, maybe."

"When did he leave?"

"Dunno exactly, but Walter come askin' me where he was, it must've been sometime before two, 'cause there was some people comin' in from the bars just after that."

"What did Pinkerton and Faulkner want out here this morning?"

"I ain't sure."

"What do you mean, Chumly?" I continued to stroke his arm. Under the thin cambric of his soiled shirt the nipples had firmed out.

"I heard Faulkner askin' Walter about Owen Keller, but Ted sent me out before I could hear what Pinkerton showed up for."

"Faulkner was asking about Keller?"

Chumly nodded, his eyes growing sleepy.

"Has Keller or Faulkner ever been out here before?"

"Sure. They was both out here around Eastertime."

"Together?" I moved across the front of his shirt and began unbuttoning it. His eyes were glazed, staring at the glove compartment door. I have watched the same response in a frog, but a frog can't talk. "Were they together?"

"Not at first. They come out here separately, but they left together."

"What about Lawson? Does he know this Keller?"

"Sure. He met him up in L.A. But that was sometime before Faulkner and him got together."

"Faulkner and Keller, you mean."

"Yeah."

"Did you ever see Glenn Symonds—here?"

"Who's he?"

"You heard about the doctor who was killed Friday night in Eugene's trailer?"

"Yeah. I heard Ted and Walter talkin' about it to some of the people up at the house. I don't remember him ever bein' here though."

It was not the sort of place Symonds frequented, to be sure, but so much had changed in his behavior over the past year that there was a chance he had looked in for one reason or another. It surprised me, though, to hear that Faulkner had been seen in this milieu—and with Owen Keller. Even though Symonds' liaison with Owen shaped up as a recent development, my knowledge of it appeared now to be of later date than that of anyone I had talked to so far. That Keller was mixed up with Faulkner, as casually referred to by Chumly in his trance, was something else about which I had been in the dark. Putting bits and pieces together, I could remember the first intimations of Symonds' shift had come from Eugene; he had mentioned his master's new interest in that neutral way of his which was always the preface to an announcement of some new putdown he was about to suffer. I had let him know at the time that I was not interested in hearing about it. So I heard no more until Tuesday, when Symonds himself made mention of it, and there too I had failed to listen to what he wanted to say.

But now Owen and Faulkner? How many knew of this and had called it to my attention without my taking any heed of it?

I have already said that I considered Symonds' actions in regard to Eugene as a means of reaching me. It was with no feeling of pride that I assumed this; I was years past being flattered by the perseverant assaults of one I knew too well, I thought, ever to assume him capable of a direct and uncomplicated action. In his mind, the way I reasoned it, I counted as unfinished business, and it was the mind of a meticulous surgeon little given to letting loose ends go unsutured. I represented a loose end for him, but had Faulkner, I wondered, been a loose end too—one who all this time had gone unnoticed, by me at any rate? Eugene, Symonds,

Faulkner, Owen . . . The field was narrowing, and out here, now, there was the feel of them, their scent, like the scat of an animal who has been and gone, faint but discernible.

"Was Pinkerton still around when the sheriff and those others showed up?" I said. Chumly's shirt was open now. Obligingly, he removed it as he answered my question.

"They called from the gate, and while I went down to let 'em in he took off around the fence road. When he got to the gate they was already through, and I let him out. I was just killin' time there when you showed up. You must've passed him goin' the other way."

He settled back against the seat, the sweat standing out on him in droplets that gathered and rolled down his chest and stomach, dampening the waistband of his jeans. You never had to ask Chumly to take his clothes off; a hint was enough to get him down to the buff. He found excuses to parade his body as naked as circumstances allowed. I let my hand drop to his belt, and loosened it. He took over with the buttons from there on, and I concentrated on one of his nipples. They both stood out half an inch from the hairless plates of oversized pectorals. Like a woman, he shuddered at the touch as I gently rolled the cultivated dugs. ("How'd you get tits like that, Chumly?" "Oh, just sorta worked on 'em, I guess.")

"What do you know about Faulkner and Owen Keller?" I said, letting up on him a bit so as to penetrate the rapidly congealing glaze of wantonness before he sealed himself off completely in his private pleasure pod. His eyes were closed, and a smile of idiot joy transformed his usually placid features. Through a proto-suffusion of ecstasy he answered, already far away.

"Nothin' really. They was out here, that's all. They started goin' together out here."

He was breathing hard. His Levis were down around his ankles and he was fisting himself, far gone in some obscene little fantasy of—I don't know—horses maybe. He had his free hand on me now. I tried a last time. "Chumly! Do you think Eugene knew Owen Keller?"

"Sure, Scotty, sure. He knew him. Everybody knew him."

In rage and rut, I leaned over and took his near nipple in my mouth, eliciting a wallowing groan from him. The taste of salt and the musk of slept-in sweat got to me. I reached under him and grasped one of his beautiful buns that had skin like a baby's and began to knead it like dough. My head was killing me, but I went ahead, chewing on a leathery dug and working a finger into his ass until, with a cry, and a fine repertoire of contractions, he shot, spattering himself, me, and the seat with his seed. When his breathing had quieted, he opened his eyes, still smiling, as if basking in the afterglow of what he had seen in there. Then he leaned over and accommodated me.

For a long time after that we sat there, drowsed and baking in the dry heat. My head was worse, if that were possible, and the release of tension followed by the buzzing stillness served only to open up a susceptible channel in my mind to something very much like despair. The senselessness of everything that had happened hovered like mocking laughter on the dusty, heat-shimmering air. And the gnats were at us.

"It's gonna be a hot one today," Chumly said at last, breaking the spell. "You comin' up to the house?"

"No." I began mopping up. "There's nothing I want to talk to those two about, and nothing they'd tell me if I did."

Chumly got out of the car with his Levis halfway up, and toweled himself down with the shirt he had been wearing. "What're you gonna do now?" he asked, standing there, still dripping, his natural tendency to allow himself to be looked at in no way diminished.

"I've got to go," I said. "Come on. I'll drive you over to the gate."

"That's okay, Scotty." He laid the sodden shirt over a perfectly chiseled shoulder and relaxed in an antique pose: Faun After Coming. "I'll take a short cut. Drop around again soon, hear?"

I said I would, and pulled tire-crunching away, leaving him in the encinal grove, sun-dappled and smiling, his jeans down around his boots. In the rocking rearview mirror I saw him bend down slowly with rustic grace, a grin on his retreating face as he drew up his britches over a cock still half alert. He had enough foreskin to hang himself with.

15

OUT ON THE blacktop again, I headed toward the city by way of Lemon Grove. I would now get around to seeing Mrs. Carmichael. It was on my way and, being Eugene's mother, she was someone I felt I ought to at least pay token attention to. I doubted that she would be of any help to me, or to the police, who had been at her, but it was another door I meant to open and close on the principle that something might be overlooked if I did not.

Yet, even as I said it to myself, I knew the reasoning to be specious. For my line of inquiry up to now had been almost lackadaisical in its manner and all but ineffective in its results. There was a randomness to this roundabout search over the past twenty-four hours, an approach atypical of my way of doing things which was not lost on me in the aching forenoon of this second day. This kind of systemless casting about reflected a desire not to come to grips too quickly with what lay ahead, but to nevertheless keep heading in the general direction as I saw it. It was also the result of slackening will in the face of an aching tiredness. It had been over two days since I had gotten any real sleep. The narcomatous crash I awakened from that morning did not count, and the impromptu scramble with Chumly had left me more unrefreshed than before. In the present circumstances, venereal relief was no more than a baby's pacifier. I needed sleep and order badly in order to make a functioning man of me again.

Stopping on Grove Street, I looked up her address in a phone-booth phone book and searched out the little house on the split lot where she lived. It was really a garage someone had put a floor in and built an addition onto, making it into a knocked-together two-room cottage. There she made do on welfare checks and what Eugene doled out to her in spending money. She whined a lot about the stinginess of her allowance from him, and he griped about the demands she made on his paycheck, but I happened to know that he was generous with her. She carried on bitterly to anyone who cared to listen that he left her to all but starve, but this was the way she chose of throwing dust in her own and other people's eyes so as to hide the neglect she suffered from her older, and favorite, son.

Eugene had talked to me on occasion about his elder brother toward whom he remained curiously neutral considering the account he gave of him. This may have been an altogether natural result of the way this senior sibling treated him when they were children together at the family hearth which, as I gathered from his telling of it, must have been a scene of continual intramural strife between his parents who had no time left from their own squabbles to referee the imitative infighting practiced by their offspring. His father, a pulp mill worker of no more intelligence or ability than such a job demanded of him, was periodically out of work due to unspecified illnesses whose genesis was the bottle but which were blamed by him on working conditions at the mill and the living situation at home. Both sons, and a third who died in infancy, were born in the trough of the Great Depression and left fatherless during the war boom when the old man stomped out of the house one evening after a more than usually vituperous session of retributive bickering with his wife over some trifling matter of money, or booze, or it might be his prolonged absence from the house they rented in a shanty community called Samoa that hugged the skirts of the mill on the west shore of Humboldt Bay. At any rate, he disappeared into the night, never to return. Once she had determined in her own mind that her husband was gone for good, and that the mill would do no more than hand over to her the terminal pay he, in his obtuse rage, had left uncollected, Mrs. Carmichael, no more canny or talented than her spouse, settled into the lower grades of domestic service, first in Eureka, then in Arcata, finally further north in Crescent City. The two boys were left pretty much to themselves and to the school systems of Humboldt and Del Norte Counties until Larry, the elder, graduated from high school and promptly disappeared, breaking his mother's heart, to hear her tell it, and leaving her with her younger one under foot.

A series of minor disasters during childhood had accustomed Eugene to the prospect before him, and had established in his mother the ingrained notion that he never could be trusted to get his shoes on right, much less make his way in this world and properly see to the comforting of those years that were to be her prelude to the next. She lived in dread of what costly mishap might befall him in her absence, and maintained with grim complacency that she knew it, or something akin to it, was bound to have happened whenever a new calamity or embarrassment (they were much the same in her eyes) involving him occurred.

Tall and robust for his age, Eugene had nevertheless an innate incapacity to stand up to anybody in defense of his rights, and, for that matter, seemed in large measure ignorant of what his rights might, in any given situation, be. He was the target of every bully in the block and the butt of practical jokes in the neighborhoods and schoolyards he was subject to. His rangy good-natured acceptance of the badgerings meted out to him were put down to feeble-mindedness by oldsters who enjoyed, at the same time that they deplored, the aggressions worked on him by his peers. These latter accepted him as the resident sissy, although even they were somewhat puzzled by the muscular resistlessness of one who might have felled them with a single blow were he so inclined. What none of them could be expected to comprehend—least of all Eugene himself—was that he had learned at an early age the raptures inherent in bodily contact, but only through the medium of punishment. Father, mother, brother, all tolerated his presence at the price of corporal abuse and verbal degradation. There remained only bondage to complete a trinity of bliss.

When he was twelve years old, a bus returning him from school in Arcata to the sour-reeking purlieus of Samoa, the company town, ran into a ditch and catapulted Eugene against an enameled pole provided standees. (Eugene never got a seat.) He emerged from this mishap with a cut lip and minus most of one maxillary incisor, but otherwise unscathed. Indifference, delay, penury and decay combined to lose him that and several other teeth, gaps made good by the government when he entered the service.

He also had an unretractable prepuce until the age of nineteen when a Navy doctor, aghast at the inflamed stink of it uncovered during a short-arm inspection, ordered the phimotic member abridged. The medical lieutenant assigned this routine task at first expressed impatience at Eugene's shifty-eyed plea that he undergo the operation without benefit of anaesthetic. But then, intrigued by the novelty of the request and the hint of mania and possible frenzy in the softly intoning voice (and amazed at himself for giving it a second thought), he finally allowed himself to be inveigled into granting the eccentric request. An excuse was found, and the circumcision performed as asked, without fear or clamor on the part of the patient, but with some hindrance to the posthetomist's art, since the organ refused to relax from a rigid state it assumed at the first aseptic swabbing. Blood, smegma, and semen were the issue of this genital tonsuring, and the young doctor's admiration of the prodigy before him was surpassed only by his own excited reactions to what had taken place. So instead of certifying the seaman as a nut, he kept silent, and saw him as often as could be arranged after the annular wound had healed. Together they learned from each other over a period of months those things heretofore unknown to either but secretly germinated as varient potentials of the same leaning within them, leanings which had awaited only the quickening light of experience to be realized in practice. By the time the Navy bureau of personnel disbursed them on their separate ways, they were both well into the woods, and as a result of this seminal encounter Eugene worshipped doctors the way some women reverence priests.

The Navy had been Eugene's escape from home, but his departure from the service cast him back into the fold. Discharged in San Diego, he discovered his mother waiting for him on the dock, so to speak. They had maintained a sparsely worded, intermittent correspondence during the four years of his enlistment, but her presence in the southland came as a surprise. She was not there solely to welcome him back to civilian strife; she needed support. Larry, her favored first-born, had surfaced in Mexico during Eugene's absence, where he supposedly worked for an export firm, the nature of whose commerce was never made clear to her. He had sent her something every month in the way of remittance money with the understanding that she was to keep her distance, meaning she was not to cross the border or otherwise attempt to see him.

With Eugene home from the sea, both Larry and his money disappeared. One of Eugene's first acts on assessing the situation was to go down to Tijuana and confront his brother on this fiscal matter. And Larry, who apparently enjoyed connections in high places, arranged to have him thrown in jail and kept there for a while as a lesson in what he might expect from further meddling where he was not wanted. The confinement itself would not have been so bad had the *carcel*'s regime given him more in the way of fantasy to feed on and had anyone spoken English enough to understand his needs. As it was, he was released inside a week and hustled back across the border at San Isidro with a warning from one of his brother's agents never to return. Mrs. Carmichael gave no credence to his relation of this aborted errand, but believed instead that he had blown the funds from his mustering out in the fleshpots of Baja, leaving her to stew in want. The idea that the money had gone to pay for bribes and sustenance while he waited for his brother to spring him was unacceptable to her, and she railed against him ever after for the hard days that resulted from what could only be his thoughtlessness and profligacy.

He lost no time in finding a job once back in the country, however. While working for a building contractor, he went to trade school at night. After six months of this he landed the welder's job in the shipyards, set his mother up in Lemon Grove, and handed over enough money every month to keep her out of the poor house and his hair. At the same time he arranged to get her on the county welfare rolls. It was in the nature of their relationship that he neither received nor expected thanks for any of this. In her eyes he was only doing his duty, and the benefits she received at his hands were no more than her right. She got it squared around in her mind that Gene had somehow turned her beloved older boy against her. And this, in a cross-grained way, made her feel better, knowing where to put the blame. As might be expected, she and Eugene did not get along very well.

I had met her only once in my twenty-four months down there. That was on a Saturday afternoon when Eugene had been living with me a short while. He had come back to the house to pack his leather satchel with equipment to take to a referral in Palm Springs that was to last overnight. I was in my back room when the doorbell rang, and I told him to go answer it. I heard voices, and came to the front of the house to see him arguing with an elderly woman as tall as he was and twice as big around. He was trying to push her back out the front door in which she had wedged herself, refusing to budge. On seeing me, he gave up the struggle and introduced her, reluctantly, as his mother. He had taken her for a visit to the doctor, and was going to drop her off in Lemon Grove on the way to his arrangement. She, he later told me, had convinced herself that he was living with a woman who was after his money (the Phoenix operation had made inroads on all his usual expenses, and he had been understandably evasive in accounting for the reduction in her widow's mite), and had forced entry prepared to do battle with the hussy. Somewhat deflated after being introduced to the landlord of the premises, she exchanged civilities with me, entered a formal note of praise for Larry, and a passing one of disparagement for Eugene, before he told her to shut up and get back in the car. I waved them off and watched them down the hill. Her mouth was moving and Eugene's jaw was knotted in silent rebuttal as he gripped the wheel of the Dodge with both hands, the better, perhaps, to keep from belting her one. It was the only time I ever saw Eugene order anyone to do anything.

He mentioned her rarely after that, and I did not relish seeing her now. Nor did I expect to find out anything I did not already know of Eugene's whereabouts, which is to say nothing. The most I hoped for was that she might let slip some piece of information that could be fitted in with others.

Parking the car near a weed-grown driveway that led to the converted garage, I walked past the main bungalow and knocked at the hollow-core door. Mrs. Carmichael was a great wheeze of a woman, prone, as Eugene had informed me, to temper and tears, so I braced myself as the shuffling footsteps from inside shook the termite weakened frame of the building, and the door handle turned.

It was tears.

She took no more than a second to recognize me. "Mr. Scott!" She shooed me in and closed the door, then tuned up. The police, she said, had been to see her three times since yesterday, and were determined to ring the phone off the wall in their efforts to drive her out of her mind.

"No sign of Eugene?" I said, standing in the middle of the small, untidy room, sorry already that I had come.

"Oh!" she wailed, "now don't *you* start!" In between sobs she swore she had not seen him, and that if she had, she certainly would not tell those people. I tried to assure her that I understood, while she dabbed at her eyes with a sopping handkerchief.

"They've been hounding me too, Mrs. Carmichael," I said. "I know what you've had to put up with, and I don't blame you for being put out with them, but . . ."

"Mr. Scott," she interrupted. "You know as well as I do that Gene wouldn't hurt a fly, much less kill anybody!"

This speech seemed to exhaust her. She fell with an audible plop into a spavined, overstuffed armchair facing a television set whose sound had been turned down. The screen was alive with a cartoon animated in crudely filtered color. She mopped her eyes some more, then waggled her head a few times as if she were having an argument with a more than usually unreasonable antagonist, and finally asked me what I thought she ought to do were he to turn up on her front stoop.

"I hesitate to say, Mrs. Carmichael," I hedged. "Of course, if Eugene didn't have anything to do with it . . ."

"He *didn't* have anything to do with it, I tell you!" And she was off again, a mother's tears, boohooing into her handkerchief. "*You're* his friend," she blubbered. "*You* know."

"Yes, ma'am. I know how Eugene is. But the police don't know him. They're going to want to question him at least. You can see their point in that."

"No, I can't!" she snapped. "It's all crazy to me. A man murdered in my son's trailer when he wasn't even there—Gene, I mean. What if I was to come home from the store some day and find a man murdered in my house here? That's no reason to think I done it, is it?"

For a moment I entertained a mad vision of her plodding up the driveway, exhausted from carrying a bag of groceries under the hot October sun, and opening the door to find a man hanging by his heels from the ceiling fixture which, I noted with the demented, split-second attention one pays to details in such inappropriate imaginings, would not hold a body. And no, I doubted if the police would have suspected her—assuming they found her alive or coherant after being exposed to the sight.

"But," I insisted as gently as I could, "they might want to ask you about it."

"Ask!" She spat the word at me. "They want to lay the blame on Gene because they're too lazy to go look for who really done it. Oh! I wish I knew where to get hold of Larry. He'd know what to do."

"It's only routine, Mrs. Carmichael." I was still standing; I bent over to put a hand on her heaving shoulder in an attempt to calm her down. "They have to start somewhere, and they may only want to know where he was at the time, if he wasn't there at the trailer when . . ."

"He *wasn't* there, I tell you!" She shook off my hand and looked up, glaring like a snapping turtle. I saw the craggy features of Eugene, rounded, but not softened, by the fatty paddings of age, staring at me with a ferocity born of frustration. It was the Gorgon face that must have frozen him in terror as a child. As then, she was now reduced to raving, and I listened to her while she ventilated her feelings.

"You're as bad as them. . . . them policemen that was here. And they're worse than the welfare people. One of 'em talks to you while the other one snoops—as if there's anything here anybody'd want!"

And more of this. When at last she ran down, I asked her, "What exactly did they want to know?"

"Well now . . ." She looked at me as if she were suddenly dealing with a simpleton. ". . . what d'you *suppose* they wanted to know? What've we been talkin' about since you come in here?" With exaggerated patience, she spaced the words out as she might for a backward child. "They wanted to know if I knew where my son is." Then she shifted into the more comfortable swing of diatribe once more. "As if I ever know where he is. He never tells *me* anything, and I never see him more'n once a month, if that much. He calls me when he feels like it and treats me like I was a scr-vant." She added, with a cantankerous wrench of her body that threatened the chair, "An' I hate to think what'd happen to me if I *had* to get hold of him sometime."

"Did they tell you anything more than that they were looking for him?" I cast about for a tactful way of putting it. "I mean, did they say anything about what happened?"

She waited a moment, a champing look of complaint, considering, I suppose, how much she ought to say to me. Then she sighed, as if something inside her had given way. "I'll tell you, Mr. Scott. They said some terrible things; things I don't for a minute believe. It was worse than what you see on the television. They're scum, Mr. Scott, just plain scum. Comin' around to see an old lady and tryin' to poison her mind with lies about her own son! You can't imagine the things I had to listen to!"

With no effort at all I could imagine what they might have told her. And what they did tell her could not have amounted to much considering what they might have divulged had they been of a sadistic bent and had the time to waste. But even the little they revealed was enough to very nearly unhinge her. She blubbered on about it, at times nearly incoherent with indignation and rambling all over the landscape in excursions of self-pity at her lot in life. I gathered from it that a summary inventory of that trailer had been made and some choice items of interest described to her in their visits in an effort to jog her memory as to where it might be most profitable to look for her fugitive son. They went so far as to let her glimpse some of the contents of his file cabinet. I had seen that bin of smut many times: three drawers full of correspondence ranging as far afield as Australia and Sweden and running a gamut from mere randiness to blithering depravity; all the referrals and inquiries, detailed, dated, handwritten, or pecked out with smudged erasures and corrections on his second-hand Underwood. In his spare time Eugene also engaged in fictional experimentation, summoning up fantasies of horny exploits full of impossibly complicated arrangements and feats of ingenious villany, all described in penny-dreadful prose. The spelling was the most truly creative feature in the whole collection, a baker's dozen of manuscripts otherwise unrelieved in dullness. All of it was actionable, however, under the laws as they were administered then. Even the old lady must have realized that. And the local heat took them very, very seriously.

Then there was the art work, as preposterous as the writing, although some of it had merit as far as draftsmanship was concerned. Like all this collection, it was the proportions that were out of sync, and which, in turn, suggested the informing warp of mind behind them. Eugene possessed an almost complete series of Nelson Faulkner's fevered conceptions—unsigned, of course; Faulkner was no fool. (Oh! And remember, remember: this was before the Great Unbuttoning.)

And his journal. This last they apparently did not regale her with, either because they had not discovered it, or because so direct an evidence of her son's recondite behavior might have pulled the plug on her already strained sensibilities. But she might also simply have forgotten it in the welter of other bewildering items paraded before her in an attempt to soften her stubborn disbelief and mold her into a more cooperative stance in the face of the law's inquiries. I edged around the subject in the first opening her nonstop monologue allowed.

"Did they," I asked, picking my way carefully through the debris that littered her mental landscape. "Did they find anything like a calendar, a datebook . . . something Eugene might have noted things down in? Things such as where he went, or . . ." Here I swallowed hard. ". . . what he did?"

She pulled her face down in a long, sour look and said in a voice quiet with disgust, "Lord, you got as one-track a mind as them."

It must have looked that way. But the techniques of temperate interrogation are limited, and depend for their effectiveness on the art of the questioner in response to the good will, or lack of it, of the respondent. Mrs. Carmichael, having completed her diatribe, and exhausted from the review she had given me of her trial by exposure, was now considering me with a different eye, sinking, in the process, into a sullen uncommunicativeness. I backtracked, trying a different approach with her since it was possible that the indignation she exhibited at what she had been told resulted less from the substance of the disclosures than from the sudden apprehension of previously unimagined dynamics in an offspring she had thought she knew like the inside of her mouth, potentials kept from and therefore repugnant to her above and beyond any absolute moral fault which might be found in them. But while Eugene was worthless in her own eyes, she was prepared to look the other way when faced with these new and inexplicable revelations. The whole thing was no better than a plot, an unprincipled intrusion by the world on the peace of her declining years. And she was not one to take such a shaking-up in her stride.

"Anyway," I concluded, as offhandedly as I could manage, "if we could find something like that, itwould probably get him off the hook. The police want to know where he was around the time it happened."

"Well, he wasn't around that trailer," she said. "And if they did find somethin' like you're talkin' about, they didn't mention it to me. They didn't tell me nothin' but filth, and made-up filth to boot! I'm glad I'm far enough along so's I don't have much more of this vale of tears to look forward to. If I was a weaker person, this woulda done me in. But do you think they give a hang about that? What's an old lady to them? As it is, I don't know how I'm gonna stand it."

Looking at her, I could not imagine any grave physical deterioration was in progress. She was the stuff from which Eugene was made, and she would last, grimly and with rancor, for more than the alloted three score and ten. I wondered, standing there, what her husband had been like, but brought myself back to the subject I wanted discussed.

"Eugene told me once that he kept a kind of daybook."

She eyed me with suspicion. "A what?"

"An appointment book . . . sort of."

"Now, Mr. Scott." She relaxed into almost amused disdain. "What would Gene want with an appointment book, for heaven's sake? Who in the world did he have to make an appointment *with*, except me? And what good would that do anyway? He never knew how to be where he was told to be on time in his life."

That was not quite so. Eugene was habitually late, unless forced under the terms of a liaison to be on time, but he always kept a rendezvous, and he always set down the pertinent facts of what he had been up to. His neatness and system were remarkable in one so otherwise fouled up. With early direction and some ambition worthy of his organizational abilities, he might have gone fairly high up some corporate ladder of success. As it was, he expended his limited talents to insure a good issue to his adventures and a safe return to home base.

By now, it was clear to me that she knew nothing of the journal. It was also evident that her confusion was a cover for the fright she had sustained. I could not at first imagine that anything the police might have told or shown her would make much sense to her obtuse and determinedly benighted comprehension, but yet, but yet . . . One never knows about the canny American mom. She lives on multiple levels, even as do the rest of us, and she no doubt has a care to keep her own cohesive view of things in tack as a measure of self-protection—and never more so than when her sons are in jeopardy. If Mrs. Carmichael could be so blind as to the worth of her youngest in contrast to the witnessed merits of the abandoner she prized in his stead, then she was capable of any self-delusion necessary to maintain the comfort of her established beliefs.

"What are you going to do if he shows up here, Mrs. Carmichael," I said, forgetting that she had asked me that same question shortly after I came in.

She smoothed the damp hanky out on her lap, then looked quickly around her as if fearful someone might hear us. Leaning forward and thrusting her lined, saurian face up at me, she said, "I hope he's got sense enough to keep away from here, Mr. Scott," after which she sank back again and stared off out the small curtained window that offered a view of the stucco back of the house that shielded hers from the street. "If I was him," she added, almost as if she were now talking to herself, "I'd put plenty of space between me and San Diego. That's what I'd do."

"But if he didn't do it?"

"Don't make a bit of difference." She waved away my objection. Her voice was soft now, dejected. "After what they're spreadin' around about him, he don't have a chance. Anyway, they don't care who did it. They don't like my boy, and they b'lieve he done it, and they'll say whatever's needed to bring folks around to their way of thinkin', and that's all there is to it."

"They'll have to prove it," I offered.

"Pooh!" She looked away in contempt at such naïve reasoning. "They'll prove what they want to prove, and it won't matter a tinker's damn what you or me or anybody else has to say about it. And the last thing they care about is some old lady who might be dependin' on him for support. Why, even if they *was* to decide he didn't have nothin' to do with this mess—and he didn't—d'you suppose that shipyard's gonna keep him on after this here filth's been spread all over the newspapers and the radio and television? He's already smeared, just as well as if they'd took tar and brush to 'im."

"It might be difficult for him," I conceded, reminded of my own position in the face of mounting publicity. "But Eugene could get another job. Some building contractor would probably hire him again."

"They don't pay what he's gettin' now," she sniffed.

"No." I moved toward the door. "Nothing you have to take when you lose what you've got is as good as what you had."

She agreed that that was so, and got up with great effort from the chair to follow me to the door. With a hand on the knob, she leaned out and spoke to me while I stood on the little roofed stoop.

"Mr. Scott," she said quietly, "who was this doctor fella they said was murdered? Gene never mentioned to me any doctor he knew."

"Eugene has a lot of friends and acquaintances, Mrs. Carmichael." I sought for some way of saying it that would not at the same time stir up the spectre of the revelations the police had spread before her earlier. Something truthful but uncompromising was needed; scruples enter in at the strangest moments. "He was . . . much liked." But no, that was not quite it. I gave up, and said simply, "He knew a lot of people."

"Well," she sighed, "I sure didn't know any of 'em, an' that's a fact. You're about the only one." She tilted her head slightly and peered at me through a conjectural squint. "You're a doctor too, ain't you?"

"Not exactly."

She appeared to relax. "Eugene said you was. Course, Gene fibs a lot. But I'll tell you somethin'; I felt better when he was stayin' with you, Mr. Scott. I never liked him bein' down in that trailer park. That's no kinda livin' for anybody that amounts to anything, d'you think?"

I agreed with her again, and took my leave, promising to call her if I heard anything. As I turned to go, the telephone began to ring. "It's them," she said simply, and disappeared inside the house.

Leaving Lemon Grove, I headed west on Imperial, heavy with a sense of nonaccomplishment. It was as if I knew less now than when Pinkerton had called me the day before, for the little I had learned since then seemed to be fading into some shadowy background in which nothing presented a definite outline, and everything was in a process of merging. A curious stillness was settling around me, the effect of sleep loss and anxiety, no doubt, but I had the impression that all the objects of my inquiries had withdrawn discreetly into the distance, as if waiting. I was almost too tired to drive, and the conflicting questions posed by these comings and goings pursued me like the wail of distant sirens in heavy traffic: questions about questions about questions.

Nearing Encanto, I wove my way up back streets and lanes to the bare hills of Radio Heights. Across two saddles, with the twin eponymous antennae appearing on the left, I reached a crossroads with a settlement of shanty houses on it, then took the steep dive to the bottom of the gulch whence another turnoff led up again on rutted gravel to my house. Thief, knowing the sound of my car before all others, came bounding down the hillside to greet me, all but leaping into the rocking car in his loud-mouthed joy. This I found strange, because I remembered having locked him inside the patio when Ray and I took off the night before.

I pulled onto the level parking area below the house instead of going up the road further to the patio entrance at the front as I usually did in good weather. Tiredness, and some idea of putting the car inside as a way of insuring my staying home this time and getting some sleep may have been the motivation for this. I cannot say, unless the telegnosis Ray and Linda preached as an article of faith to each other was at work on me. Probably it was a subliminal alert that nudged me, seeing Thief outside. I don't know whether I believe in hunches or not, if belief means anything at all. What the living believe, the dead alone may know in truth. So, still alive, I have no way of knowing. I wandered along this train of thought as I came to a stop.

Paying some attention to Thief in return for his welcoming act, I steered him clear of some potted cuttings I was trying to establish and which his thrashing tail was threatening, and crossed the gravel to the garage door. I grasped the handle and tugged. The seldom-used panels ground on unoiled tracks as the door rolled, protesting, up across the cobwebbed ceiling which formed the floor of a sundeck. And there it sat, revealed in a flood of noon light: Eugene's white Dodge sedan.

And up top, my bike was gone.

16

EXPECTING MOMENTLY to be collared by the police, I checked the Dodge only to see that it was empty, then yanked the garage door down and walked up to the front of the house, saw the place where my bike used to be, and went inside, where I lay down on the couch—the same one the two lawmen had sat on the night before—and let my heart quiet down. The living room still smelled of their stale smoke, and on an end table where I had left it lay the inspector's card. "Don't waste any time in getting hold of me," he had said. I felt time wasting while I lay there, held in perilous torpor, listening to the silence all around me, a familiar but now alien quiet in which my option of choice was ticking away.

It was entirely possible that they already knew the car was there and were watching the house from one of the hills, waiting, knowing now I knew, timing the minutes wasted by me before I should have gotten hold of them. Or, if they did not yet know, there was Eugene, liable at any moment to snag himself in the net seined out for him. All that was needed was for them to pick him up somewhere astride my Harley and I would be implicated. Never mind that I had not known his car was left in my garage in exchange for my bike. I knew it now, and still was loathe to rouse myself and call them.

I closed my eyes to ease the throbbing toll of last night's folly and the two fool's errands I had wasted the morning on. I would call the detectives, but not quite yet. I could not trust myself to talk to anybody about anything for a while; in my present state I would have spilled everything I knew.

This new fear had, after a few minutes of hard breathing quieted my heartbeat, a strangely soothing effect, like a vein-injected soporific, and I found myself drowsing and awakening, only to drift off again. Half-dreaming, half-imagining, I watched a parade of phantoms in the buzzing afternoon, faces and names dropping like living creatures out of the warm sunlight streaming in through dust-blemished windows and investing the creaking house with their distinctive and oppressive presence.

Last night's folly. The dregs of it disquieted me more than I wanted to admit, which may have been a portal of wisdom—leading in or leading out, I could not tell. Half-awake, I dwelled on these people's roundabout approach to something most of us came by in less stylized contacts and with easier manners. The distinction of "these people" and "most of us" was blurred for me now. I tried, half asleep, to concentrate. Was it all merely the lavage of sick minds, or truly the exaltation of technique that most who practiced it maintained it was? Or was it the hermetic art a few like Symonds tried with such expenditure of thought and effort to realize—the vestiges of a cult as dark and fraught with meaning as the Attic corn mysteries reeking of butchered pigs and rotted sperm? Was it the mind-blown yearning for the meat of things that Nelson Faulkner discovered in his drawings and lived for in the hampered fastness of uncultured reverie? Or was it, after all, only another special madness, a bully's dream venerealized to a point where the press of bodies and the clash of wills inherent in the act had to be made explicit to achieve any satisfaction, so that, in accord with the law of diminishing returns, it needed ever a further push along the paths of excess to keep it going, until there came that forking of the road where one had either to abandon it totally or drop over the edge into total abandonment? Certainly the psychosexual decathlon in which Eugene competed, aiming always toward something newer, bigger, farther out, had been shoving him, arrangement by arrangement, further along a line that pointed ultimately to his being scared to death. Nothing for as long as I had known him had ever done more for him than raise a few goosepimples on his glabrous skin, but then, there were not many that I had encountered who would acquiesce in even half of what it took to get Gene up and off. Maybe Glenn had finally succeeded in getting him there, and this resultant scurry was evidence of the unhinging that must logically have followed it. In which case where was I who had heretofore refused to make any personal comment, as did these people, on the cruelty of choice made necessary in a world where none is equal in the flesh, or the mind that reflection of the flesh? Inclined as I was to dismiss such faultlines as inconsequential, I had observed with equanimity how lying, fraud, and killing were embraced as forms of self-discovery—of salvation even.

I woke with a start, as if someone had just touched me. I sat up and looked around. The pervasion of Eugene's presence hung about the house like the cloying fragrance of a cut melon lingering in the room where it has been eaten, and where the rind still waits to be thrown out. What could the shallowbrained fugitive from sanity have been thinking of to go leaving his car in my garage and then to steal my bike? The answer had to be that he was not thinking, or that somebody was doing his thinking for him.

I sat up, shivering in the heat of the closed room. Everything which earlier had appeared to me in the light of careless violence and puerile extravagance gone astray was now etched in suspicion and malice. A lacy encrustion of betrayal ran like tattered embroidery through the happenings of this weekend, revealing a gangrenous fringe to wounds unclosed, still festering; wounds hidden and left too long unhealed. Could there be accident in any of this?

Through the length of the small house I made my way to the bedroom I had converted from a sun porch by weather-stripping the glass and caulking the dry rot-ridden seams. It caught the morning and a good part of the afternoon sun and, with its half-glazed roof, gave me the stars at night or the rain splashing overhead in late winter and early spring. The sun, now on its wintering angle, was streaming in, overheating the plants that hung from every casement seam and rafter, making it a kind of greenhouse as well. I unlatched the windows to let in air and then got out of the clothes I had not changed since returning from Crystal Cove. In little make-work trajectories I set about watering the thirsty pots, all the while thinking . . . thinking . . . trying to guess at hidden movements and sorting out ostensible motives, knowing I must make that call, and soon.

Instead, I picked up the phone and dialed L.A. information for the number of Symonds' secretary, Jessica, without any idea of what I would say to her. I needed someone to talk to who had been close to Glenn. There was no answer at her apartment in Santa Monica. She might have been at the clinic, but so might a lot of other people who would have been interested in knowing whom she was talking to, although the law up there had probably already combed through his office with or without her help.

I next dialed Santa Ana information and got the number at Crystal Cove. Yet I hesitated to contact Louise Silver, although I owed her a call. While holding my finger on the cradle button, undecided, I studied the deck and the side of the garage where it emerged from the slope of the yard beneath it, bordered in tall hibiscus. Its familiar shape was now gravid with the secret it contained. I put the telephone aside and went back to watering the plants as if I had no better thing to do than household chores.

Well, I loved this dilapidated house, the sanctuary it had been for me in a long time of trial. The buying of it had taken everything I had, and keeping it cost almost more than I could afford, yet it was mine, and I could do with it as I pleased. Here the sense of alienation I felt in the exile my work and the paucity of positions open to me elsewhere forced upon me let up somewhat, and I could feel secure, wrapped around and protected by the cover of these familiar walls. But now Eugene's invading it with his contaminating trace had destroyed the comforting illusion of peace that used to reign here.

Why?

In neither the private nor professional sectors of my life was I ever wont to indulge in the fashionable despair whose ritual utterance consists of that interrogative *Why?* The sincere "Why this?", the plaintive "Why me?", the churlish "Why him?", or the dreamer's "Why not?" were entirely alien to the way my reasoning was trained. I was looking for what and when and, most important, *How?*

There was nothing more to do for the plants. I went to the

kitchen and threw some things into the blender, then to the worn-out bathroom of falling tiles and crumbling plaster to draw a tub for soaking. The sunflower-headed stalk of an ancient shower attachment had ceased to function long before I bought the house, so bathing was the style here. The calcified pipes, sclerotic victims of the county's alkaline water supply, took nearly half an hour to produce enough soap-congealing liquid to sit in.

Back through the kitchen to the bedroom again, this time with the blendered brunch in hand, I dined, staring across the broad gully to the bare, antennaed hill crisscrossed with rally tracks worn through the dry grass, thinking . . . thinking . . . while I watched a lone rider practice on the slope with a sound like a chain saw.

If, as Chumly reported, Eugene left Stallion Oaks a little before two on Saturday morning, it was reasonable to assume that he put his Dodge in my garage some time before daybreak; the risk of its being seen and remembered by someone in the wake of the news alert following the discovery of Symonds' body was too great for him to have come around the Heights any later than, say, five-thirty in the morning. This being so, there remained the question, among others, of what he had done during the nearly twelve hours of daylight that ensued before I returned on my Harley from talking with Pinkerton at around six in the evening to play host to those investigators from National City. If he were hiding in the house, or in the hills nearby, then why had he not come out after they left?

I thought back. In the house, possibly in the spare room, he could have heard Don's interrogation and my responses. Maybe that scared him off, or at least gave him second thoughts about seeking me out. If that was not the case, then a margin of caution might conceivably have held him back for a while before he felt it safe to show himself after they left. Then too, I did not hang around long; feeding the dog and telephoning Ray, I was probably out before he could make up his mind to appear—*if* he was in the house. My guess was that he was hiding in the garage, maybe even asleep in his car. Hearing Ray and me talking when we returned to pick up my car to go to The Club, he would not have thought it prudent to emerge then either, no matter how much he may have wanted to establish contact with me.

Of two things I was sure: I had shut Thief in the patio when Ray and I left, and my bike was in the upper driveway, unchained. Yet if he were here, then why had he fled?

Eugene and Symonds, Symonds and Owen Keller, Keller and Faulkner, Faulkner and . . . The pattern made no sense, for the circle would not close. The vision I had of what was and had been going on remained opaque, and was clouding the transparence of my own actions. Could, for instance, someone besides myself and the police be after Eugene?

I stretched out on the bed and observed the shadow line of afternoon crawl slowly across my legs, remembering how, ten months before this, while I was napping here just so, Symonds had come around to see me, a week or so after his liaison with Eugene had been cemented—the master couchant, the slave lodged. Throwing something on, and wondering as I walked the length of the house why Thief had not alerted me, I thought it might be Ray. The sight of Glenn, when I opened the door, left me stunned for an instant. His head was backlighted by the sun in that radiant halo effect the almost peroxided albino gloss gave to the brushcut hair, as if the setting had been arranged and the angle posed. I stared at him, trying to cloak surprise and anger as he bent down and petted Thief with a proprietary air, asking me with a smile if I were going to invite him in.

Being confronted by a former lover you've not yet cleaned out of your system is at best embarrassing. Time, which can so quickly cement an affair, will eventually disjoin it, and, left to work at it, will furbish the scar that remains after the hurt is forgotten. I was not long enough away from Symonds, given the years we spent together, for this last to have taken place, but I could fake it, and I did. We played a game like the other games we had for so long tried our skills at, and in celebration of the preliminaries to this rematch (which were the finding and taking of Eugene) he must have felt, talking to me in the sunny front room where only our voices disturbed the Sunday quiet, that his reach had scored sufficiently to risk a confrontation by which he might establish some new rapport. He was certain I would behave myself if he could once get to me.

But what did that mean?

Those who have never been in psychic bondage, addicted to the potency of another, cannot know what I am talking about. Those who have never had to save themselves by breaking such a habit have no idea of what is entailed in calling if off. Try leaving someone when the doing's at its best, even if the living with them is at its worst. You must either have character equal to your feelings, or have no character at all. And afterwards, when the cleavage is done and the disengagement accomplished, there develops a terrible tropismatic exclusivity that passes the limits of time and separation; the nick of the beast you quit remains long after its teeth can no longer fit the imprint they once made, and serves as a pattern against which all newcomers' bites are measured and, usually, rejected as something short of the mark.

And I did behave. We had a very civilized conversation. In the time, almost, that it takes to write it down, I heard how much money he had made in the year just passed, what taxes he would not have to pay on the amount, the newest additions to his ever-swelling list of celebrity patients, and the added distinction he now enjoyed from a faculty appointment at UCLA Medical School.

No one changes very much in sixteen months unless great good fortune or an elemental catastrophe has intervened. The ravages of emotional woe are something else, of course, but these inner afflictions more often than not enhance the lamina of surface appearances, and I wondered if this could have occasioned his remark that I looked the same. I had not, after all, looked very good when he had last seen me. Still, the lean and hungry look has much to commend it to those interested in what it was that brought the bearer to so barren an estate. Whatever Glenn may have gone through, the long abeyance had improved, if that were possible, the sinister incandescence he projected. That knocked-together house I lived in, with its flea-market furnishings and peeling paint, was transformed by his appearance, and I felt myself a stranger in a place he had, by his mere presence, taken possession of. His slightly mocking voice, subtled with accents of tenderness and concern, was laced with the triumph of one who was certain of himself, but careful not to overdo it. What he really thought I could only guess at, and I guessed at the time that he still wanted me, because that was the gist of what he proposed in the roundabout way our conversation took. And, while on my guard against him, I marvelled, studying him, at the maintenance that kept him in such good array. I found myself recapitulating all those assured qualities of voice and manner, and the surface gloss, still polished, that had captivated me that first time so long ago.

And I was struck too by the linearity of the emotions I felt and had been feeling; how like a ladder or a spiral ramp they were, or—better—how much they resembled a meandering trail which I had once followed across an accidental countryside. The passage of time revealed a progression along an unbroken path which I had trod, but my attempts to retrace the course thus covered revealed a prospect at once the same yet different, because the point of view was changed. What I looked forward to before, I was gazing back on now; what was dimly forming was now clearly done, perfected, dead.

It was during those intervening months of separtion that I had made a fundamental error. I saw suddenly that during all that time when I had been rebuilding my existence free, as I imagined, of him, I had as conscientiously undermined my resolve by yearning for the feel of him and grieving over the loss of what I thought of as the best in him, dwelling in unguarded moments on what might have been. In this way what I thought was the building of a defensive wall against the past turned out to be a hidden place of unacknowledged worship. I had come to sanctify my enemy and made an object of impassioned meditation one who formerly had answered to the more manageable stirrings of an ardent and open lust. So that now the physical presence of him, like the appearance of a jinni, or an angel, was unsettling in the extreme. His was a pressure against which I had forgotten how to contend.

Well, Glenn always had the element of sheer push going for him, as well as most of the other attributes the world and the flesh put value on. Yet he could never leave well enough alone; excess was his undoing. Even I, in a way, was one of his excesses. And sitting there, wrapped in the mesmeric folds of his velvet suasion, I suddenly came to my senses hearing a proposal he set forth to soften the blow to what he took to be my dented pride and broken heart. He offered to share Eugene Carmichael with me; to, in effect, use Gene as a means to bring us together again.

Hearing it, I could not keep from laughing. The perceived strength in him that took all my force of mind and character to resist was almost in a single stroke demolished by the obtuse assumption on his part that Eugene was the object that stood in the way of any reconciliation that might be worked out between us. I think he thought my laughter was an expression of relief.

It was, but not as he imagined it. Some grasping faculty in me had collapsed; the captive feeling for him I had ransomed from the first severance drained away that afternoon and was lost in the instant after he proposed his little scheme, and my heart was almost light again.

A pleasant kind of boredom filled the field between us after that. No more longing; no more grief. What I represented to him, how I affected him in this again-seeing, I have no way of knowing aside from some idle thrusts he made from time to time, testing me, I suppose.

It was in this way, with an offhand remark, that he let me know that it was he who had gotten me the curator's job in San Diego. Questioned, he would tell me no more, and I did not push it. His point, I know now, was to let me know that my leaving him was of his choosing as well as mine—not that I did not understand that too, and from the first. It was an old trap I had fallen into: I had left him and he had let me go, therefore he had abandoned me. This was heart's cure meted out to the soul, but I had pondered and thrashed over this through too many sleepless nights after leaving him to want to resume it again. This stale evidence of his impenetrable egoism lessened him in my eyes, but I welcomed the emotional relief that accompanied such a sorry revelation.

Even so, the emotional sewage in which we (or at least I) had wallowed threatened at times after that to rise up and engulf me. How often in the lonely, sun-drenched afternoons of the past summer when he was abroad, had I lain in this quiet, glaucous receptacle of light which was my bedroom and onanized an aching vision of him as he used to be, summoning up again the allokinetic turning of his body to receive me as I touched his ghost. Those cowardly reveries indulged in when there was no chance of reaching him were the closest approach I would allow him in my self-imposed state of reverence for the past. It was a ghost I used to purge the living creature. And the phantom battened on my weakness as the living image could no longer do. Even when others were hard by, handy to use as body and warmth whereby to work off the regret and bitterness I let build up, I would as often as not dispense with them and use instead myself to rid myself of what it was that rode me, riding in my imagination that body and spirit who had, without my understanding it, mastered me so long ago by submitting to me.

17

"SOMETHING'S running over."

The voice came out of the dream. I had fallen asleep, thrashing out these chaffing thoughts, and woke to find Ray standing over me by the bed and Thief upon it, laving my face.

"Get that damned dog out of here," I said, reaching down for my Levis on the floor and drawing them across my body, suddenly the prude that Linda had accused me of being. Ray, perched on the window sill, looked at me with a smirk of insolent familiarity.

"Don't cover up for me, doll," he said. "Where's the water coming from?"

"Christ!" I leapt out of bed and ran through the sopping kitchen. I had drowsed well past the plumbing's capacity. For the first time in my stay there I had a full tub. The bathroom was afloat in an inch of water, but I could not face mopping-up operations in the still sleep-numbed state investing me. Instead, I opted for the washing I had planned on, pulled the plug, lowered the water level enough to avoid further slopping, and sank with a groan of weariness and satisfaction into the tepid soak.

Ray, having put Thief outside, came in and sat down on a metal stool, hooking his feet on the rungs, clear of the shallow puddle in which floated, undulating softly from the stir of his entrance, the bathmat.

"Are you just going to leave this mess?" he said, the fastidious housekeeper clucking at his neighbor's indifference to dust, disorder, and now damp.

"Later," I sighed. "What are you doing down here anyway? I haven't called you yet."

"I know, but I couldn't wait. I saw you come back over an hour ago. I didn't mean to wake you though."

I looked at him closely. "How did you see me?"

"I was outside, watching a sparrow hawk with the binoculars . . ." He motioned vaguely behind him. ". . . on the road."

From the road leading from his house to mine he could not have seen the doors or the inside of the garage. But while I did not doubt that he was bird-watching (it was a hobby of his), I also knew that it was more an excuse than a reason for him to be on the hither side of the hill with lenses at this particular time.

"Something on your mind?" I said.

"Not anything important for me, I guess. Dale Pinkerton called and talked for a while."

Just like that. I began quietly soaping, waiting for him to get it out.

"He said you really didn't do anything down in Chris' basement last night."

"I told you that."

"I know, but he also told me who doctored your drink when you weren't looking."

When I wasn't looking? I almost laughed: I couldn't remember ever looking. I continued soaping myself as well as was possible in that calcimined water. Ray was beginning to squirm with impatience.

"Don't you want to know?"

"Sure. Who?"

"Faulkner."

"That's interesting."

"Dale said he did it so you wouldn't leave there before he did."

"Pinkerton!" I laughed. " Did he have any opinion as to why Nelson didn't want me to leave before he did?"

"No. But do you know what else he said?"

I scrubbed away, wondering if I ought to attempt my hair while I was at it. I reached for the shampoo, which had water softener in it. All this was about to drive Ray up the wall.

"Scotty! This may get the police off your back! Don't you understand?"

"Not exactly. But go on. What were you going to say?"

"Dale thinks Eugene was in on it too."

I submerged for an imperfect rinse, then stood up.

"Ray, what do you know about somebody named Owen Keller?"

"Oh, *that* one."

The name had never come up between us before. I looked a question at him.

"Well, he's Faulkner's friend, but everybody's supposed to be after him. I mean, he's supposed to be divine looking." He fell silent, looking at me with his mouth open. "Scotty, do you think *he's* the one?"

"How did you get all this together, anyway? I hadn't even heard of Owen Keller before this week."

"Oh, I don't know. He's supposed to be into that awful leather thing like all the rest of them, and he's a doper too, so who cares?"

"What do you mean, he's a doper? Who said so? And what's he supposed to be on?"

"I don't know, but I guess it's more than pot or acid."

"How do you know that?"

"Would anybody mention it otherwise?"

It depended on who was talking. What bothered me in hearing it, however, was that I could not imagine Glenn having anything to do with an addict; he disapproved of grass even, preferring absolute clarity of perception and function in everthing he did, and demanding the same of anyone he dealt with.

"Did Dale tell you Keller's an addict?"

"No, but Linda thinks he is."

"Linda!"

"Well, I don't think Linda would say a thing like that if she didn't believe it was true."

"Does she know Owen?"

"No, of course not. But she hears a lot."

What he was telling me smacked of gossipy Chris exchanging tidbits with his sleepovers at breakfast that morning. Hearing now who was the source of Ray's information inclined me to discard it as trash. Which, while it erased an impediment to Owen Keller's connection with Symonds, left me about where I had been before. But I was reminded of something else.

"Did Dale happen to mention anything about the police looking for a drug connection in all this?"

"No, but the news this morning on the telly did. There's nothing definite though; they just said they were looking into all angles."

He glanced with distess at the sopping floor.

"Scotty, this water is going to lift up the linoleum if it sits here like this, and you'll never get the smell out. Let me get a mop started on it, then you can tell me what you're going to do."

"I'm not going to do anything, but you're welcome to mop. There's one in the broom closet, I think."

I pulled the plug, and through the drain's rattle I heard him call from the kitchen.

"Why do you really think they doped you and Lawson?"

"Lawson was probably a mistake. I gave him a drink from the same glass I was using. And I guess Pinkerton wasn't far wrong. When I got out to Stallion Oaks this morning, I found he *and* Faulkner had been there ahead of me."

"That place!" Ray's round face appeared in the bathroom doorway. He had the mop. "You don't suppose Dale had anything to do with Glenn Symonds' getting killed, do you?"

"What do you think?"

"Lord, I don't know! I mean, *cui bono*?"

"Or, in this case, . . . *cui malo*?"

Without comment, he set to work while I dried off. As he backed out the door, making figure eights with the mop in between wringing it out in the toilet, I was warmed by a renewed fondness for Ray and his genuine concern, which was the other side of the worn coin of his curiosity. Absorbed for the moment in his work, he did not catch my smile as I watched him. Then suddenly the warm feelings faded when I saw the black tongue of an evaginated sock flick from a corner into the the center of the floor by the lick of the mop. Bending over to retrieve it, he shook his head sadly at what he assumed to be further evidence of my sloppy habits which were, in truth, a poor comparison to his obsessive neatness.

The sock, however, was not one of mine.

"Let me have that!"

I snapped it out of his finicky grasp.

"You can't wear it now!" he protested. "It's wet."

"Never mind. There's the kitchen yet to do."

"For heaven's sake, Scotty! I didn't come over here to clean house for you, you know."

"Right!" I stepped out of the tub and wrapped the towel around me, still holding the sock. "I'll finish up here and see you over at your house as soon as I can. You want to have dinner together?"

He leaned on the mop handle, aware that something had happened to divert my attention into channels he could not follow. He protested, but I hustled him out with promises of seeing him later.

"You're just going to get into more trouble," he said, then turned and left.

As soon as I heard the front door close I turned the sock outside-out and examined it. Living with someone for any length of time, you come to distinguish their clothing from your own, even (or especially) if they happen to wear the same size. What I held in my hand was Eugene's.

I went through the house from front to back. A second look through the spare bedroom turned up a wad of dirty clothes crammed in a corner of the closet: a T-shirt, Levis, the other sock, and a denim shirt. The shirt and Levis had a few rusty smears on them that could have been blood. Everything reeked of soured sweat.

After throwing the stuff back into the closet, I sat down on the bed and carefully reviewed my movements in another attempt to mesh them with Eugene's. He must have returned to San Diego with Symonds sometime between six and seven o'clock Friday evening if, as he said, he was somewhere in Hollywood when he called me at four-thirty. He was at Stallion Oaks around midnight, and was gone from there before two. That left nearly thirty-six hours to the present time unaccounted for. To him, waiting, it must have been an age.

When I caught that hurried glimpse of Symonds hanging upside down in Eugene's trailer with someone holding him, talking to him, I never for a moment considered that that someone might be

Eugene—no more than did the thought of Symonds being dead, or dying, cross my mind. Now, although I still could not buy Eugene's killing Symonds, the idea of complicity began to cloud earlier convictions on that point. It could, after all, have been Eugene I saw.

Then I recalled the telephone ringing on my desk at the museum when I left to go see Pinkerton. That also could have been Eugene calling from my house. For, where else was he to go? Who else could he have turned to in his fear and confusion, if he knew Symonds was dead, but me? Once more the notion of trust betrayed took hold of me. At the same time a rising exasperation at the foolish panic he had succumbed to both in hiding and, now, in running away added anger to the guilt I felt. He had been in or around the house waiting for me to show up while I was stringing myself out all over the county in an attempt not to come back here where I belonged, and where he might reasonably have expected me to be.

I searched through the Levis and shirt, but there was nothing in them. Something . . . someone . . . some missing element in all this kept nudging me. I got up and sloshed back through the sodden kitchen to my own bedroom and dialed Los Angeles information. The operator had a listing for a Keller with the initial O in South Gate and another on Silver Lake Boulevard. I took down the latter. Then, after pulling on a pair of trunks, I went outside to my car and fished out an L.A. street map from the glove compartment and made sure of what I already knew: Silver Lake was on the western edge of Hollywood. I went back up to the house, to the phone again, and dialed the number the operator had given me. There was no answer.

Someone . . . something . . . I knew now what I had to do. I called Ray and said I could not make it for dinner. He protested, but I told him I wanted to be alone and was going out and get drunk. It was not like me, but it calmed him somewhat. He offered to come and fetch me back home if I found myself unable to drive later on. That done, I swallowed another dexadrine and dialed the number on the card the detective had left. I reached him at his home. I told him what was in my garage. After hanging up, I went down and reparked the car in front of the house, then I shaved, got dressed, and went out on the deck with the dog as they came barreling up the gravel road from Weaver.

18

TWO HOURS later I was headed north on Highway 99 with the dressing-down I had taken from the law still ringing in my ears. I had gotten off easy, to hear that detective Don, tell it. And he was easier on me than I expected him to be. While uniforms and plainclothesmen from two cities and the county turned my house upside down, he and his sidekick Pete and I sat out front in their car and talked. He listened to what I had to tell them about my movements since I had last seen him at the morgue, and what I had come home to here, then he began to pick at it. I think it rankled in their minds to imagine Eugene might have been in the house when they were there questioning me. We went back over it a number of times, and I was hard put to convince them that I did not remain inside the house after they left, and also, that Eugene might have been outside somewhere, afraid to come near the place until they were all gone.

I suggested that they check with Chuck Lawson and Ray about my timing, and, even as I mentioned them, I saw out the car window where, among the scattered neighbors and stray rubberneckers who had been drawn out by the appearance of so many marked and unmarked patrol cars, Ray and Linda were standing underneath one of the pines by the road. As if it were a sporting event and I a contestant, they waved to me. I was holding Thief by his leash through the open front door; he was getting hard to control with all the strangers hanging around. I asked Don to let me call Ray over to take him off my hands. They both turned around to look, Don beside me in the front, his partner, Pete, in the back. Don assented, and I leaned out, calling Ray. Linda started to follow.

"No, Ray," I said, "just you."

Both of them had idiotic smiles of embarrassment plastered on their faces. Ray bent down and peered in, bestowing an ingratiating grin of trepid curiosity on each of us. "Is everything all right?" he said.

"Would you take care of Thief for me?"

"You want me to keep him?" Ray was playing the simpleton.

"Just take him," I said. "He's a little upset with all these people around here. Keep him for the time being."

He looked bewildered. I closed the door and turned back to Don, who asked me for Ray's last name. Then he took up where he had left off. He wanted to know why I had not telephoned him when I learned about Eugene being at Stallion Oaks.

"It must've been three o'clock in the morning," I said. "And I was sick."

"How do you mean, sick?"

I told him it might have been from staying up that late. He and the one in the back seat exchanged looks, then he told me that withholding information could make me an accessory.

"I wasn't withholding anything," I protested. "By the time I got out there the sheriff, you, and Pinkerton had all got there before me. And anyway, I told you just now, didn't I?"

They ignored the mention of Pinkerton and took up the matter of my bike. I explained to them again that I had not noticed it was gone until after Ray left, and that I had gone down to see if it was in the garage, which, I said, was when I discovered the Dodge. They found no fault with this, other than wanting to know why I thought the bike might be in the garage if I had left it by the patio.

"I told you, I haven't been feeling well, and I wasn't thinking very clearly. I wasn't sure but what I had put the thing in the garage and forgotten I did it."

"Do you usually forget things like that?"

"Not usually, no."

Twice some plainclothesmen came out of the house, once with Eugene's clothes, and talked to Don & Co. The two of them got out of the car to confer so that I could not hear. More people were showing up, standing around gawking and talking. A reporter arrived, and a photographer. They were down the slope in front of the garage where a towtruck had dragged the white Dodge out on the parking gravel. They had found something inside it, but nothing was told me. One by one the cars pulled out and left, and the truck towed Eugene's car away to be impounded. Finally, there was nobody but my two questioners and myself left with the diminished crowd of people crowding closer now, trampling my grass. Linda and Ray were nowhere to be seen.

"Do you want to go back into the house again?" Don said, for which I was grateful. Inside, he wrapped it up with an admonition and a warning. I was not under arrest, but I ought to be. They considered Eugene dangerous and my efforts in his behalf misguided. They wanted me to stop meddling. I protested.

"And," Don went on, brushing the objection aside, "whatever you hear, or think you hear, let me know—right away. And you'd better stick close to home too. If Carmichael tried to get in touch with you once, then he's liable to try again."

"Does that mean I can't leave the house?"

"Where do you want to go?"

"I don't know." I looked around me. They had ransacked the place searching for further evidence of Eugene. "I'm going to need a drink before I start putting this house together again. And I don't like drinking alone."

They, of course, preferred me glued to the phone, but they could not force me to stay there. "Just remember," Don said, turning at the door, "if you hear or find out anything . . . *anything* . . . call Pete or me."

As they went out I asked them if they could move those people off my front yard. They said they would do what they could, but it was a half hour after that before the last of them wandered away and I could get out of there.

Crystal Cove is about seventy-five miles north of San Diego. It lies between Laguna and Coronado del Mar on a piece of the old Irvine tract, most of which is now taken up by suburban Santa Ana, industrial parks, and another campus of that sprawling boondoggle, the University of California. The Silvers' house was next to the last one down the gravel road along which the properties were generously spaced at just about shouting distance from each other. It was basically a frame summer cottage of '20s or '30s vintage which sat on the shallow shelf of sandstone facing the sea. Through the years additional rooms and another story complete with decks and balconies had been added onto it, and now, besides the luxury of space, there was a muddled charm in the heterogeneity of its design—or lack of it. A fenestrated wall of brick and adobe plugged with geraniums, succulents, and shrimp plants screened the first story from the road and surrounded it on three sides. Jutting out above the beach was a broad deck supported by pilings through which the surf combed at high tide.

Morris Silver's wife, Louise, owned—or rather, leased—the place. The lease had been in her family for over forty years. All the places at the cove were leased; it was a way of maintaining a restrictive covenant without falling into the tangle of civil suits ("Impeach our Chief Justice!"). Silver, who, while very rich, had not always been so, thought the place was a joke, and did not come there very often. He felt uncomfortable, I think, although the families in the Cove had only their leased interests in common and saw as little of each other as possible. Silver owned an impressive spread on the Rancho Santa Fe tract between there and San Diego where they lived most of the year. Louise spent part of the summer and fall at the Cove with the two children, and Morris and friends came up to visit on weekends. This was to have been the last one of the season.

I pulled up before a double iron gate that opened onto a flagstoned patio. Louise was standing there as if expecting me. Their dog, a black Lab named Molly, had roused her from a portable chaise, barking at my approach. I got out of the car as she came forward, smiling and extending a hand.

"Scott!" she said. "What a pleasant surprise! I'd given up on you."

I apologized for not calling. "A lot of things have been happening. I thought I'd drop by and pick up my trunks. I left them here yesterday."

"I'll have Gertrude look. She's cleaned your room. Would you like a drink?"

"If you're having one, Louise, thanks."

We settled for gin and tonic, and she went to check with the maid. In loose cotton slacks, with a matching jacket fastened across her shoulders, she did not look like a mother. A handsome woman past her prime, but gliding on the updrafts of luxury and the attention she paid to the upkeep of herself, she was, I guessed, in her early fifties. She returned shortly with the drinks, shaking her head so that the last of the afternoon sun glinted on the short cut scallops that framed her face in a blond cap. There was some gray there, but it looked as if her hairdresser might have streaked it in for effect. Her fine skin was without wrinkles; the skin of her hands smooth and well cared for, the fingers long, manicured, and adorned with a few stones in fine settings.

"Gertrude hasn't seen them. There's a pair of green ones by the shower under the deck." She looked amused. "But I seem to remember you wearing something more colorful. And these are a bit large."

She was right; I had been wearing an almost-bikini of red and blue nylon, something I had picked up in a men's shop in L.A. earlier in the summer. She and Morris had remarked on it Saturday. "Maybe Molly got them," I said.

She motioned me to sit down on one of the cushioned, iron garden chairs while she half reclined on the chaise, setting her drink untouched on the glass-topped coffee table between us. "Morris and the boys are off somewhere down the beach with a friend who came in last night from New York. They ought to be back soon. Will you stay for dinner?"

"Thank you, Louise, but I can't. I've got to get back to San Diego. I thought I'd check about the trunks, but I mainly wanted to thank you for having me up Friday, and explain my not calling."

"We looked forward to it, Scott—the children especially. I had hoped to have you here for the whole weekend." Then, without changing her voice or modulating her smile in the slightest, she reached for her drink and said, "The police came by this morning to talk to us . . . from National City."

"I expected they would." I set my drink down and watched her slip a cigarette out of the pack on the table. I picked up the lighter and held the flame out to her. "Do you mind if I have one?"

"Please. I didn't know you smoked."

I shook my head while I lit it, then picked up my glass, and said, "They talked to me too."

She nodded and lay back with one arm behind her head, looking at me, her clear blue eyes inquisitive, patient. There was the hint of sympathy in her unhurried pose. I went on.

"I meant to call Morris and alert him to the possibility they might be checking with you . . . about me, I mean. But I've been, well . . . preoccupied since I left here."

She nodded, but offered no comment.

"It was something of a shock—what happened."

"You were a friend of Dr. Symonds."

It was not a question. I nodded, and looked down at the glass frosting undrunk in my hand, wondering what she might be intending, and worried about my eyes betraying the unease in which I found myself. I heard her sigh, and looked up.

"It's horrible, something like that happening. I told them you were here all night with us." She waited. I did not move. "I think they were on a routine check, that was all. Was he a very close friend, Scott?"

"I knew him pretty well some time ago, Louise. I was still in school then."

"He was very successful."

"Yes, very. Tops in the field."

I suddenly wanted very much to get away from there. The stop-and-go quality of our conversation boded no good, and the commonplaces we were uttering indicated she was fishing in troubled waters. The question in my mind that had been the reason for my stopping by was answered, and there was only one other thing I had to touch on. I toyed with the idea of mentioning Eugene's car and the police again, but decided not to.

"What does Morris think of this?" I said.

"About the police?" She looked surprised.

"The police . . . the, what happened."

"I don't know that Morris is upset, if that's what you mean, Scott. He likes you, you know. And the fact that you knew somebody who was murdered wouldn't make any difference, I'm sure." She laughed suddenly, and set her drink on the table. "He never tires of telling that story about the sheriff and your salary."

It served to break the tension for a moment. The story she referred to had grown into the proportions of a legend around the museum, and there were some, among whom might have been Morris Silver, who believed I had thought up the scheme as a way of getting a raise the board had twice refused me on the grounds of there being insufficient funds. Their salary scale was way below par as it was, but my own pay was stuck down where it had been when I was first hired on a provisional basis pending the completion of my thesis. In less than a year there the thesis was finished and submitted, and the doctorate was mine, but the board, made up of businessmen and retired Warbucks, kept stalling on my request, and Silver did not appear to like me enough to build a fire under his cronies who had appointed him to manage the affairs of the institution. I fulfilled my part of the bargain, and in anticipation of higher earnings had racked up an impressive total of outstanding bills for clothes, household furnishings, and money borrowed on the car for some dental work that needed doing after years of graduate and undergraduate neglect. It got so bad I finally had to call a halt to most of the payments except for the house, and had hunkered down in the face of increasingly strident dunnings from the other creditors. Car, house, and food were essential, and what I made was more than eaten up by these. I tried token payments—interest only—but this only made them howl louder for payment in full. One afternoon after a staff meeting, a sheriff's deputy appeared with a writ of attachment on my salary. The confusion this caused in the director's office was such that the flustered secretary ushered the peace officer directly in to see Silver who was attending to business that day instead of walking the golf links with one of his banking and yachting buddies. I was called into his office shortly afterward to face a bored deputy and an indignant boss. Asked to explain this state of affairs, I recalled to him my reasonable expectations and recited a short-form résumé of what it took to keep me alive, clothed, housed, and transported to and from work—barely; then put it to him that someone with the delegated responsibilities of a curator of invertebrate zoology in even so modest an institution as the one he headed could hardly subsist on the salary a junior lab assistant might expect to earn in any other place worth naming. It took him back a notch hearing it tolled out to him like this. He fumed for a few minutes, stuttering about how he had never before considered that a person of respectable background and splendid training might be working for less than he needed to live on. It embarrassed him mightily to have the deputy listening to my bill of complaints. He wrote out a check on the spot for the amount needed to get the county off my back, and told me the salary increase would be along in my next paycheck, adding that he would try to talk the board into making it retroactive to my award of the Ph.D. This last he did not get done, and probably did not push too hard, having thrown it out as window-dressing for the stolid witness to our exchange. After that, however, he warmed considerably to me, and took an increased interest in me and my department to the extent of approving minor expenses for badly needed equipment and part time help for cataloguing. The office and most of the staff were scandalized and for a long time vaguely embarrassed by the incident. The mentality of San Diego wage earners being what it was, the idea of a foreclosure would have shamed most of them into the woodwork and out of sight forever. They took it amiss, I suppose, that I did not appear undone by the experience. I was thereafter regarded as some kind of crank. Intelligent perhaps, even a good curator, and certainly astute, but odd—possibly a crank.

"Scott." Her voice was serious again. "Tad told me you weren't in your room last night after we all went to bed."

Tad was the younger of the two children. He was six, and his sister was seven.

"I couldn't sleep," I said.

"He said he woke up too and couldn't get back to sleep, so he decided to go in and see you. He shouldn't have done that, but he likes you so much." She smiled; it was like an appeal.

Tad and I had got along fine. I played with him and his older sister during a large part of the short time I was up there, teaching them to body surf and arbitrating the jealous fights they got into. "I went out for a swim," I said. "It must have been about one or two o'clock. I don't know. Probably that's where I left my trunks. I showered off under the deck so as not to wake the house."

"Tad went looking for you. He said your motorcycle was gone too."

I told her I had put it around the side of the house and covered it while I was out. "I was afraid of the salt air getting at it."

She nodded, her gaze noncommittal. I knew she did not believe me. On Saturday morning the bike was standing where it had been the night before, just inside the patio enclosure. She would have noticed it when she came outside after she got up. I waited, studying the remaining chips of ice in my glass. She took a deep breath, as if she had decided on some course of action and was about to put it into effect. She reached for her drink again and said, quite softly, "I didn't mention any of this to the two policemen who were here."

"That was thoughtful of you, Louise. And thank you. But it probably doesn't matter one way or another."

"Scott . . ." I could hear it coming from a long way off. There was no way she was not going to touch on this. ". . . they said that you were Glenn's . . . that you lived with this man for several years. With Dr. Symonds, that is. And that you had seen him often recently. Is that true?"

This all came out so fast I half-believed I had not heard what I manifestly did hear: that she had said "Glenn." I stalled, trying to think of some way of diminishing the effect of what the law had laid on her in the light of what I might say in response to her question. I had not been mistaken; she had said "Glenn," for sure. I told her it was more or less true.

"When I was in graduate school in Berkeley I used to come down to L.A. a lot, and I stayed at his place. We knew each other that well, I guess. Except when I was in the field, I spent most of my vacations there. We took some trips together—Mexico and abroad. He had friends in Tangier, France, the Middle East. . . ."

She listened, the cool, almost almond slant of her eyes lending an impassive cast to her expression of patience and discernment. But behind the careful, almost imperceptible makeup, the shadings of disappointment, like cloud shadows, darkened for a moment the surface of control. I felt as if she had me squirming on a label, ready to affix the pin, but unsure yet if it were the right one. I ran on.

"I haven't seen much of him since I joined the museum. He was down here for a medical convention and called me up. I've also seen him now and then over the past year, I guess."

She smiled, as if to reassure me. "Morris' only concern is to keep the museum's name out of it, Scott. Otherwise, of course, it's none of our business."

"Was Morris upset?"

"No, of course not! Don't misunderstand me, Scott. He has a great admiration for you, and there's no reason why you shouldn't see whom you please." She was losing some of that perfect calm she had displayed up to now, aware, I suppose, that she was doing variations on an embarrassing theme. "People have their own lives to lead, and we have no reason to expect . . . well, you know what I'm trying to say."

I nodded, depressed by the platitudinous turn her remarks were taking. She hurried on with it.

"The only thing Morris ever worries about is adverse publicity whenever anyone at the museum gets his name in the paper. And," she hastened to add, "there's no reason to suppose you should be mixed up in this horrible business."

"I understand, Louise." I put my drink down and stood up. "But under the circumstances I think I ought to wait and talk to Morris at the museum; not up here. And I do have to be getting along. It'll be dark pretty soon; I'd like to get back to San Diego early enough to get some sleep. I haven't had very much in the past few days."

"Yes, of course." She rose too. "I'll go see if Gertrude has turned up your trunks."

"Oh, don't bother," I said, tired of the pretense. "I've got another pair, and if they were out on the beach, the tide probably took them. That, or Molly buried them."

We both smiled. We both knew—or thought we knew. I wondered what she would say to Morris, and what they had said to each other, and to the police, about me. From the sound of it, my job was safe as long as my name did not hit the papers. Walking in silence to the gate, I stole a glance at her face, looking for the fine, telltale lines that would indicate that she had been under the knife, lines hard to find if done by an expert and the skin, like hers, was the kind that lent itself to such adjustments. In the dusk I could not tell, but it would have explained a lot if they had been there. She caught me looking at her, and said, "Scott, do you understand anything of what it was all about?"

"How do you mean?" I walked around to the driver's side of the car.

"Oh, I don't know." In the gloom I could not see, but she sounded as if her forehead were creased as she spoke. "The newspaper and the radio are so vague, and the police merely wanted to know if we knew you . . . if you were here with us last night. I get the impression that it wasn't anything to do with robbery, or the usual sort of . . ."

"No, it was unusual, Louise." I got into the car and closed the door, the leaned over to talk to her from the opposite window. "That's probably why they're so vague. It'll take a while to put it all together, from the looks of it."

"Yes." She turned her head, and it caught the late, red sunlight in a way that silhouetted her profile for an instant. There was something unbending there; a proud, almost masculine line to the nose and chin. "It sounds as if . . ." She shook her head and smiled. ". . . I don't know."

For a fact, she did not know what she was talking about, but I could have told her what she meant.

19

DRIVING ALONG the gravel road that skirted the cliff on the way back to the highway, I looked out at the ocean; a frayed, gray curtain of rain had lowered far off nearly to the horizon where a fading orange slit announced the day's final scene. It would be fully dark by the time I reached L.A., and dawn before I was back in San Diego.

It was apparent to me that Louise Silver knew I was not in my room on Saturday morning after the rest of the house had gone to bed. She must also have realized, if she gave it any thought after I left, that I could not have hauled better than a quarter-ton of motorcycle over an inclined patch of grass and sand to stash it beside the house in a more exposed position than it was in in the patio and then have dragged the thing back up again early in the morning before anyone was up to see me do it. A Harley handles smooth as butter when under way in a long run on dry pavement, but resists with mulish and often embarrassing obstinacy efforts to manhandle it cold over uneven terrain. A slope can take it away from you no matter how strong you are, and drag you down with it like a falling horse. It had almost been more than I could manage to wheel that monster Hog along the level roadbed to the turn that cut through the cliffs to join up with Highway 1, a distance of perhaps five or six hundred yards. The damned thing did fall on me once, and nearly brought me down twice again before I got a safe enough distance between me and the house to where I calculated I might stomp on it without waking all of Santa Ana County. As it was, some of the other residents cottaged along the strand must have heard me, because their lights came on. I counted on the social isolation enjoyed by the Silvers to guarantee no comment about it from their neighbors. But now two, and possibly three, people knew that I was gone from Crystal Cove early Saturday morning. None of these had yet talked to the police about it—as far as I could tell.

These and allied thoughts preoccupied me as I made my way north along the badly chopped-up freeway. The going smoothed out at the Long Beach turnoffs, but it was night when I made the Santa Monica junction. I kept to the outside lanes then, taking my time as the sparse lights of the Baldwin Hills swung slowly around on my left and those of Hollywood glittered far off to the right.

Louise catching herself over the name "Glenn" kept tugging at me. I was not even sure it was a slip on her part. She may in some way have been testing me. But whatever else it might mean, it certainly meant she knew Symonds. Ordinarily it would not have been very surprising for me to learn this. She was rich, good looking, and well into that period of life when beautiful women who are neither performers nor professionals (both of whom come earlier to this sort of thing) take to one or a number of rejuvenation schemes in an effort to stave off the inevitable ruin that is harder for them to endure than for most because they had it so good to begin with. Such was the meat that Symonds fed on, and she would give herself to the best. I would have been willing at that moment to bet my six wheels that her name was in his files in that pastel-shaded complex in Beverly Hills where he performed his miracles of integumental uplift. It would at least explain the teasing claim he had made of having gotten me my job at the museum.

Still . . . "Glenn." Doctor's patients who call them by their first names know them either socially or from having gone to bed with them. Bed was out, so she had known him socially as Louise Silver, or by whatever name her first marriage conferred on her.

Thinking which put me in mind again to call Jessica, who might know something about her. By now, I reasoned, Jessica ought to be home with a drink in hand, unwinding from the strain she must have been under these two days and fortifying herself for what lay ahead on Monday when the tax people would come to take over her books. Her telephone had probably been ringing off the wall. The permanent cancellation of Symonds' crowded surgical schedule would send tremors of dismay into unstable pockets of vanity throughout the southland. His many clients would find themselves shunted to other scalpeled luminaries for the peeling, lifting, smoothing, chiseling, and implantations they had to have. Most of them would wonder until their own dying day if Dr. Glenn Symonds might not have given them a finer job than the one they got had he lived to practice on them.

I had only met Glenn's secretary once, and knew her mostly from mini-conversations of a salutational nature over the telephone, conversations which took place on the few occasions when I called the clinic to reach him. She knew her boss (or at least understood him) better than anybody other than myself—not that either of us really understood him, considering the amount of knowledge and insight our long associations with him might have been expected to yield.

This rambling train of thought took another turning, and spurred in me a desire to make a detour past Symonds' house in Coldwater Canyon. It was an urge as compelling and as feverish as a voyeur's need to roam and peer. I had, over the past twenty-four hours, grown so accustomed to giving in to the slightest notion that followed in the wake of each new unexplainable stimulus that it was almost a normal part of my behavior, and I hardly gave it a second thought beyond reflecting that I at least ought to approach the place with a certain amount of caution in case the Beverly Hills police had it staked out—for what reason I had no idea. But then, I was so much in the dark as to what was going on that this minimal amount of prudence was no more than residual good sense put in the service of improvident compulsion.

I took Overland turnoff and drove north to Santa Monica Boulevard and the Mormon Temple, all lit up and topped by a moronic angel that looked like the bonnet ornament of a Rolls-Royce, then turned east to Beverly Drive. Symonds' house was about a quarter of the way up the canyon and halfway up the ridge behind Franklin Lake. It was typical of a lot of the spreads up there: roomy and expensive; a four-bedroom, split-level, Mediterranean-style, frame-and-stone house set in the hillside about twenty yards from the street and screened from it by oleanders. It was built to an L-shape that embraced a swimming pool in the back, and a covered walkway overgrown with purple bougainvillea joined it to a double garage. A rail fence and brick gateposts kept the dog in and strangers out. Although ample, it was no more than average for the area. Symonds had not been given to public show, if one discounted his leather moments and a taste for expensive wheels. Polished but understated assurance was more his style. Most of his money went into investments: real estate, bonds, safe stock, and Swiss and Lebanese bank accounts. He bought the place I was looking at from a television producer a few years before we split up, and I knew it well from having lived in it during that last depressing period of our commitment to each other.

I parked under an overhanging oak several car lengths from the gate and doused the lights, puzzled, now that I was there, as to why I had made the fool excursion in the first place. The lamp over the gate was out, but some of the windows of the house were lit, probably on timers. Symonds had a Mexican houseboy, a former male nurse who had worked at the clinic for a while, but the authorities would have shooed him out if they had sealed the house, which was likely. Carlos was devoted to his master, to hear his master tell it, but I do not think he played his master's games, or even kept score for him. Symonds no longer did that kind of thing at home; he had another place laid out for that. Anyway, Carlos did not know me, and I had no intention of getting to know him now, assuming he might still be there in the house.

One I did know though, and who came to mind as I was looking at the house, was Fagin, Thief's litter mate. I got out of the car and approached the driveway gate. There was no bark, no growl, no whine of greeting from behind the fences. Fagin knew me well, and Glenn never let him in the house, which meant that he was gone. This confirmed, for the moment, my guess that the place had been sealed. I walked back to the car.

The house, the houseboy, and the Rolls-Royce Glenn had twice come down to visit me in were all a far cry from the the basement in Topanga Canyon. He was nine years my senior, and he already had a solid practice when we first met, but he had come a long way since then through country verdant with money and reputation, each nourishing the other, while I was still barely started. I can remember being asked, in substance, on more than one occasion when Glenn and I were out together, how I managed to hold onto a beautiful thing like that? It was not the underlying and unflattering supposition of my unworthiness in being paired with the likes of him that irritated me. Social subtleties run to other modes in the intense, yet jaded, undercurrents of speculation encountered in West Coast parties tinged, as they are, with ambiguous sexual connotations. What they could not know was that I had asked myself that same question too often to be surprised or offended or anything but bored at having it posed to me by yet another, for whatever reason. The answer I could have given, but more often than not did not, was that it was a simple matter of hard work, patience, and a great waste of time. The real question, one no one ever asked, was "Was it worth it?" I eventually answered that question, and acted on the answer I came up with.

But now, here I was, come back to the scene of the unindictable crimes we had practiced on each other. It should never have come about but when I completed graduate work in Berkeley and found myself with no job to accept, a thesis still to write, and nowhere to go, I came down here to live with him. He insisted, and I needed no great amount of persuading. Spending parts of summers, a week or two of vacation, and most weekends with someone you love is one thing; settling in with them is another. The experience quickly proved unsettling in the extreme; a crashing demonstration of the self-evident: We were not meant for each other. Or, at least, we were not meant to live together. The sudden, intense throes of antipathy that came and went over trifles and were quickly forgotten in those short periods we spent together before I moved down had gone unheeded, and after I was settled in we neither of us knew how to contend with the situation but by making it worse.

It was evident in a few months of clashing tempers and mounting rancor that we were trapped like cats in a bag, and the factor that exacerbated it was the realization that we were no different from so many others we had seen this happen to. Unable either to let go or to take loving hold again, we flayed each other daily. The cause of it was in our heads, so, as is the way in situations like this, we sought release through the body: our own bodies at first, then those of others. Sex was a fix, a temporary relief from the confining horror of boredom and the pettiness spawned by the inescapable familiarity of this new arrangement. The reentry phase of an erotic addiction found us uncomforted in our disappointment and confirmed in mutual distrust until we were soon unfit to address a civil word one to the other.

There resulted from this enforced proximity a delumination of Glenn Symonds in my eyes from the far heights of the dulia I used to render him to a sullen regard darkened by contempt, which was nothing but the inverted scorn I felt for myself, trapped as I was. This deepening resentment began to lay waste to every aspect of him and his works in my eyes. Where before I had felt the presence of a magical surrogate of the superessential, one who had come into my keeping and into whose service I had been drawn, a kind of sacred king whom I alone was able to assault with a tenderness and familiarity denied to others, I now felt only increasing weariness and hostility, and began to take a new and resentful view of what he was and did. And gradually I came to regard that other life of his in a different and threatening light. And worse still, the very excellence of his well-kept body (what he referred to as his physical integrity, much as a lecturer in art history might speak of a structure's architectural unity) and accipitral good looks, which had for so long remained a basic attraction when quarrels and temporary fallings out threatened the mutual fascination which kept us tied to each other, became, in our close living together, yet another source of aggravation. It was as if he were parading them simply to annoy me with the thought of how much others valued what I had in hand and was even then toying with the idea of letting go of.

He harbored, I am sure, no more charitable feelings toward me at that time. Jealousy, hatred, esteem, goodness, evil, cruelty, madness, courage, submission . . . Trying to apply Occam's Razor to a relationship that made no sense anymore, if it ever had, was like cutting linoleum with a paring knife. I gave it up at an opportune moment when the job from San Diego was offered me like a *deus ex machina*, and one bright fall afternoon Symonds came home to find I had cleared out that morning.

But then I remembered, during the last disruptive months of our living together, an attendant phenomenon began to intrude on the sexual action at its height: a finely wrought incongruity of ritual utterance. More and more, as it became evident that we were not going to make it for much longer, a new, almost unbecoming voice was heard from behind that facade of unshakable determination and potency, a voice crying to be spared. On the bed, or pad, or floor, or ground, wherever, beneath me, he would ask over and over in a kind of singsong, the query of a sick child or an old man, plaintive, trancelike: "Will you miss me when I die?"

And, "Yes," I would answer him, almost offhandedly, as one importuned while engrossed in work requiring all his attention. Yet my own calculus of weaknesses and strengths tuned in, alert to this new voice in the conjugation of our familiar and still satisfying act.

"Will you remember me?"

The wording varied, often spinning off into passionate, eulogistic, ass-clutching gibberish, but the vapid substance of the query remained the same. And always, never pausing in my labors, I would answer yes to what he asked, giving brutal assent to a question which indicated some protest, or at least an allaying counter-question, was in order. In baldly assenting to the formal commitment demanded of me, I denied his hidden plea and left him the prey to what it was that was bothering him. It was at once hurtful and delicious—to both of us. The unfamiliar but erotic rhetoric was also a clue to something I did not, at the time, understand: a dark self-doubt that was eating away inside him. Not only did I fail to grasp the significance of this new departure in his behavior, I did not care enough even to acknowledge that I noticed it beyond the curt, dismissive "yes" I offered in reply.

So, it was with these nocuous expostulations and replies that we made it in those final hate-flared days and nights. Our twin saltations spanned only momentarily the betrayal that always remained, when it was over, to cloud the glimmer of affections that never proved strong enough to last past the borders of reestablished quiet, separating us like rooms whose doors are closed again. Each retained the bare basis of attraction that was, in the final analysis, insufficient ransom for the other, and this last coded message of intimate confession on his part, and the indifferent acceptance of it on mine, served only to tear through the weakened seam of loving to produce one more unpatchable rent, one which, had I bothered to think about it afterward, would have revealed to me the putrilage eating away at his personality, but which, with traitorous good sense, I wanted no part of, and from which I was fleeing for my life.

For in leaving, I left him to his rot and sought to reclaim my own peace of mind in anchoritic solitude in the wastes of San Diego. I opted for lysis over the ever-recurrent crises our nearness to each other spawned, and exchanged the fraudulent struggle of living with a lover I no longer loved for the more rewarding exercise of earning the peon wages paid me by a second-rate institution where, for the sake of sanity and the future, I tried to do a first-rate job.

There is a sublime, a sensuous feeling of personal transcendence that accompanies an act of abandonment. From the vantage point of distance, the wounded but still swaggering heart surveys the imagined fearfulness and grief left in the dark behind it like landmines set in sequence to go off whenever the bereaved survivor will have traced his way along the paths first of knowing, and then of feeling, tearing himself to pieces as he goes. My act, I knew now, though wise, was informed by vengeance, for, viewed in a different light, leaving as I did was a counterfeit of suicide, as sleep is said to mimic death. And the terrible thing about such deception is its indefiniteness, for in the back of the mind lies the cunning possibility that you can always go back. And that was exactly what I did in Symonds' eyes—even though he had to come and get me, as it were.

The means by which I left him, which I had regarded as ironical, since it came about, as I believed, through the friend of a friend of his, now established itself as a double irony in that that friend of his friend might well have been Mrs. Silver, which indicated the possibility that Glenn had been as eager to be rid of me as I was to get quit of him. And this intermediary "friend" was no more than a sop to what he may have imagined would be my wounded sensibilities had I discovered how willing he was to let me go. In any case, he gained his object, and that other one as well, the promise now fulfilled that he had extracted from me with his babbling intimations of mortality. I would remember him all right, and I already missed him, although I could not for the life of me figure out why. But with this final, improbable outrage, he had seen to it that I would have him, a ghost over my shoulder, for the rest of my life: a judgment and a sentence. Glenn Symonds could not have planned it better had he set out to accomplish his own unlikely end himself, which, with unconscious craft, he did, although the realization of the goal he had set himself may not have become apparent to him until he had very nearly reached it.

Back on Sunset, I turned left and drove easy with the traffic clear across Hollywood, and took the jog towards Silver Lake. It was now around eight-thirty, and as good a time as any to call Jessica. I pulled to a stop in front of a telephone booth by a liquor store.

Jessica Fenton was in her late forties, divorced, and settled into a life of routine oriented around her management of Symonds' affairs and the care of a small apartment building she owned and lived in in Santa Monica. As with most who worked at the clinic, she had undergone some of Glenn's stitchery as a fringe benefit ("I just let him start at the bottom and work up!"). She was devoted to him to the point of possessiveness, but had no illusions about what they meant to each other. She ran the clinic like an efficient carwash, bossed the staff of eight, which included an assistant surgeon, buffered Symonds from the more importunate clients, researched and advised him on the peculiarities of important ones, absorbed the abuse and passed on the praise that was aimed at him, and saw to the collection of what was owed him before and after his work was done. Behind the liquid eyes and the artificial sweetening in the brisk, contralto voice was an accomplished bully, very savvy, and utterly unimpressed by anyone who could not afford the scale, no matter who they were or what their story amounted to. She knew how close Glenn and I had been; how she felt about it I could never tell. Our paths crossed only on the most impersonal and amiable terrain. Glenn had let her know that a call from me was one he would always take, and, although I had never tested to find out, I imagined that directive was still in force all the time since I had left him.

She was home this time. There was no mistaking the voice, although it sounded less formal and less cordial than I remembered it during office hours. I identified myself and asked her if she remembered me.

"Of course I remember you, Mr. Scott! Where are you?"

Not *how* are you, but *where* are you?

"I'm calling from out of town, Jessica. I've been concerned about what happened to Glenn."

"Concerned isn't the word! Do you know, I must have been the last person to hear about it? His houseboy called me yesterday evening and said the police were there wanting to close up the house. The little imbecile was practically out of his mind. I didn't know what he was talking about. I've been there and back at the clinic nearly the whole time since. I just got in. Doctor's been gone since Wednesday, and I can tell you things are in a hell of a state. Where are you calling from?"

"I'm in San Diego. You say Glenn was gone Wednesday?"

"No, he was in Wednesday, but there's been only one message from him since, and that was Thursday, just before noon, telling me to cancel his whole schedule through Saturday. He didn't give any reason; he just said do it. I didn't like the way he sounded. Did you see him before this happened?"

"I had dinner with him last Tuesday in San Diego. That's all." I repressed the urge to tell her about the viewing at the morgue.

"Well, for God's sake! What a thing to happen! It doesn't make any sense, and the police won't tell *me* anything. What do you know about it?"

"They won't tell me anything either, Jessica. But I called you to find out about someone I think may know Glenn."

"You mean somebody who might've done this?"

"No, no! But I'm checking out something that may have a bearing on it."

"Who is it you're talking about?"

"Well, I know you usually don't give out information on Glenn's clients, but . . ."

"Yes, that's right. Doctor doesn't talk about his clients, and I can't either. Who is it?"

"It's a woman."

"A *woman*?" She sounded indignant.

"Her name's Louise Silver. I was wondering if you could remember anyone with that name ever being a patient of Glenn's?"

"Oh, for God's sake!" She sounded exasperated. "She's never been a patient, no. What're you asking about her for?"

"You mean you do know her?"

"No, I don't *know* her! But I know who she is."

"I don't understand." And I didn't. The Silvers were loaded, and Morris was a wheel in San Diego banking and real estate circles, but the name never cropped up in the papers, and Morris was certainly not anything tremendous enough for Jessica to have heard of his wife unless it was through Glenn. "She's my boss's wife," I said. "Glenn never mentioned her to me."

"I don't know that he ever mentioned her to anybody, except me. They don't have much to do with each other—or they haven't for many years, I'd say."

"I don't understand, Jessica. What's the connection?"

"Louise Silver is Doctor's sister."

All I could find to say to this was, "Oh."

"What's this all about, anyway?" Her voice was becoming strident. "What made you think she was a patient of his? Did she put you up to calling me?"

"No, she didn't. Jessica, listen!"

But she was off and running about her life and hard times, and how people were plotting against her, and what did I know that I wasn't telling her. I left her screaming at me on the line as I put the receiver back on the hook.

Symonds had never told me much about his family or his childhood. It was as if he had not had any, but began life fully grown out here in California, which, in a sense, was the case. When his father died a few years before this, he went back East to settle the estate, what there was of it, but exhibited no sentiment and recounted no history concerning that event when he returned. He might as well have had no mother for all he ever told me of her beyond her name, which he bore, and which was Rose. Of siblings heard I nothing.

As I thought of it, sitting in the car after hanging up on Jessica Fenton, I remembered that he found children a bore, and people with families who insisted on talking about them even greater ones. Childhood was something he seemed to have shucked as soon as he could, walking away from it the way a suddenly endowed mendicant might leave the shameful rags of beggary behind him.

I never questioned him on these matters; I was as little interested in his family as was he. Needless to say, my own parents never formed a topic of conversation between us either. My mother and father had died before I met Glenn, and since I recalled them only with a kind of fond indifference, his silence on this aspect of his life never impressed me one way or another.

But knowing now who Louise Silver was made me feel somehow unclean, used, in the sense that all concerned in my coming to San Diego, except myself, seemed to be party to a grubby little secret whose only mystery, as far as I could see, was that it had been kept a secret. Had they (Glenn? Louise? Morris?) imagined I would refuse the job had I known it had been arranged as a way of getting me out of Glenn's hair? Or that I would have made a nuisance of myself by presuming on what they may have regarded as their charity? In my anger and bafflement, it did not occur to me that Glenn might have been the only one who knew, and had used Silver and his sister (through another) as the agents of my riddance. It needed neatening up to make it rhyme.

And Louise Silver: very cool for one whose younger (?) brother had just been heinously done in. "The only thing Morris ever worries about is adverse publicity when anyone at the museum gets his name in the paper." And . . . Morris.

By now I was in no fit mood to see anybody else, but I was too close not to make the attempt to peel back what might be the remaining layer of this kernel of doubt I had been picking at for two days. Before leaving the phone booth, I checked the book for Owen Keller's address, wrote it down, then returned to the car and found it again on the map by flashlight.

The Silver Lake district was a seedy genteel neighborhood that appealed to a lot of singletons of both sexes because the rents were not exorbitant and the people in the area were pretty relaxed about what was beginning to be called "lifestyles" other than their own. It was near Hollywood and not far from the downtown area. Finding Silver Lake Boulevard, I picked up the address within a block off Sunset. It was a flimsy, pitched-roof, single-storied duplex, unshaded and squeezed in among others like it on a street picketed with a line of tall goon palms. There were lights on in the back where a concrete driveway led to a garage and what might have been a backyard. I parked in front of the place and got out to have a look. The driveway served as a walk to the shallow front porch, but at the steps I stopped, having noticed something at the back of the house where the concrete faded into shadow. In the foreground of the darkness back there I caught the telltale glint of chrome and plexiglass reflecting the curtained illumination from the neighboring kitchen window. I walked on back along the side of the building, and even before I reached it I recognized what and whose it was. Blue, with a studded Stanley saddle-seat, thin, red-line trim, and conchoed saddlebags: I would have known it anywhere. It was all coming together.

Bending over to touch it, I sensed, or heard subliminally, the nearness of someone behind me. As I straightened and turned, there was just time enough to catch the light on his face. And it was the look he had in that split second of recognition—the terrible glary eye, like a startled colt or a stallion rearing—that told me I had made perhaps another mistake.

"Faulkner!" was what I might have cried out, but the syllable, "fal . . . !" was probably all that was uttered. I only remember the crash of stars and enveloping blackness, and the descending, tristisonous echo of that faltered name, because I never would have imagined him coming on like that with me.

20

ONCE, a half dozen years before this, when I first knew Nelson Faulkner, we had been playing around on his bed in San Francisco, and I had mocked him, quite innocently, on the subject of the things he did with other people and did not do with me. It was an affectionate scoff, not meant to be taken seriously. I was in an antic mood, but in no way intended challenging his devotion to these other interests of his. Faulkner had been, during the brief span of months when I used to come over to his place, a good performer, gentle (if somewhat silent and intense), and in no manner rough. It was as if, with me, he put aside all those other procedures, just as did Symonds. Only, what we did together was different from what Glenn and I were wont to do. I thought we understood each other past his taking offense at a little friendly teasing thrown out in the relaxed pillow talk of those hours when I found time to come across the bay from Berkeley and see him. It was a mistake on my part; as I would not have done it with Glenn, so I ought not have with him. Faulkner and Symonds were of the same race in this.

Now, Nelson's history included specialized service in the CBI theater during World War II. He had barely been of an age to make it in that contest, and mentioned it to me only in passing. A perceptive assessment of his induction tests had channeled him into the OSS; that was all I knew of it. At any rate, what I said in jest triggered in him some mechanism that did not want bothering, and, in what seemed less time than takes the telling of it, he had me immobilized, manacled, and recovering in sickening pain from where he had reached me, somewhere in the neck.

I was frightened, even though I knew he meant me no real harm.

He sat in a chair tilted against the wall of his studio, staring at me, expressionless, while I came around. It was the first time I had ever experienced restraint like that, and if I did not like the idea of it, the idea made real appealed to me even less. As consternation mounted, then modulated into resentment, and resentment, frustrated, blossomed into panic, I questioned, pleaded, cursed, and with sweat rolling off me, attempted reason to make him let me loose. He simply watched me get through these predictable paroxysms until, having resolved something in his own mind, he came over to the bed and undid me, who was already morally undone. He told me curtly to get dressed and get out. And I did, as angry and as silent as he, but at myself for what I had said more than at him for what he had done about it. It was not until after I slept on it that I realized I had embarrassed him, and that he had compounded his own confusion by reacting in the way he did. I was, after that, a kind of keeper of his conscience in much the same way that I guarded Symonds' secret, immolatory yearnings. Both of them saw windows in their soul opened at my casual prying, so that it might appear I had been not unpurposefully tampering at the locks.

I saw Faulkner after that, but, unlike Symonds, he was never at ease with me again.

Now here I was on replay, this time bound hand and foot on a couch in a place I did not recognize, and with the pain in my neck far worse than the effect of the first blow had ever been. And there again sat Faulkner in a chair tilted against a wall, looking at me. He wore Levis now with his boots, and a leather jacket frogged with chains. A mesh T-shirt beneath it revealed the dark tone of a spare, but well-defined chest. The lines of wear about his mouth and eyes were more pronounced in the shadows cast by the light of an old stand lamp with a parchment shade that shone between us. It may have been the filtering of that shade that gave a liverish look to his face. He was tired, but not, I guessed, as tired as I. He saw I had come to, but no sign of recognition acknowledged it. He was watching me; nothing more. I moved my head, rubbing a cheek against the scratchy pile of old plush.

"Where's Owen Keller?" I said.

"We'll get down to that right away." His voice was as expressionless as it had been had we been going over an invoice. "What are you doing here?"

"I was going to ask you the same thing."

"You're not in any position to ask anything. I put you there for answers, not questions, friend."

I rubbed my cheek against the plush again. "If this is how you treat your friends, I'd hate to be your enemy. Untie me, Nelson."

"You're not tied. What were you poking your nose around here for?"

"What if I came to see Keller?" I moved my hands, straining against the bonds, which, I felt now, were not ropes or anything that was tied. Something professional, as the newspapers might say. "What difference does it make? Do you go out and deck everybody that comes up the driveway, for Christ's sake?"

"You don't know Owen. And he doesn't know you. What are you doing here?"

"Nelson, let me go. This isn't amusing me the least bit, and I'm going to start raising hell in a minute."

"You do, and I'm really going to waste you."

"Nelson!" I cocked my head up to get an untilted view of him. "It's me—Scott. Remember?"

I twisted around to lie on my back, then gave that up as a bad move; with my hands bunched together behind me, it was more uncomfortable than staying on my stomach. I compromised on my side.

"What were you coming to see Owen about?" he said.

"I thought Eugene might be here."

"You can do better than that."

"No, I can't. Eugene was up here Saturday, and I thought he might have come up here again."

"Why would he be up here?"

"I don't know. It's something to do with Owen Keller. That's all I've been hearing is Owen Keller ever since Symonds got it, but everybody still thinks Eugene did it."

"And you think Keller did it?"

"I don't know."

"How did you get Owen's address?"

"Easy. He's in the phone book."

"Why didn't you call first?"

"I did. There wasn't any answer."

"When?"

"Two . . . two-thirty."

He looked at me in silence, as if checking his own time references, then he asked, "Where's Eugene?"

"Hah!" I flopped back over on my stomach. "He changed his clothes in my house and then took off on my bike. How about letting me loose?"

"Were you there?"

"Are you simple? Of course not! It looks like he was hiding in my house some time between the time he left Symonds in his trailer and the time I got back home after your party last night."

"Why was he hiding at your house?"

"I don't know!"

I could have cried with frustration.

"And that goes for questions about everything and everybody connected with this shitty business. I didn't even know you and Owen had a thing going until this morning when somebody down at Stallion Oaks let me in on it. Christ! You people have been playing all kinds of games I've been happily unaware of until now. And, by the way, why in hell did you doctor my drink last night, you sonofabitch?"

"Did Eugene tell you to come up here?"

"Nelson!" I spoke as one might to a backward child. "He did not *tell* me anything, do you hear? I found his car in my garage, his dirty clothes in my closet, and my bike gone."

"How does that lead you to come up here?"

"It was only a hunch. I'm looking for him, remember?"

"So was I."

"What do you mean, 'was'?"

"I want to know where Owen is now."

"When was the last time you saw him?"

"I called him Thursday night from San Francisco and talked to him, and I was supposed to see him here Friday. He said he'd been over to see Symonds Wednesday."

"Yes." I failed to completely suppress a smile. "I understand Symonds was moving in on him."

"You understand shit!"

"That may be. I've had a good deal of it slung my way the past two days. What was he doing at Symonds'?"

He did not answer, but the look of concentrated fury gathering on his utterly still features cautioned me I was going too far, or too fast. It was the look of the *Würgengel*, the image of the destroying angel Symonds had worshiped and thought, in his wildest dreams, to have assimilated in himself.

"Maybe," I said, attempting to diminish the insolence of my question. "Maybe he's still over there." I knew how foolish it must sound to him even as I said it, but hoped it would confirm for him the almost complete darkness in which I was moving where knowledge—real knowledge—of what was going on was concerned.

He relaxed at that with a scoffing sniff, reaching in his shirt pocket for a cigarette. After lighting it and exhaling the smoke my way, he took it out of his mouth, looked at its glowing end, righted his chair, then stood up and came over to me. He leaned above me and put it between my lips. Relieved, I drew on it, and thanked him in a gust of smoke. He returned to his chair.

"Have you been to Symonds'?"

"Just to look," I said. "Before I came over here. I think the law's closed it up, but his secretary gave me to understand the houseboy may still be there. They would've sealed everything but his quarters."

"You talked to his secretary?"

"Between here and there, yes."

"Did you tell her you were coming here?"

"I didn't even tell her I was up here. I didn't tell her anything. I wanted to find out what was going on at the clinic and at his house."

"Did she say when she last saw Symonds?"

"Yes. Wednesday. He was at the clinic then."

He nodded, as if that confirmed something he had been wondering about.

"Owen wouldn't have been over at Symonds' place," I said.

"I didn't say he would."

"Was that why you were down in San Diego yesterday—looking for him?"

He nodded.

"What did you think he would be doing down there?"

"If I knew that, I probably wouldn't be looking for him."

"Do you think Eugene was with him?"

"Maybe. What do you think?"

"I hadn't considered it."

"What *have* you been considering?"

"Eugene. Where he is, mostly. How about untying me?"

"I told you, you're not tied."

As if to illustrate the point, he got up again and approached the couch. Putting the half-smoked cigarette to my lips and leaving it there, he unsnapped a ring of keys depending from a belt loop of his Levis. He reached behind me, pulling me away from the back of the couch. There was the brief bite of metal against my wrists, and then my hands were free. He pulled me up in a sitting position, put the handcuffs in his back pocket, then sat down beside me, took the cigarette out of my mouth, and had a final drag on it while he watched me rubbing my wrists. After stubbing the butt out in an ashtray on the end table near him, he offered me a fresh one from the pack, lit it for me, and leaned back.

"How about my feet?" I said.

"One thing at a time, friend. I'm not through with you yet."

"Nelson," I said, "do you really think I had anything to do with this?"

He looked at me, his face a yard from mine. There was the ghost of a smile there. "I'm trying to see how you couldn't."

"Would I be running all over creation looking for Eugene, or Lawson, or Owen Keller for that matter—somebody I don't even know—if I did? Would you?"

The smile faded.

"Did you?" I persisted.

He said nothing. I adopted a more reasonable tone. "Listen. The police will fit this together no matter what we think or what we find out. They're already onto everything, or almost everything, we've been doing."

"Then why are you moving around so much?"

"For the same reason you are, because I can't sit still. We're both of us clean because of that, don't you see? And Eugene's clean because *he* can't sit still." I was a little amazed at my own inventiveness, but I went on. "Find the one who's being so quiet and you'll find the one who did it."

He moved slightly away from me so that he could reach the back of my neck which he grasped, not untenderly.

"Does it still hurt?"

"It's felt better, I can tell you."

"Sorry about that. I've been under a strain."

"Join the club."

"You really are in the dark about this, aren't you, friend? What I can't figure out is why you keep sticking your nose in it. As far as I can see none of it had anything to do with you. I'll admit I wasn't sure about that last night, and it's possible I got some wrong ideas about you a little while ago when you came sneaking back here to look at my bike."

"I wasn't sneaking. And I thought at first it might've been mine."

"Yours?"

"I told you: Eugene took mine—or I think he did."

He appeared to weigh this in his mind.

"All right," he said at last. "I can buy your looking for your bike, but why were you so interested in finding Eugene Carmichael before this morning? You weren't anything to each other any more—if you ever were."

"Do you know, I was asking myself that same question all day yesterday and half of today. Maybe it gave me an excuse not to think so much about Symonds. I saw him, you know."

He looked at me sharply, as if he found some hidden intent in the remark.

"You mean you saw him yesterday, or the day before?"

"Yesterday."

"Where?"

"Pinkerton took me up to the morgue. Last night. I identified him."

"That worthless hick!"

"Who, Dale? If he's so worthless, why did you let him into that session for last night?"

Faulkner turned his head and gave me a real smile. I had never seen it so full-out; not quite ingratiating, not quite a sneer, it was lubricious, taunting, utterly charming. "That was Lady's Night, couldn't you tell? You were there."

"Very funny."

"What did he look like?"

"Symonds? Pretty bad. The pathologist had already carved him up, and besides that, he looked as if he'd drowned."

"God didn't mean people to hang upside down by their feet," he said. "Otherwise he'd have made them bats."

"How many have you strung up like that?"

"Not many. I don't like looking at people upside down. They look like they're making faces at you. What did the autopsy turn up?"

"He was killed with cyanide. And then there were a few other things."

"Such as . . . ?"

"I'd rather not talk about it."

I felt my neck squeezed, less gently this time.

"Talk."

"Somebody had done an apprentice's job of castration on him."

"What else?"

"It looked as if Glenn patched it up himself, or tried to."

"What would you do if you found Carmichael?"

"I don't know. I'd be relieved to find him, I think. I'd probably talk him into going to the police. What else could I do? He's mixed up in it some way."

"But you don't think he did it, do you?"

"I told you: I saw Symonds afterwards. Eugene isn't capable of that kind of butchery, and cyanide isn't within his range of competence, or knowledge, for that matter."

"It probably isn't within range of many other people's ideas of killing either."

"So . . ." I turned to look at him, and it hurt my neck. I winced, and his fingers, firm and warm, eased the pain. ". . . he couldn't have done it. You know that."

"No, I don't. Anybody can do anything, given the right circumstances."

"You believe that?"

"I believe a person will do something you least expect from him under enough pressure, or just to get out of an embarrassing situation. This session Symonds set up might have all been a big mistake."

"All your sessions are big mistakes. And what session are you talking about? The only one I heard he set up was supposed to be Saturday night. Which reminds me: You haven't told me what you put in my drink."

"I didn't put anything in your drink."

"Then who did?"

"I thought it was Pinkerton, to tell you the truth. That was my glass."

"Your L.A. friend insisted I drink out of it instead of the bottle."

"No . . ." He shook his head. "Rublatz is clean, but Pinkerton and I had a race to see who got out to Stallion Oaks first."

"Old Longhorn won."

"Yeah. He knows the roads." His shoulders jerked in a disgruntled shrug. "I used to think Pinkerton never had sense enough to pour piss out of a boot with the directions printed on the heel, but he's not as simple as he acts."

"I'll tell you something about him," I said, "if you undo my feet."

"Tell me anyway."

"Feet first."

"That's about how you'll go out of here if you try my patience any more, friend. Talk."

He stood up, maybe to distance himself from me, shook out another cigarette, and gave me one. I lit it myself. He stood in front of me, waiting. "Well, what about him?"

"I have to tell you first that I went down to National City Friday night—early Saturday morning really. Eugene had asked me to check on him. I phoned him, but when there wasn't any answer, I was worried, so I went down."

There was no reaction on his part to this confession. He stood there in the middle of the room, an elbow cradled in one hand while he smoked, the cigarette held close to his face with the other, waiting.

"I've got a key to the trailer, but I didn't have to use it. The door was unlocked. I let myself in without making any noise; I know the layout in the dark. There was a night light on in the back room, and there was somebody talking—arguing or coaxing, it sounded like—almost in a whisper. I knew it was Symonds strung up. Whoever it was was half-holding him, talking to him. I couldn't recognize him in that light. And when I saw it was Symonds like that, I didn't want to hang around. I got the picture and got the hell out, and back up to Crystal Cove."

"Where's Crystal Cove?"

I explained about the Silvers.

"You think it was Pinkerton you saw?"

"I couldn't tell. He didn't look that tall, but then, he was bent way over. Also, he might've been wearing leather—a jacket anyway; Pinkerton doesn't have any leather."

"What about the voice?"

"It could've been anybody's. Whoever it was was whispering, hissing, as if he were angry, upset at something. It looked to me then like some kind of talk between the action. That's why I cut out."

"Doesn't sound like Pinkerton."

"It could've been anybody. It could've been you."

"Or Eugene."

"Or Owen Keller, whoever he is. But there's something else. After Pinkerton took me to the morgue, we had a talk. And in the course of it he asked me where I *really* was Friday night."

"What did you say?"

"I told him the police had already talked to me, and we had covered that. But I thought then that he knew something, that he might've known somehow that I had been away from Crystal Cove early Saturday morning."

"You said that when you got the picture, you got the hell out of there. What kind of a picture did you get?"

"It was an arrangement, a session as far as I could see. And Symonds was it."

"You didn't think maybe he was in trouble and could've used your help?"

"On the contrary. If it was a session, I was an intruder. And, more than that, I didn't want to see Glenn in a situation no one had ever seen him in before. He certainly wouldn't have been overjoyed to know I was looking at him. How was I to know he was in trouble?"

"You still love him, don't you?"

"He's dead."

Faulkner looked at me, studying me through the smoke of his cigarette for a long moment of silence. Then he asked, "If it was so dark you couldn't make out the other one, how could you tell it was Symonds that was hanging up? Could you see his head?"

"No. Whoever had hold of him had his head cradled out of sight. It was weird, but I knew the moment I saw them that it was Glenn he had hold of—the line and shape of the legs, I suppose. The way he was dressed, too. I couldn't mistake the kind of outfit he had on. Glenn never bought leather off a rack, or if he did, he had it cut and changed to his taste. Lenny did all his nighttime tailoring, you know that. Nobody could wear leather like that man. Not even you, babe."

"You know," he said, disregarding my last comment, which in any case would be taken as a compliment by most people in the circles in which he moved, "if you hadn't cut and run right then, all this might've been different."

"You think I haven't considered it?" I nodded to him. "Just as you said, people do strange things when they're embarrassed."

"Were you embarrassed or scared?"

"I was scared *because* I was embarrassed."

"At what?"

"At seeing a degrading performance on the part of someone I found I still cared enough about not to want to be caught watching. How's that?"

Nelson laughed without smiling. "You always did have feelings that were too delicate for your own good, Scott." He shook his head. "I don't understand it, and never did."

"What?"

"Why it is you like to be in the neighborhood when the party's going on, but you never want to join in. Is it really true that you used to fuck Symonds until he was silly in the head?"

Now, as I've indicated before, people who knew Symonds and

me could no more than have guessed at what we had going between us in bed. His view of himself, warped as it was, was something so important to him that I would never have lasted beyond our initial bout had I not respected and, by my silence, sustained it. To be sure, I could never see what real difference it would make to anybody else but him. He could have carried off a public appearance in drag, I believe, and not diminished one shimmer of the *baraka* he commanded with these less talented worshippers—Faulkner here excepted. Men get fucked, like to get fucked, get fucked repeatedly by other men (and this not figuratively), and the experience does not undo them. Faulkner on top of me—I looked at him, remembering—had neither jarred my sanity nor affected the way I looked at myself before or after the action. But then, if I had learned one thing from seven years with Glenn Symonds and as many months of visiting Nelson Faulkner, and all the rest of the ruck that followed in their wake, it was that what might ordinarily be viewed by most people in the light of pleasure or discomfort was twisted around into an impossibly complicated pattern of stimuli and reactions in which anything might mean anything, and nothing was what it would appear to be to an outsider viewing it from the cool perspective of sweet reason. I was not going to get my legs free for a while, it seemed, so I decided to probe some myself.

"Are you and this Owen Keller lovers?"

Faulkner rubbed his cigarette out and reached for another one, as if he had not heard the question.

"All right," I said, trying to keep the impatience out of my voice, "were you worried about him seeing Symonds? I'm not trying to pry."

"You are prying."

"I heard he and Symonds started seeing each other after Glenn came back from his vacation this summer."

"You heard shit!" He dropped into a chair, the same one he was in when I came to. "They knew each other, but they . . ." He gestured angrily with one hand. "Owen's never touched him."

"That you know of. What's he like—Owen?"

Faulkner sighed and looked away. It was not sentiment that elicited this response, but rather his inability to describe anybody, even somebody he was interested in. He could probably have drawn me a picture without any difficulty.

"What does he look like?" I persisted. "Blond? Brunette? Tall? Short?"

"Blond. Medium, I guess; about like me. I don't know. This thing with Symonds had gone too far."

"Was there something going on?"

"He was interested in Symonds."

"Interested how? Was he an M?"

"No, he wasn't."

I digested this first course for a minute while I nursed the butt of my cigarette. It was laid out plain from where I looked at it: Faulkner was deeply involved. I was seeing one gripped in the talons of jealousy and feeling the bite of that bird for the first time. He was not the type to profit in any prudent way from the pain of the experience. Faulkner did not like pain. Like many of his kind, he did not really understand it. It was that that held him fascinated by it.

"Listen, Nelson. I don't know from A about Owen Keller. Do you care to fill me in? It might take my mind off the cramp in my legs."

He got up and crossed the room, settled on his haunches in front of me, and again unclasped his key ring. In a second my feet were free. Without any preliminaries or further hesitations, he began telling me about Owen, all the while massaging my calves and ankles, having pulled up the pant legs and shoved his hands into my boot tops.

21

OWEN KELLER had gotten out of the Marines two years before this, and before that he had managed to get in a couple of years of college. He was born in southern California, but had been raised in Seattle. His father was dead and his mother had remarried while he was in the service. After he was discharged he came back to the southland where she was now, and laid around, living off her, not knowing what to do with himself, until he got on everybody's nerves, including his own, to the point where his stepfather asked him to leave.

He drifted after that, and found himself up in San Francisco where he began hitting the bars and the bottle, supporting the habit with a job on the loading platform of an electrical supply house south of Market Street. In the course of this regime, he ran into somebody who was mildly into bondage and discipline, and who turned him on to it—mildly. Then he met Cusped through the same channels, and Cusped, that epitome of blabbermouths, fed his mild fantasies until they grew into a more and more fevered interest in Leather, cycles, shackles, and all the rest of it. By this time he was living in L.A.

"Was he on drugs all this time?"

Nelson looked at me narrowly. "Who said anything about drugs?"

"It's been mentioned," I said.

He ignored this and went on to say that a little over a year ago Cusped put him in touch with Symonds.

"Is that when he met Symonds, then?"

"No, he didn't meet him then."

"Well, what do you mean by saying Cusped put him in touch with him?"

"Just that. Symonds had already heard about him, and he began leading him on. I told him to keep away from him."

"Wait a minute," I said. "You knew him then too?"

"Yeah, I knew him in San Francisco."

"How were they in touch—Symonds and Keller?"

"Mostly through other people, I think. You know Symonds was out of his mind, don't you?"

"We're straying from the point, I think. Anyway, the last time I talked to him he didn't appear to be any different than he's always been."

"He's always been crazy."

"That can be said about a lot of people we both know. There are some who might say the same thing about you."

He showed no interest in this observation; just stared at me, waiting. So I came back to the subject of Eugene, and asked him how he thought he fitted into the picture.

"You tell me," he said. "You're the one who knows about that. You expected to find him here."

"All right. Let's backtrack. Symonds called Eugene Friday and told him to get up here, that he needed him for something. It didn't sound like an arrangement to me, but I may have been mistaken. Eugene hadn't called me to report on his activities in a long time. But the real point is, he sounded worried."

"Worried?"

"He gets that whine in his voice that's a plea to be questioned closely."

"And you didn't question him."

"I didn't have time, and I wasn't interested. The only thing about it that stuck with me and kept nagging at me while I was up at Crystal Cove was that he asked me to call him in National City later on and check to see if everything was all right. He'd never done that before. Now, you say Owen Keller wasn't an M. Suppose—just suppose for a moment—that Symonds and Keller arranged Eugene for something."

Faulkner shook his head and semaphored "no" with his hands. "Come back down, Scott. You're way off. Owen's already worked on Carmichael, and he's not interested in going any further with it."

"When?" Owen taking Eugene was another piece of news to me.

"At an Easter bash down at Stallion Oaks. There was a lot of people there. He was just one who worked on him."

I let it lay, and kept on with the hypothesis I was putting together. "Maybe he wasn't interested in Eugene, but Symonds and he, *plus* Eugene, would be something new, wouldn't it? That could make it jell, if, as you say, Owen hadn't met Symonds before."

"Wouldn't Eugene have mentioned that?"

"Ordinarily, yes." Not only would he have mentioned it, given the fact that he called me at all, but I realized now that what had not sounded right, coming from him, was the lack of any specific details about anything other than his worry, and the information that he was calling from a post office in Hollywood.

I shifted gears.

"You met Owen at Stallion Oaks, didn't you?"

"No. I'd seen him around the bars up in San Francisco and I worked on him once, last year, before he moved down here."

"Just once?"

He nodded. "He came on strong at a bar one night, and I played along. He was a good looking kid, but he talked like he had his head locked up his ass. I gave him a surprise party when I got him back to my place. He didn't like it, and he wasn't very good. I didn't see any more of him until Easter."

"You didn't know he was going to be there?"

"No, it wasn't anything planned. He'd matured, you might say, between that first time I'd seen him and when I saw him at Ted and Walter's. He's all right now. He's a little like you, only he digs all the action you don't care about."

"And that's when you became lovers?"

He looked at me with that crinkling around the eyes; a look of burgeoning distaste. I made a gesture of amendment; the list of terms that might offend the truculent sensibilities of Nelson Faulkner and his buddies were too numerous to remember in the heat of trying to make sense of this mixture.

"You said he was living in San Francisco when you first met him. What's he doing down here?"

"You mean for work? Nothing. I think his mother pays him to stay away from her."

"He could do that as well up there. Why move down here in L.A., if that puts him nearer to her?"

"To tell you the truth, I think he moved down here to be nearer to Symonds."

"I don't understand. Why 'nearer'? Why didn't he just get together with him and be done with it?"

"It was a game they were playing. I told you, Symonds was leading him on and Owen was swallowing it all, point, barb, shank, and eye."

"What do you think Glenn had in mind?"

"I'm not sure. Didn't Symonds ever tell you about the lovers he had before you?"

"No. He hardly ever mentioned anyone he did things with except the anonymous ones he bragged about when I was in a mood to listen to it. Some of them were very funny."

"This isn't funny. Both of your predecessors killed themselves." He looked at me with the trace of a smile thinning those dark lips. "I wonder how you missed that rut."

"I may not have, if things keep going the way they are. The last time you talked to Owen was when?"

"I told you, Thursday night. He called me Wednesday, just after I got home from work, and said he had it set up with Symonds. I told him not to do it, to wait until I could get down here this weekend. But he said not to worry, that he had everything under control."

"What did he say when he called you Thursday?"

"I called *him* Thursday. He sounded like something was wrong, but he still wouldn't tell me what was going on. All he said was that he might have gone too far and hurt Symonds worse than he ought to've."

"What else?"

"What d'you mean, 'what else?'? What else did he say? He said he had to go back and see him."

"*Had* to?"

"He meant he wanted to, or felt he ought to, I guess."

"Did he say he had seen him again since Wednesday?"

"He didn't see him, but he'd heard from him. I pushed him on this and he said he'd talked to him, and that Symonds was in a bad way. He wouldn't tell me what he meant. I told him to keep away from him until I got down here. I offered to take a plane down that night, but he said not to, that he would wait if I thought he ought to. I told him I'd call him as soon as I got off work Friday and then come down, so he said he'd wait."

Faulkner offering to fly down from San Francisco was a measure of how strongly he must have felt in this, for he had an almost superstitious fear of air travel and had not left the ground to go anywhere since his discharge from the service twenty years before.

"Didn't you find out anything specific about what had happened?—what he'd done to Symonds?"

"I told you, he didn't want to talk about it. And it's not my style to push anybody over a phone. You know that."

I did not know, never having been pushed by him before this evening, but I understood his attitude, his "style."

"He sounds to me like a kid who wants to make a reputation fast by being the one who gunned down the gunfighter."

"No." Faulkner sat back on the floor in front of me with his arms around his legs. "It isn't that simple. If it was something like that I wouldn't have worried because Owen wouldn't have had a chance. He had to've already been close, otherwise Symonds wouldn't have let him get to first base, much less cut his balls off."

I let my breath out in a long sigh. "Your boy sounds like he's got a real vicious streak in him, Nelson. Do you know him well enough to say that he's got all his marbles?"

"He's not vicious and he's not crazy."

"No? How many balls have you cut off in your line of work? I can tell you Symonds never has, and he was a surgeon and might've gotten away with it. If Owen Keller isn't absolutely insane, then I'd say he's topped both you *and* Glenn.'

"Take it easy, Scott. There's still a lot you don't know about this."

"I can believe that, but I already know more than I can handle. And I'll tell you something else. I know Symonds didn't die from anything Owen Keller did to him up here. At the most, he may have made life not worth living for him, but Glenn left L.A. alive. And I believe now that it was Keller I saw down in that trailer. I also think that the reason you weren't able to get hold of him Friday after work was because he was on his way down to San Diego with Symonds still alive, in tow, accompanied by Eugene."

He shook his head, closing his eyes in emphasis and pursing his lips. "No, it won't wash. Owen hated Symonds. He wasn't about to enter into anything aimed at him that included Eugene either."

"Hated him? If he hated him, then what were you so worried about?"

"I didn't want him messing around with Symonds. It was more than he could handle."

I studied him for a moment, then I said, "You really had it bad, didn't you, Nelson?"

He moved across the room and stood for a moment with his back to me as if waiting.

"All right," I said. "I'm telling you Eugene was here and then down in San Diego, in that order, and that Keller was in the picture—physically. I'm telling you, and anybody else, that Eugene wouldn't, couldn't, didn't have anything to do with Symonds' death. That's why I'm looking to find him and why he was waiting to get hold of me. But the reason you want to find Keller is because you're afraid he was the one that did the killing."

Faulkner just stood there, staring at the dingy, worn carpet and biting his lower lip. I got up and stretched.

"Is there anything to drink in this house besides water?"

"You stay here," he said. "I'll get it. And don't touch anything."

He left the living room and returned shortly with two cans of beer. Opening them, he handed me one and pocketed the pull tabs. "Fingerprints," he said, in answer to my puzzled look. "You might not want them to know you were here."

"Them" could only mean one thing. I smiled and glanced at my watch. It was already ten-thirty. Faulkner sat down on the couch again and lit up. I remained standing, easing the cramp of the long sit I have been through.

"Did you tell the police everything you've told me?"

Faulkner never said "cops," or "fuzz," or any of the other designations people used for his heroes. It was a measure of the veneration in which he held them. I knew that what he was going through now must be something in the way of an emotional contretemps. His involvement with Owen Keller was contending with his older bias in favor of those who might soon be hunting down this one he had come to care about if what I said proved to be true.

"I didn't tell them anything about you because I wasn't asked. The same goes for Owen Keller because I didn't know. And I said almost nothing of importance about Symonds because . . . well, because I'm worried about my job."

"What about Eugene being up here?"

"Just that—that he called me from up here."

"And about your going down to National City Friday night?"

"It was after midnight—Saturday morning. I didn't mention it."

"How did you get away with that?"

"Luck, I guess. They didn't press the point about my spending the night at Crystal Cove. They checked it out with the Silvers."

"Who?"

"Louise Silver. She's the museum director's wife. Morris Silver's my boss."

He looked at me, frowning, as if he found some discrepancy in what I had told him. "Are they friends of yours?" he said.

"Not really. It's the first time I've been asked up to their place in the two years I've worked under Morris. And something else: I found out tonight for the first time that Louise is Glenn's sister. I didn't even know Glenn had any family other than his father. Jessica, his secretary, told me."

"His sister didn't tell you?"

"No."

"She knows about Symonds."

"Of course."

"How was she taking it?"

"The only thing that seemed to be bothering her was some concern over anything that might connect the museum—and Morris, her husband—with this. 'Anything' means me. She and Glenn must not've been very close. She asked me how close I was to him though, but she didn't push it. I got the idea she didn't want to know very much. Anyway, she told the law I was at their house all night."

"Does she know you weren't?"

"She doesn't know anything for sure. She's a smart woman though. She'll probably do whatever's necessary to guard her own and Morris' interests. And, by the way, the police haven't talked to you, have they?"

"No."

"How can you prove you weren't down in National City Saturday morning or Friday night?"

"I was with somebody out in public until the bars closed, and then I spent the night at his place."

"Convenient." I sat down in the chair facing him.

"I didn't do it for that. I wasn't going to sit around here waiting for Owen to show up."

"Yes," I said. "Somehow I can't imagine you sitting around waiting for anybody. When did you go down to San Diego?"

"Saturday afternoon. Rublatz and I had brunch over at a place near Glendale after I checked in here to see if Owen had gotten in during the night. He hadn't. I guess we were in San Diego by about three o'clock. I sacked out at Chris' and he started making the rounds of the bars."

"You stayed at Chris Hanson's place?"

"That's right. Rublatz—you remember him, he was in the basement last night—he used to be one of Chris' wards, I guess you'd say. He stays with Chris when he comes down there. I hadn't even gotten to sleep when Chris came in and told me about Symonds. Then Rublatz called me to tell me too. I called Owen again, collect, but there wasn't any answer."

"Was that when you planned that session?"

"Shit! It wasn't planned. Guys were coming in from all over after Carmichael's telephoning the night before. Rublatz ran into Lawson down at The Club and listened to him shooting off his mouth about what he claimed he knew about Eugene and Symonds, and about what was going on."

"He didn't know anything."

"Did you know that yesterday?"

I admitted that I did not.

"Neither did anybody else until we checked it out on him. Lawson was just looking to profit from the situation and get what he never had a chance at before."

"I don't think he'll want it again for a long time," I said. "You very nearly had a third killing on your hands."

"Yeah, I know. I was in a hurry to get out of there, and I left it to Cusped to undo him. I guess he thought it was funny to leave him like he was."

"Cusped lacks a sense of proportion. And while we're on that, why do you think Pinkerton doctored my drink, and why did you leave me there like that after he did it?"

"I told you. In the first place I didn't know you'd been doped, and in the second I was in a hurry. Anyway, I still didn't consider you completely clean in this—although I didn't give a shit what you might have done to Symonds. I was still not sure but what Owen might be down there; that you and he might have gotten into some screwy arrangement involving Carmichael and Symonds."

This suppositive arrangement seemed to grow by another member every time he mentioned it. I had to laugh. "You knew better than that."

"Did I?" He looked at me, amused in turn. "What've you been doing with Carmichael for the past year?"

"No more than what you and I did when I used to see you."

He thought about it a moment, then said, "That's believable enough now, hearing it from you. But you've been getting yourself talked about. I hadn't seen you in a long time, and people do change, usually for the worst."

I told him it was flattering to know I had been talked about as far away as San Francisco.

"It was only because of your connections, babe," he said. "The things I heard mentioned, when I heard anything at all, sounded as if you'd been converted."

"Did you believe it?"

He shrugged. "You were a long way away and a long time gone. Like I said, people change. It was worth a laugh, hearing it. I didn't care one way or another."

"Look," I said. "If Glenn Symonds couldn't bring me around to his way of doing things in seven years of close living, do you imagine someone like Eugene Carmichael could change my mind?"

He almost laughed. Then, putting a hand on my shoulder, he rocked me slightly while he said, "It's probably never dawned on you, but I've known it for a good long time: you're probably one of the most accomplished second-hand men in the field."

"What's that supposed to mean?"

"Kick it around a while," he said, and let his hand slide off my shoulder in a motion that finished by fingering another cigarette out of his pocket. He must have used up two to three packs a day. I shook my head when offered another.

"One more thing," I said. "Why did you let Dale Pinkerton into that arrangement last night?"

"No way of keeping him out. He knew about it, and he wanted in—as an inactive observer, you might say."

"But somebody—anybody—there must've known about his connection with the law. You're lucky we weren't raided."

He pursed those fierce vermilion lips and waved two fingers holding the smoking cigarette in front of my face. "That's the main reason he *was* there, babe: so we wouldn't be interrupted. The police know about it. I don't doubt Pinkerton laid it all out for them after it was over."

"If not before."

"Probably after. He stuck around and talked to them after I got nowhere with Ted and Walter out at Stallion Oaks."

"What about Chris?"

"Oh, he's bit all his fingernails down to the quick, I guess, but Pinkerton and Buddy—Rublatz—quieted him down. He owes Buddy something; that's why he didn't interfere with what went on downstairs in his building. And Pinkerton being there made him feel better about it. But he's got a higher estimate of that canary's pull with the police than I have.

"Did you hear about the screaming fit he threw at The Club before they dragged Lawson out of there last night?"

"Yeah. That was more or less planned roughhouse to give Chris a chance to proclaim publicly his aversion to violence. The truth is, he wanted to watch."

"If I had seen him down there when I came in, I would've figured it was a setup. As it was, I was suspicious of Dale's being included."

"You almost walked in on the end of it anyway." Faulkner did laugh this time, a snort of scorn. "Lawson had already told us about Carmichael being out at Stallion Oaks. Then your San Fernando cowboy friend misfired, and it looked like a real dud. I wasn't getting anywhere in finding out where Owen might be in any of this, and Buddy was bored. Cusped was the only one still hot to go ahead. He brought you down, and you got me convinced Lawson really knew something and was holding back. I decided to let you get it out of him because I didn't want Pinkerton to know I was looking for Owen. It was one big misunderstanding. Lawson probably thought you were crazy; you scared the shit out of him, insisting he tell you something everybody already knew. If you didn't have a reputation then, you do now. He'll see to that—he and Cusped."

"You've made my day," I said.

We sat there facing each other in silence for a while. I looked at my watch. It was approaching eleven. It was hard to believe I had been nearly seven hours absent from Radio Heights.

"What are you going to do now?"

Nelson shrugged. "I'll head back to San Francisco. I was ready to leave when you showed up. I've got a job now, you know."

"Doing what?"

"Illustrator."

"What's new about that?"

"Well . . ." He did not even smile. "I get regular pay and work eight-to-five. Most of it's advertising layout."

"Ichabod," I said, "how the glory has departed!"

"Yeah," he smiled faintly. "I felt that way too in the beginning. You heading back to San Diego?"

"First I want to make a telephone call. Okay if I use the phone here? It'll be collect."

"Who's it to?"

"Ray Reynolds."

"Aren't you worried about them tracing you up here?"

"I suppose I should be, but I'm not. I'll tell them when I get home. I just want to check to see if there's going to be a reception committee down there waiting for me. And anyway, there's no reason why I shouldn't have come up here to see you . . . or Owen for that matter, except that I didn't tell them beforehand. And now . . ." I opened my hands in a gesture of emptiness, ". . . what's there to tell?"

He motioned to the hall. "Go ahead. It's in the bedroom."

Ray's blossomy voice accepted the charges with no questions asked, but after the operator told me to go ahead and we had the line to ourselves, he began to bleat.

"Where *are* you?"

"What's the matter?"

"Those two policemen you were talking to this afternoon have been back up here looking for you. They were *here*!"

"Did they give you a hard time?"

"They're perfectly awful! They want you, and I'm supposed to tell you that if I see or hear from you. That's the way they order people around! They found Eugene."

First things last. I hardly took time to swallow what felt like my heart coming up in my throat.

"What happened?"

"I don't know anything except that he's dead."

"Dead?"

He started to babble again, and it took a direct order to quiet him. Faulkner must have heard me; he appeared in the bedroom doorway and stood there, monitoring me.

"Ray, now calm down and tell me what you know."

"All I know is what Pinkerton told me over the phone about an hour ago. Eugene got shot, or smashed up, or both, somewhere along the border, between Tecate and Mexicali, or in one of those places. I don't know exactly."

"What did he mean, 'smashed up'?"

"Eugene ran into a ditch, or a tree, or something. He was shot, or getting shot at, Dale didn't know for sure."

"Was he on my bike?"

"Yes. Dale said there isn't much left of it either. He also said he's got to see you. What are you doing in L.A.?"

"I'm coming back down—right now."

"Do you want me to tell *them* if they call again?"

"Not unless they ask you."

"But they said I was *supposed* to tell them if I heard!"

"All right then, tell them!"

"Well, Scotty, I just don't want to get you in any more trouble than you're in already."

"Do whatever you think best, Ray."

"Oh!" he wailed, "I don't know what's best. I feel absolutely terrible!"

"Well, stop it! I'll see you later."

After I hung up I stared at the telephone, trying to make sense of what I had heard. When I looked up at Faulkner, he was still regarding me with that expressionless stare that told me something more than silence was expected of me. I related what Ray had said. He received it without comment. It was not until we had returned to the living room that he spoke.

"You think this is it?"

"It depends on what Eugene had on him."

"What do you mean?"

"His journal. It'll blow the whistle on everbody, including Owen, if he was a party to any of it. I imagine he noted down every name he called Friday night."

"What is this journal?"

I explained it to him. "It's pretty cryptic," I concluded, "but all the same it's not hard to figure out once you get the hang of his hieroglyphs."

Nelson fetched up a big sigh, then said, "Do you think the police are going to be around here?"

"It shouldn't take much time if, as it appears, Owen is mixed up in it. Right now, all I see is two down and one still unaccounted for."

"Are you going back to San Diego now?"

"No. First I'm going to try to get inside Symonds' place."

"You said it was sealed."

"That's his house. I'm not interested in where he lived. They've already been there. I mean the other place."

He looked for a moment as if he were going to ask me "what" other place, but then something in him relaxed, or gave way. He looked me straight in the eye and said, "There's a guard dog there, and it's locked."

"Been there already, have you?"

It was plain that both of us had had the same idea a few minutes before Ray's call when we were blithely prepared to part—Nelson ostensibly destined for San Francisco, and I for San Diego. I put a hand on his shoulder and, with a calculatedly condescending smirk, said, "I know the dog and I've got the key."

I started to turn, to go, but he pulled me around. There was that glary look in his eye again. I put a restraining hand on his arm.

"Easy, Nelson," I said. "Trust me."

22

THE EVIDENCE of that key had bothered me for longer than the period of events covered thus far. The sight and feel of it among the others on my ring had for two years been a reminder that the way was open should I choose to return to Symonds. It was a palpable symbol of that option. And it was through no forgetfulness on his part that I still had it, nor the result of an oversight on mine that I had never handed it back to him on leaving. Things are sometimes left undone through inertia born of unadmitted scruples. And the flawed moral crystallization of my second thoughts was a truer index of what was going on inside the shell of resolve behind which I had been hiding than any public gestures signified. There was an unconscious stealth at work, nourishing the potency of these unadmitted qualms which masked their meaning even from myself, as if I, like those who observed me, had nothing but my words and posturings to go by.

There were seven keys on the ring I used to clip to my left belt loop. This what to some might indicate a sinister buckling was, in intent, nothing of the sort. I dare say I learned the convenience of carrying a bunch of keys this way from Symonds and the others who formed his entourage or circled the outskirts of his nighttime retinue. They may have believed that they invented this sartorial conceit, but it was plain to any who looked beyond their present state that they had only appropriated it from the vestural customs found in the precincts where worked the blue-collared paragons they were to copy; which precincts also, but later, became the setting most congenial for the doggeries they established and frequented in the evenings, transfigured then as extravagant counterfeits of the super males they emulated in queer and stylized fantasies.

Their tasseled keys were an affectation that drew upon unnumbered and unacknowledged corroborating myths full of talismanic and erotic implications. For these bright, swinging indicia of andromorpholatry were at once a badge of office and a proxy for the concealed weapon a night's work would disclose, a promise or a threat of what might lie ahead—and of where a head might lie.

Keys in clusters, signaling as they do that the wearer has something worth his locking up, carry about them overtones of wardenry, the custodial and correctional significance of service and severity: the verger's function. Dandies and dullards in expensive skins, mechanics, clerks, and hairburners in artful denim, the cozener costumed as a cop, portal keepers and taskmasters of every trade and calling, sly winking clavigers who finger keys and work at penetration, riddlers with the knowledge and the means to open, close, bind, release, punish, or withhold as they please, or as they are told . . .

Well, and where is the predicate of all that?

Just this: that keys grow on who would wear them, for, being worn, they tend to multiply, for show or use, no matter. Once used to them, I came to feel somehow unmanned and vulnerable without their jangling weight, like a sabretache at my side. Symonds said he felt naked without them. Gloved, bonneted, and masked in the operating amphitheater, he wore them underneath his gown. He removed them only to go to bed or bathe, and even then he kept them near at hand.

All who wear keys walk with a common chinking sound, but each man's ring rings different from the next. Thief, his sharp ears, knew at a distance my keys' brassy tattling from all others. Just so, each cluster, ring, and clasp hanging full in view are deputies and surrogates for the collared, caverned, acorn-headed rod whose potency they danglingly proclaim for all to know who care about that sort of thing.

The keys on my ring, like battle pennants or ribbon clusters, represented certain developments, turning points in a past that was now catching up with me; a past I had tried to deny while, as with the keys, still holding on to what unlocked it.

The oldest of the seven was the one to my car. The car was a gift of Glenn's in the form of a sale (the money it took to have the pink slip transferred to my name) made to me when I was still living in Berkeley and coming down to see him almost every week. It was that same Oldsmobile, now trundling east on the Hollywood Freeway, that Symonds was driving the night he plucked me out of the bar on Melrose and took me out to Topanga Canyon. The fine black leather of the seats was ruptured in places, with the herniaed foam rubber popped out here and there. The blacktop had been replaced twice since then, and needed replacing again. Glenn had bought a Cad and had no use for this. At the time he took over Eugene he sold the Cad for good money, I guess, and bought a Rolls. When Symonds still owned the Olds, black and chrome were all that shone on it. Now it was black and faded red, and, needless to say, the chrome was pitted and rust-stained where it was not missing altogether. I used the car more than I did my bike, although the car cost more in upkeep and was on its way to becoming a collector's item.

The next oldest key fitted the ignition on my bike. And my bike was a Harley that had also once belonged to Glenn; that same bike I brought into Page Street years before to have the cams adjusted, and that resulted in my meeting Faulkner. Symonds also sold it to me for a token price the year I moved down to L.A. He bought himself a new one at the time, forewarned that the last of the stompers were barely to be had. No Electra-glide for him! Mine now, it was a vintage '55, the same as Faulkner's. Black, with blue and orange-lined trim, it had a chrome-edged windshield, a great, butt-polished solo saddle with cochoed fringes on the skirt, and black leather saddle bags like fat maniples over the rear fenders. The news that that classy Hog had been trashed by Eugene meant that the car was now my sole means of rolling support. But it was not the bike, but Eugene, I would miss.

I will skip the next, to return to it last, and take up the one just past it in time. This key unlocked the service entrance of the museum as well as the glass-paned door to the department of life sciences (my cubicle of an office had neither door nor lock: we were a poor and understaffed institution). This key was given to me on my first day at work, and represented independence and freedom and what I had believed to be a new life clean of the snares that had entangled me in the past.

The next to it symbolized much the same thing in that it commanded the front door of the creaking pile of asphalt, stucco, and timbers that I called a home. It opened the front door to the house. All other doors and windows locked from inside.

Then there was a thin one, somewhat like a car key, that unlocked Eugene's trailer. He had presented it to me when he repurchased the thing, and I joined it to the circle, never imagining that it might some day be regarded as incriminating evidence.

I bought the seventh, and two Schlage locks it opened, after I bought the house. One was for the tempered steel chain that secured the bike, and the other fastened the garage door shut during the rainy season when I stored the bike in there. The garage was otherwise usually empty and I seldom locked it.

Which brings me back to the third and now most important one.

When I moved in with Symonds, he gave me a key to his house in Coldwater Canyon which was also a master key to several other properties he owned around the city and county. My coming to live with him was made complicated at the outset by the stickling nature of my reluctance to enter into so undeniably a client relationship. With an advanced degree in reach but no job in sight, I had grown touchy with the pride of one who considers himself worthy but unappreciated. Specialized knowledge often temperamentally unfits a person for doing anything other than what he's spent his time and effort in perfecting. Glenn must have understood the slough I was in, and probably had as many second thoughts about my joining him under the same roof as I did, but he softened the way it rubbed against my self-esteem by establishing the fiction of a job that I might do in return for keep and enough money to make me feel that I was in some degree independent.

I had, and have, only a general idea of what Glenn did with the money he netted and none at all as to how much it amounted to, but a sizeable chunk, as I have mentioned already, was invested in property. Some of this was parcelled out in pieces of real estate bought as the opportunity arose and whim suggested, then held as tax shelters for a while before being sold for a profit as his agent, a partner in a firm with branches throughout the state, directed. Among these were modest items of transient interest: small apartment buildings and now and then a single family dwelling. There was also a beach house at Malibu that he rented out in the expectation of using it himself some day when he had the time. The master key was his idea of personal convenience, and at the time I lived with him it opened the four doors that gave ingress to his house, the entrance and six apartment doors of a white stucco job on Los Feliz, a large Culver City house near the Fox Studios that he sold again while I was living with him, and a five-unit job on Crane Drive near the top of Mount Washington in Highland Park. This last he had owned for at least five years, and it was here he had his workroom.

Faulkner wanted to know how I still had the key.

"I never gave it back to him when I moved down to San Diego."

"How did you get away with that?"

"It wasn't a question of getting away with anything," I said. "I didn't think of it, and he didn't ask for it back."

"You must've thought about it sometime after you left."

This was true, but, such being the rigid nature of my resolve to be quit of him, I saw no reason to make any special effort over a key, an action that might be misconstrued by him as a covert overture aimed at resuming relations of one sort or another. That I did not remove the key from the ring and throw it away was a result of inertia—perhaps. It is hard to tell sometimes where fish-bellied sentimentality is likely to disguise itself.

"Didn't he miss it?"

"If he did, he didn't miss it enough to ask for it. He didn't have to worry about me having it. Who knows? Maybe the idea of my holding onto the key to his house and the other places amused him."

Faulkner sniffed in disbelief. "It wouldn't amuse me. What are you going this way for?'

"It's shorter." We were headed south, and I was maneuvering to get into the turn lane for the Pasadena Freeway north.

"Thought you said you'd stayed away from this place."

"I had to go up there on tenant business a few times. I've never been down to that bottom one though."

"Why not?"

"For about the reasons you'd expect. I wasn't interested. Or, you might say, I was *dis*-interested."

"He tell you to keep away?"

Faulkner had been watching the freeway traffic most of the time, but I glanced over and saw he was looking at me. The mockery on his angular features took a slightly mad look with the lights from passing cars flashing on those wide-glaring eyes.

"Glenn never gave me orders," I said. "And he never told me to keep away from anything. I kept away from it for the same reasons I never hung around his clinic. His work and his play were no business of mine. And I don't know but what I'd have gone ahead and done it anyway if he had told me I couldn't."

"Sounds like you had a close relationship all right, if the only things you didn't have in common were work and play."

"I meant his work and his play. We worked at other things together, and what we played at was private."

"Why did you leave him if it was so good?"

"I forget."

Saying I would have done something anyway if Symonds had forbade me to was not meant as a boast, but as a measure of what I would have considered so unusual in his largely permissive behavior with me that I would have thought something was amiss enough to have poked my nose into it, as much out of concern as from curiosity. Eugene's call and my reaction to it was of much the same gauge.

And anyway, the real sense of Glenn's handing over his master key was that it indicated his complete trust in me. For the record (and his tax returns, to be sure), I was listed as his property manager, but what I really did was spend most of the time putting my thesis together. The few rents I collected, the occasional plumber or electrician or carpenter I oversaw, and the half dozen tenants I contracted with in his name hardly cost me a full day's work a week in the period I lived with him and performed those chores. Glenn knew I was aware of what he used the bottom unit of that hillside ziggurat for. He may even have thought I took a look at it at some time or other, although I had not. But we were by this time so fixed in the well-worn grooves of established behavior that any deviation on my part into areas I had for so long spurned would have been a cause for anxiety, if not protest, on his. He had grown to like me too well the way I was.

"How did you find out about this place?" I said, after we were headed toward Pasadena.

"After I got in yesterday morning from being out with Buddy all night, and I saw Owen hadn't come back to the house since I'd been gone, I shook it down. His bedroom looked like a bomb had gone off in it—he's usually pretty tidy—but I found some names with addresses and phone numbers. One had an address and a number, but there wasn't any name on it, only a doodle that looked like it might be like that tattoo Symonds's got on him. There wasn't any answer when I tried the number, so I rode up to take a look at the address before I had to meet Rublatz for brunch, and there was this goddamned monster dog behind the door trying to tear it apart."

I laughed. "That would be Fagin."

"Symonds'?"

"He's Symonds' all right, but he's my dog's litter mate. I did the obedience training on both of them, and both of them heel better to me than they ever did to Glenn because he never had the time to spend working with them."

"There was some guy that came out of one of the units halfway up those steps when he heard the dog going crazy. I asked him if there was anybody in it, and he said he didn't think so, but that the dog had been barking. He didn't know whose it was. Don't the tenants know their landlord?"

"I told you, Glenn never had anything to do with his renters. If they ever saw him there, they probably thought he was a tenant too."

"Think you can calm that mutt down?"

"He won't be any trouble. He was inside, wasn't he?"

"Yeah."

"Did you ring the bell?"

"Yeah. It turned him into a wolf."

I thought about it a minute, then said, "I don't think we're going to find Owen there."

"I'm not expecting to. But there's a chance we might find something that has to do with him."

". . . or Symonds."

I slowed down and took the dangerous pocket turnoff at Avenue 43, then swung left across the viaduct.

"Did you train him to be a guard dog?"

"He's not a guard dog, Nelson. He's an intelligent, well-behaved shepherd, that's all. And he'll take your arm off if he thinks you don't belong there."

"Fine!" he said. "Let's hope he thinks you belong there."

Fagin was not what was worrying me. Even had I not seen him in two years I would not have been afraid to approach him, but Symonds had brought him down to San Diego on one of his visits after we started seeing each other again. He only did it once, and I made friends with him again as if I had never left him. But we had to keep him away from the house and Thief. They were strangers now.

"He shouldn't be there," I said. "It sounds like the poor sonofabitch hasn't been fed in the last two days."

"Does that make him meaner?"

"Wouldn't it make you meaner?"

This seemed to amuse him, but I saw he was nervous, fretting about what lay ahead. I went north on Figueroa a few blocks, then took a turn in front of a blacked-out supermarket and began the climb up Mount Washington along the curving web of streets that pile the houses one on top of the other. Most of them were hidden behind screens of bamboo and mock orange and pines, with only their garages presented to the road like the gaping butts of gargoyles squatting out over the steep slopes, disdainful of the groping traffic behind them. Crane Boulevard became a narrow lane near the top. I was going by memory now, and Faulkner, the directions fresher in his mind, had to direct me when to slow down. It all looked alike at night.

He spotted it first. There was just an open carport and a rustic-sided closed garage at the curb with an overhang of maytens trees. The number was on a peaked wooden housing that sheltered five mailboxes. We parked, out of caution and necessity, over a block away, around a curve in the road and out of sight of the place. Then we got out and walked back.

"Aren't you going to lock it?" Faulkner saw I had left my window down.

"It's a convertible," I said. "Besides, we may want to get out of here in a hurry."

"What makes you say that?"

"The same reason that made me suggest we bring gloves: You never can tell."

The gloves were an afterthought. I had none with me, but Nelson loaned me his riding gauntlets. He had drawn on the skin-like pair of kid he used in the session the night before.

"What if he changed the locks?"

"Then we're out of luck. I'll know before we go down."

"How?"

"Wait and see."

Like most people, Symonds was a creature of habit, but unlike most, his habits were formed and reinforced by a conscious effort to conserve time and direct his energies in the most economic way. With his schedule and his interests, he had to. He was also a man more secretive than most. Who besides myself would know that one master key unlocked everything but his clinic? My guess was that no one did, unless it might be Owen. I did not suggest this to Faulkner; he was in a somber enough mood already.

"For all we know the police may have been here already," I said, "but it's not likely. The county offices are closed on weekends, so the police won't begin to sort out his holdings until Monday. I doubt if his business agent's moved on anything yet either, even if he knows what's happened."

"What about that broad you talked to tonight?"

"Jessica? She doesn't know his business agent, just his attorney. And he's the one who may already have gotten things rolling, but he won't know offhand where it's all located. Symonds played his cards close, particularly where money and property were concerned. Sort of not letting the right hand know what the left one's doing. The only thing I don't like about this is the dog."

"I thought you said you could handle him." Nelson had stopped in the middle of the street. I never realized before that he had a fear of animals.

"Come on. Fagin's not what worries me. It's his being here instead of at Glenn's house. One of the other tenants may have already complained if he's been making much noise. They howl, you know, when they hear a siren, and this place overlooks Figueroa."

"Don't you guess the other people here know their landlord's had it?"

"Don't think so," I said. We were almost there, and instinctively we slowed our steps. "The rents I used to collect were made out to Glenn Rose. He used his middle name. It was his mother's."

"Why did he do that?"

"I never asked him. Here," I stopped in front of the mailboxes. "Let me check something first."

I stepped over to the garage door and found the lock. The key slid easily in. This time it did not take me by surprise when I lifted the sectioned panels. By the pale light of a street lamp some distance away I saw all I needed to, and dropped the door back shut.

"What is it?" Nelson said.

"His bike." I lowered my voice as I rejoined him.

"Now I know how he got up here."

"It's Symonds'?"

"His very own." I motioned to the steps. "Shall we?"

The property was a series of five separate but adjoining cottages, boxes that stepped sharply down the steep hillside from the street, the floor of the one about level with the roofline of the one below it, all set in a row beside a long flight of steps. The garage and carport were at street level above them all. The wooden stairs connected the units like an open, scandent corridor. At the top, a small deck jutted out between the garage and port, and from there the steps dropped at a forty-degree angle. To the right, as one descended, were landings at the entrance to each unit. On the left-hand side patios opened out, bricked and cemented, one to every cottage, each backed by a ten-foot retaining wall and fronted with wooden railings that overlooked the one beneath it. Each patio had border plantings of mock orange and bamboo. They gave a sense of settled permanence and seclusion to what were really flimsy crackerboxes set on pilings nailed into the slope. The tenants, as I remembered them, were short-termers, mostly single occupants who seldom remained long enough to get to know each other very well. Besides their funky charm, these little dwellings possessed the added attraction of being within easy reach of the freeway, and so commanded pretty high rents. Ordinarily, Symonds would not have held onto the units as long as he had, but the one at the bottom of the slope was reserved for his own use and never rented out. The others covered expenses on the property, and then some.

We made our way quietly down the steps. The three upper cottages had lights on inside; the lower two were dark. Before the fourth landing, Fagin heard us and began to bark. By the time we reached the bottom he sounded as if he were going to tear the place apart. I stepped up on the stoop. "Fagin!" I called softly, putting my head against the heavy door. *"No!"*

The snarls quieted to whimpers, but his pawings shook the panels. I turned to Faulkner, who was leaning against the guard rail across the end of the landing watching me. "Stay there," I said, "until I see if I can get in and quiet him down."

He was not about to move. I continued crooning to the frenzied animal while I unsnapped my ring and felt for the master key. But when I grasped the handle and turned it, the door opened easily under my hand. Faulkner, I realized, had not really tried the door, or, trying it, had refrained from opening it out of fear of the beast waiting behind it. That did not surprise me, but I wondered that Symonds had not bothered to lock it when he left, assuming he was the last one out.

Fagin was out of the door in a bound and all over me in a second; I nearly went down with the force of his affectionate onslaught. I hugged him, petted him, talked to him. He left me briefly to rush over and sniff at Faulkner who remained perfectly still, then he threw himself on me again. When I judged we had had enough of his joy, I took hold of his choke collar and led him out to the edge of the landing and told him, "Okay!" He took off along the rocky slope, sniffing among the low-clinging juniper and acacia shrubs. I joined Nelson at the rail and watched him nosing out spots to deposit his own sign.

"Nice dog," Nelson murmured.

"Yes," I said. "You, of all people, ought to have one of those, of all dogs, Nelson."

"I don't need a dog," he said.

Fagin rushed from clump to clump, slipping on the rocky incline. The hillside dropped precipitously from the point where we were standing, but gentled into broad shoulders on either side. Below us the tangled lanes up which we had come were traced in the berylline glow of street lamps; further down there spread out a multicolored map of lights to the southeast over the Highland Park shopping district along Figueroa. Beyond this glimmered the moving beams of the freeway, the dotted sparklings of Rose Park Hill, and, far off, the softened smudge as of an afterglow that was the far lights of Alhambra fading into the smog-sullied night. To the northwest, I knew, the San Gabriel Mountains rose hidden in the dark; they stood out on clear days in bare, blue escarpments lifting up from the broad flatlands of Pasadena in the foreground.

"Not a bad place," Faulkner said.

That was the way I had felt when I first saw it. I pointed behind us. "I wanted to have the one next to this one when it became empty, rather than living in Coldwater Canyon."

"He wouldn't let you, huh?"

"You still don't get the picture, Nelson. He would've 'let' me, but I found out what this was used for and changed my mind. Let's take a look. I've never been inside it."

"What about the dog?"

"It might be better to leave him out here for a while. He's probably been cooped up in there since Symonds left. He won't go far."

Faulkner grunted assent. I knew he was still weighing the possibility that I was keeping something from him. I was more in the dark than he was, but the dog's being there indicated to me that the place was empty. Only Symonds could have brought him there; no one but I could have opened that door after Symonds left. As we moved to go in, Fagin bounded up on the landing. I told him to stay, then we went inside. Faulkner closed the door behind us while I felt the wall for a switch.

These attached cottages were little more than expanded studios. A shallow entry, flanked on one side by a bathroom and on the other by a kitchen, opened into a single, very large square room. On one side it extended out in a broad alcove made porchlike by an expanse of windows through which was seen the same view to the east that we had just surveyed outside. A counter and a deep valance above it accentuated the division, at the same time giving a feeling of light and extended space to the windowless main room.

A small archway connected the kitchen with the alcove and, as I remembered, tenants used this narrow, glassed-in space as a dining or sleeping porch.

Symonds neither slept nor dined here, but the place was redolent of him, permeated by an unmistakable olfactory clue, the way an artist's lodgings are imbued with the smell of turpentine. I recognized it at once: a compound of leather and Lexol, of iron and other metals, of furniture polish and a faint sweetness akin to sandalwood, and other, unidentifiable components. I was prepared for the sight of his eclectic, not to say eccentric, tastes in furniture and impedimenta that I knew made up the accessories of his covert operations, but the ordered clutter and variety of the objects that filled the room made us pause, stunned like intruders on the threshold of an ancient tomb.

"Holy shit!"

Nelson's unwittingly apposite expression of awe delivered in a whisper over my shoulder was a tribute to Glenn Symonds, for here before us was displayed all the paraphernalia of his calculated dementia, the amassed accessories of a perfectionist gone mad with striving to realize the pure, if limited, vision by which he had been for so long obsessed. I turned up the dimmer higher than Symonds would ever have done in any session he might have arranged here, and I almost physically recoiled against Nelson from the resonance of foreboding the sight of that room roused in me.

A quick glance revealed an almost exuberant excess; closer examination discovered the wide-ranging but singleminded direction of the collector's eye. Like a gallery of sacred instruments, the cork-lined walls were festooned with a heterodox display of artifacts swearing at each other from disparate modes and eras of devotion. From ceiling to baseboard they cluttered the mulberry-colored expanse: a variety of sickles, scythes, and baling hooks; an ivory wand, a winnowing fan, an old hay fork and a harrow disc; a mirror shattered into a perfect sunburst; a silver trencher, a golden hook, a crook enameled like a coral snake; a hinged flail of wood and brass, a leather sapper and an iron mace; a brace of whips ranging in size from a French, brass-butted crop of pizzle to a coiled Durango bullwhip five yards long and forked at the end; balance scales and beam scales, a dozen axe heads of different shapes, an obsidian knife; a stiff leather apron of the sort that slaughterhouse workers used to wear; a fan of daggers with a spray of swords surrounding them on a piece of tapestry depicting Saint Hubert and the Stag, a pewter basin notched for the barber's and the bleeder's art; odds and ends of armor; kepis, busbies, shakos, a Uhlan's cap, casques with spikes and helms with plumes, a cuirasse, gorgets, and bucklers; a cutibière, vambrace and gauntlet all pitted with age but polished of rust and newly joined with fine leather and velvet, ready to be worn; a score of spurs and stirrups in a line; brass knuckles (made of iron). Mounted to form a border just beneath the ceiling was a line of masks old and new, some of which he must have imagined and commissioned himself. And there, ranged along the wall encompassing half the room, boots, boots, boots of every kind and age and use. Four dozen pairs the papers later noted.

Like children in an ogre's cave, we gaped at the jumbled panoply of that chamber of parade. Much of what I saw assembled there I had seen at times before as separate entities, hardly noticed or remarked upon before they disappeared unmissed, forgotten until now. But the imprint of their combined assault on my senses still comes clearly to mind in almost full detail. I remember that a very fine, old oriental, worn in places, but still bright with its pile, covered the hither half of the floor. An obelisk cabinet of polished marquetry and ormolu stood against the wall just inside, near the entry. On its glass shelves was a further display of sacra: an ivory and silver aspergillum standing in a bucket of enchased silver, a silver sistrum with ebony handle and gilded discs, a silver incense boat cast in the form of a nautilus' shell in which sat a tiny golden scoop, and a bronze, baroque thurible plated with silver even to the rings and chains. Above it were scattered a collection of abraxas stones in clay, carnelian, ivory, lapis, and jaspar. A paten, chalice, ciborium, and communion spoon in gold, silver, and brilliant enamels filled another shelf. On the topmost stood a radiant monstrance of vermeil within whose hinged crystal lodged a diminutive, but common key whose use we knew. Two small, exquisite reliquaries flanked it, also containing similar keys. Symonds' profanitory tendencies had always amused me, but the amount, the value, and the connotations of this sacred treasure were disconcerting.

But there was more.

Against the counter stood a brocaded sofa in the style of Louis XV. Two matching chairs faced it across a walnut tea table on which a silver service sat. Behind the sofa, on the counter's end, a tall lamp made from a Celadon vase stood shaded in silk and velvet. A great Biedermeier wall clock with bronze weights and brass pendulum shaped like a sunburst ticked away at nearly midnight. Depending from a rafter by a rusty chain there was a tinted Venetian glass chandelier with candles. And a lot of other fancy stuff, castoffs or borrowings from Glenn's forays into auctions, junk shops, and overpriced antique dealers on La Cienega cluttered this space.

It was as if this epicene and overwrought elegance were arranged as a malicious fillip to the algolagnic play enacted here, an exhibition of the antithetic tendencies his personality was prey to. For in such a setting, was he not countervailing in blatant imagery what he loved, and despised himself for loving, for what he worshipped, and sought to have all others worship in him? This foreground setting was a joke, a riddle within a riddle, and the lights I had switched on on entering directed the eye of the visitor immediately to all that really mattered here.

Those lights were spots housed in recesses just under the roof. The ceiling proper had been removed, exposing the shallow crawlspace, leaving the rafters free. The shafts of light shone down on a table bolted to the floor at the opposite end of the room like an altar stripped of cloths and frontals. Its single base of enameled metal rose in sharp concavity before swelling again to support the steel surface which was covered, I knew without having to examine it, in leather. Crossing the old Sarouk, we stepped onto a field of gray gym mats cut and sewn to the dimensions of the far country which was the other end of the room. They curved snugly around the base of the operating table and had button-holed slits here and there through which anchor rings gleamed brightly against the neutral canvas. I recognized a laminated piece of furniture whose design Symonds had asked my opinion of some years before, before he had it made. It was a kind of sawhorse, but resembled its prototype the way a scimitar resembles a butcher knife. I had tried it out on Glenn just for fun to see how it worked, and saw nothing more of it after that, until now. It lay overturned on the matting in front of the table. Straps and clips were still attached to it, strung out in tangled confusion as if hastily undone and abandoned. Other ropes and chains hung loose from tackle attached to the rafters directly above the table in festoons like lianas, compounding the look of carelessness and sudden departure.

In a corner beyond the perverted altar, where might have stood a credence table, was an iron and glass instrument cabinet set on cabriole legs; a heavy yet graceful piece of medical furniture not often seen in doctor's offices anymore. Its door of beveled glass hung open, and on the glass shelves a collection of instruments caught the light with the shine of chrome and nickel. A few were spilled out onto the mat below.

Brushing the pendant straps and chains aside, I walked around to the other corner where, in compliment to the cabinet, was a maple pharmacist's chest with shallow drawers rising from the base to a milk glass top at shoulder height. Here too was evident an untidiness foreign to Glenn's habits. Several of the drawers were open and their contents spilled. Two were filled with nothing but gloves, all kinds of gloves. On the floor nearby lay his branding iron. I had never seen it, but I knew it for what it was without having to verify the rose-and-whip in high relief on the iron. I had sense enough to keep my hands off everything.

And there was the blood. Towels soaked with it were piled on the mats behind the table between the chest and ambry. Stained pads of surgical gauze lay about like waste paper, and a spattering of drops, dried but still red, covered the canvas between the wall and table. The table itself had been wiped clean, but a dull smear the size and shape of legal bond showed under the harsh lights. A blanket and pillow, likewise spotted with blood, were gathered in a rumpled heap near the towels, and a telephone sat beside these. Stooping down, I looked closely at a spancel and a set of fleece-covered pony hobbles under the table; they also had smears on them, the color of rust. The inventory grew as I poked about the debris, wondering at the soul seizure that must have possessed him to result in havoc like this.

A cry from another part of the room nearly jerked me out of my skin. It was Faulkner who had gone exploring on his own. I heard it again—"*Jee-sus!*"—in a rising wail, and, looking over the edge of the table, saw him disappear through a doorway that connected the windowed alcove to the kitchen.

I got up and made my way at a sprint around the clutter of furniture to the end of the counter. On the other side I saw legs and part of a torso clad in brown boots and chino slacks. The counter shaded the rest of it from the ceiling spots. I turned on the lamp and got the whole picture while hearing the sounds of retching from the kitchen. It was the body of a darkly blond young man. He had been shot through the chest from the look of the hole torn out of his back. He lay sprawled on his stomach with one arm twisted under him, as if he had tried to rise after he fell and rolled back over it after losing consciousness. What bleeding he did must have soaked into the Hamadan runner beneath him, further darkening its reds and browns. A trickle of blood had dried where it issued from the open mouth. It was difficult to tell much about him; the tissues had begun to settle. The eyes, not quite closed, shone with the dry gleam of death. A curl of hair partly covered a design charred on the forehead, a burn that would never slough and scar: a loop and a rose. I bent down close and touched the lock of hair to make sure, and suddenly tasted the bile rising in my throat. The body was beginning to give off an odor.

I looked up from the ruin at my feet and beheld a desolate Faulkner leaning against the archway. The tears in his eyes were from the bout he had had with his stomach.

"Is it Owen?" I said.

He nodded. The muscles of his jaw tightened in spasms. I could tell nothing from the unblinking stare he riveted on the body. Standing up, I ticked off three fingers with my thumb.

"Three out of three."

He looked at me, then back at Owen, but said nothing.

"There's something familiar about him," I said, "but I can't remember ever having seen him before."

"Symonds was his uncle," Faulkner said, and made a dash for the door.

23

WE WERE outside again and for the moment I had too much to do to think about it. I was holding Faulkner's hard, washboard belly while he puked into a border of agapanthus in the adjacent patio. At the same time I had to keep Fagin back—Fagin, who did not understand what I was doing and wanted to help me discipline this stranger. I barely kept from throwing up myself.

When the three of us had finally quieted down, I went back inside, turned off the lights, and, as I came out, locked the door. Then, thinking better of it, I unlocked it again. Faulkner stood on the landing watching me, and Fagin was poised behind him, waiting to see what was expected of him.

"Let's get the hell out of here," I said.

Without a word, Faulkner fell in beside me and we climbed the long flight of steps with Fagin heeling on my left. There were no more lights on in the other units, and I doubt if anybody knew we had been there. At the top, we paused by the curb and, as I stepped into the street, I had to take Nelson by the arm and all but lead him, so like a sleepwalker had he become. But it was not until we reached the car and I had put the dog in the back seat that I could give something like my full attention to this man who always addressed me as "friend," and then I marveled at the remarkable change the last half hour had worked in him. He leaned against the front fender, hugging himself, brooding and volitionless. I took him by the arm again, steered him around to the other side, and opened the car door for him. He got inside, and I went back around to join him in the driver's seat. His zombie-like passivity had me worried. I wanted to get out of there, but I did not want to start the car until I knew what he might do. He resolved the dilemma.

"What now?" he said in a voice that was almost a whisper.

I looked at him, trying to gauge the curious torpor investing his usually tense body. He looked straight ahead; his lean, lined face, all peaks and hollows, showed the profile of a Bedouin in which was set an obsidian eye, shining now with hatred or with emptiness. It was impossible to tell which, but it made me uneasy to see him this way, vulnerable, undone. And it came to me then that I had been thinking and deciding for both of us ever since he released me back in Owen's bungalow. I reached over, unsnapped his jacket pocket, and fished inside for cigarettes. He accepted one from me without moving his head. The matches were in his shirt. I got them and lit his for him.

"I'm going to drop you off at Owen's," I said, making up my mind as I spoke. "You can call the police from there and tell them about this."

"About you too?" he said.

"Yes. Tell them the whole damned thing. I'll head back down to San Diego. They're expecting me, and I'd rather talk to that bunch in National City than some new ones up here. Just give me a half hour to get out of the county before you call."

"You think that's what we ought to do?"

"It's the only thing to do. There wasn't anything wrong with us coming up here looking for Owen, and that door was open to anybody. Nobody else even seemed to know he was missing. And if it's anybody's fault for not alerting the law to him, it's mine, although I didn't begin to suspect he had anything to do with Symonds' killing until you told me about him."

He went on smoking, as if I were not there. I put a hand on his thigh and said, "I'm sorry, Nelson . . . about Owen."

He accepted this in silence too. I started the car and backed into somebody's carport to head us back down the way we had come. It wasn't until after we were on the freeway once more that I spoke again.

"Why didn't you tell me he was Symonds' nephew before we went up there?"

He took a deep breath and appeared to come awake.

"I thought he was still alive," he said.

"What difference would that make?"

"He didn't want me to tell anybody. He hated Symonds."

"Why?"

Again he drew deeply for air, and rolled the window down a crack to throw his cigarette out. He lit another immediately.

"You remember I said that Symonds had two lovers before you? Well, I don't know anything about the first one: he was some foreigner. But the second one was his nephew too—Owen's older brother."

"He'd be the one who committed suicide."

"That's what they called it."

"Are you saying Symonds murdered him too?"

"Not in any way it could be proved, or even suspected. But he worked him into doing it."

"How?"

"Talked him into it. Shamed him into it. Drove him to do it. You know Symonds."

I wondered now. "Did Owen tell you all this?"

"No. He found out about it from me. I put it together for him because I knew the other one—his brother. Not well, but I knew him. And I didn't want to see the same thing happening to him."

"Did you see the other one happen?"

"I watched it happening over a period of time. Shit! Everybody who knew Symonds knew what he was aiming at."

"Why didn't everybody stop it? Why didn't you?"

"You can't stop a man from jumping off a bridge if that's what he's got his mind set on doing. Besides, I wasn't interested in him."

"Is that how he did it—jumped off a bridge?"

"No. He did it with a rope. But Symonds kicked the fucking stool out from under him—or he as good as did it. It's not anything anybody can prove; it's something you know, and other people don't."

"Why in hell did you tell Owen this?"

"I liked him." He waited a moment, as if weighing what he had said. "I more than liked him. I didn't want him getting mixed up with Symonds, and I knew he was. I thought I could talk some sense into him. If he hadn't moved down here this wouldn't have happened." He waited again before going on, drawing deeply on the cigarette. Then he added, "He was a lot like Symonds."

"You said a little while ago that he was a lot like me."

"He was a little like both of you, but there's more of Symonds in him than you."

We crossed the overpass on Los Feliz; there was almost no traffic. While I waited for a light to change, Faulkner stirred, crushed out his cigarette in the ash tray on the dash, and sighed.

"Did you see the way the bastard burned him?" he said.

"I saw it. He was already dead when that was done, Nelson."

"That's what I mean."

"Glenn was insane. You've said it yourself, and I realized what was happening to him a long time ago, only I wouldn't admit it."

"Why wouldn't you?"

"I don't know. Pride, maybe."

"So, we're all crazy, but *he did it. He* actually did it. Twice! And this time it was the end of the line for him, and he knew it."

Hearing him say this, I knew that Nelson, at least, would never have to have the significance of Glenn's suicide explained to him, no matter how many others would not buy it.

When we got back to Owen's place, I pulled up to the curb and left the engine running. "You call them," I said.

"I'll call them." He was relaxed now, more like his old self. He looked out at the shabby duplex, then down at his gloved hands that held the gauntlets I had pulled off and given back to him when we got into the car. Then, as if at a signal, he moved with that old suddenness, and with a crackling of leather reached across the top of the seat and pulled me to him. Those gloved fingers wrapped firmly, but gently around my neck which still hurt from where he had brought me down earlier, and I felt the close rasp of his cheek against mine, his voice in my ear. "You call me when this is all over, friend."

He got out and then leaned back in to say he would wait thirty minutes before he telephoned the police. "You don't want me to say anything about you being in that trailer Saturday, do you?"

"Not if you can help it," I said. "But if you have to, I'll understand. And remember, that door up there was unlocked."

Before I reached the on-ramp of the freeway, I stopped and unringed the car key, then let Fagin get in the front seat where he could stick his head out in the wind. At San Clemente I halted for gas and something to eat: three hamburgers for him and one for me; then I telephoned Ray. I asked him to go and bring Thief over to his house before I got home. He wanted to talk, but I said I was too tired to do anything but drive, and that barely. After that, I called the National City Police Department and told the cop on duty to tell those detectives I was on my way in. Just past Oceanside I rolled the window down and threw the clip with the ring and six keys still on it into Buena Vista Lagoon as we sped over the bridge.

It was two in the morning when I pulled up in back of the municipal building in National City. I let Fagin out for a quick run on the grass in the little park across the street from police headquarters, then put him back in the car, swallowed my last pill, and went inside.

After

24

THE FINAL accounting stretched itself out over a period of weeks, and after all this time since then there are still patches and discordant rifts in the fabric of intent the police of two cities pieced together, but an acceptable pattern was laid out and the case sewed up in the few hours preceeding and following my return to San Diego County.

Not that my being there or elsewhere had, after all, much to do with it. Nelson and I got away from Mount Washington just before the L.A. police moved in. They were already on their way up to Highland Park when he contacted them, as we agreed he should, and a car was detached from the detail to pick him up at Owen's place, after which he went through a long night of questioning like mine.

My own interrogation was in the nature of a final sweeping-up by the National City bunch. It amounted to a reprimand with variations, and was as demeaning an experience as I have ever gone through. Don and Pete were waiting just inside headquarters, behind the glass-enclosed communications center, when I walked in. Without a word, they ushered me behind the counter and down a hall to a bare, aseptic room painted an institutional beige. We sat ourselves around a steel table of the same shade and thrashed it out. Another plainclothes detective joined us, and a cop in tan uniform stood by the door through most of it, leaving only to get us coffee when we paused in our work and took a silent break, which was not often.

"*Mr.* Scott." Having learned the distinction, Don liked to emphasize the difference. He did not offer to shake hands this time, but motioned me to sit down. "Do you want to tell us about this?"

I did not, but I did. We went over everything that I had done, and not done, since I saw him on Saturday over Symonds' body up on Kearny Mesa, although we had already covered much of it the afternoon before. The only thing I omitted were the details of Saturday night's session in Chris' basement and the motions Chumly and I went through in the front seat of my car Sunday morning. They took up again the matter of my going to Stallion Oaks, of why I had not informed them of it afterwards. To which I responded as I had when they asked me on Sunday afternoon. "You were out there checking on it Sunday morning, weren't you?"

"What about Owen Keller?"

"Yes?" I said. "What about him? I could've been calling you every hour on the hour all night Saturday and most of yesterday bothering you with the crap I've been hearing from people interested in putting forward their theories as to what was going on. Very little of it made any sense to me. I'd never heard of Keller before, and by the time I did, so had you."

Don said, "Let's go back a little. Why didn't you come clean with us in the first place when we talked to you Saturday night instead of pretending you didn't know anything?"

I told him I did not know what he asked me that I had not answered as best I could, but they saw through that. It was my equivocation on the matter of Eugene's and Symonds' sexual orientation that he was getting at.

"Look," I said. "When you talked to me then, I had no idea I was going to be mixed up in this. I told you about Eugene calling me, and I figured that was all that was important. I figured you'd have it wrapped up pretty quick, and that I could get back to minding my own business and go on as before. Why should I spread out anybody else's private life for you if what they did might lay my own out in public as well? I couldn't see that it would do you any good unless you had a prurient cast of mind."

I stared at him. He looked over at Pete, who was sitting at the end of the table, then back at me. I forged ahead, trying not to squirm.

"I thought it would get cleared up without me being involved, and that as soon as you found Eugene, or he found you, he'd be off the hook and I wouldn't be bothered. I didn't want to drag myself into it if I didn't have to, or hurt anybody else in the process."

Don let me know that they were the best judges of who ought to be dragged into an investigation, and that nobody would have been hurt who was innocent.

I found that naïve, and I told them, "I've probably lost my job because of this. How's that for getting hurt?" I was too roaring tired to care about the lip I was giving them.

"How do you mean, 'drag yourself into it'?"

"I'm talking about my private life."

"Private lives is the name of the game in this case, wouldn't you say?"

"Let me put it this way." I tried sounding reasonable. "I can't suddenly dredge up every conceivable association, or hint, or hunch, or lead, or bit of gossip about somebody I know on command from a stranger, even if you are a policeman. To put it mildly, it was unnerving. I was under a lot of strain and worry after hearing what the radio and Dale Pinkerton had to say. Over the last couple of days it's been coming back to me in bits and pieces after a lot of mulling over it and talking about it with other people."

"Why didn't you let us know when these bits and pieces came back to you?"

"What did I come up with?" I protested. "Most of the time it was the kind of idle reminiscing you hear at wakes. It was depressing. But when I *did* find something solid, I hit that phone right away. There was the car in my garage and there was Owen Keller's body up in L.A. I had my friend who was with me call on that one since I wanted to get back down here instead of being questioned up there. I also telephoned Ray Reynolds on the way down to let him know. And I called you to let you know I was on my way to see you. How's that for cooperation?"

It was not good enough for him, but I had not expected that it would be.

"Why didn't you tell us about Owen Keller before you went up to L.A., or tell us you planned to go up there?"

"I wasn't going up there to see Keller," I lied.

"Why didn't you call the Los Angeles police about the place up in Highland Park before you went in on your own?"

"Frankly I didn't think there was any need to ask anyone's permission to go look for him. It was only a hunch anyway, and we were lucky to find the place wide open. And," I added weakly, "we didn't expect to find a dead body when we got there."

"Why did you leave here and go to Los Angeles?"

"I went to Crystal Cove. L.A. was an afterthought."

"Why did you go to Crystal Cove?"

"I wanted to get away, just away. From here, from people coming at me with questions, from all these . . . associations. I could as easily have gone up to the mountains, or to the desert, . . . anywhere. But I went up there, I suppose, because I wanted to find out how this was sitting with Dr. Silver. He's my boss."

"What do you mean?"

"I wanted to know if I ought to bother coming in to work today or not."

"What's your work got to do with it?"

"Come on, sergeant!" I reminded him that this was San Diego County and not San Francisco or L.A. "The museum is privately endowed; I'm not on civil service. The stain of this kind of thing rubs off on me. Anyway, I didn't see him. He wasn't there, so I spent some time with Mrs. Silver."

"Did you know she was Dr. Symonds' sister?"

"No more than I knew Owen Keller was his nephew. You see," I added, recalling his remarks on Saturday night, "you *can* live with someone and be completely ignorant of important facts about them."

If it made any impression on him, he did not show it. Instead, he asked me why I went from there to L.A. I thought for a moment, then said, "Two reasons, I guess. First of all, I wanted to drive past Glenn's place."

"His home? Why?"

"Who knows? Sentiment, maybe. I wasn't looking for anything. Just to look *at* it. And I'd gotten the idea of talking to Jessica Fenton. She's Symonds' secretary."

"Was that the second reason?"

"She was the answer to the second one. Something Mrs. Silver said when we were talking at Crystal Cove made me think she knew Symonds. Jessica told me she was his sister."

"Did you think this had anything to do with his death?"

"No. Why should I? Something else less important but more complicated than that bothered me, and it didn't have anything to do with what you're after."

"Try me."

He reminded me of Faulkner then, except that he was no friend. Nelson would have approved of his questioning.

"It's just something very personal," I said. "One more thing I was finding out that I never knew about him—about Glenn."

What I could not make them understand, because I didn't want to go into it and, anyway, it would not interest them, was that because Glenn's influence was both more direct and more purposive than I had ever suspected, my breaking from him had been less of a salutary act of will on my part than I had believed at the time.

"If he was your friend, why shouldn't he talk his brother-in-law into giving you a job?"

"Symonds didn't know Dr. Silver, and I don't believe he was on close, or even speaking, terms with his sister. He exerted influence on someone else, or maybe on several others, on my behalf. I don't think Louise Silver had any idea that I knew her brother until after you went up there to talk to her."

He let it drop, and asked me the big one.

"If you never knew anything about Owen Keller, then how did you know where he lived?"

It was so easy, mentioning the phone book.

"And when did it occur to you to call him?"

"After I finished talking to Jessica Fenton. Only I didn't call. I just got the address and went over there."

"Why didn't you telephone first?"

"I thought about it. But I decided on surprise. I half expected Eugene to be there. When I found the place, I caught sight of what I thought was my bike in the driveway in back. It wasn't mine."

The rest we had already gone over, so they took a break after making me repeat some details having to do with what Nelson and I did after I got there. I omitted the circumstances of Nelson's violent reaction to my unexpected appearance. It would only have confused the picture further.

The uniformed cop asked me if I took cream or sugar, and left the room. Don went out with him, leaving me to Pete's sour scrutiny and the bored presence of the third detective. I guessed that Don was going to check with the Los Angeles team to find out if anything in my story failed to correspond seriously with what they might have gotten out of Faulkner. He was a long time gone, and it was a bad wait through which I tried not to fidget by downing the coffee slowly. When Don returned, he looked as if he still did not believe me, but he had apparently come up with nothing that could get me booked. Faulkner suggested to me later that I need not have talked as much as I did. He, cop lover though he was, was not one to palter, and he had not blown my cover in this or in anything else. It must have cost him a lot to tell his idols he did not know what he manifestly did know.

As for these who surrounded me now, I really do not believe they knew quite what to make of me, or what to do about what they held against me. Don sat on the edge of the table and offered me a cigarette. I had refused several earlier, but took one from him now. He lit it, lit his own, and adopted a musing expression, as if he were not quite sure how he would phrase his next question.

"Mr. Scott," he said at last, "you're a scientist." He spoke with the emphasis one might use in dealing with someone who was being unnecessarily balky over a matter on which he might be susceptible to flattery. I waited.

"I'm wondering how a logical mind like yours might see this. What do you think happened between Dr. Symonds, his nephew Keller, and Eugene Carmichael?"

I shook my head and tried not to sound stuffy.

"It's not really within my competence, sergeant. I expect you're the more logical of the two of us when it comes to speculating on that. What you're dealing with is emotions, and I'm not trained or even naturally inclined to sort out things in that area. And after what I've found out about Glenn Symonds over the past two days, I think I know him less well than I did before."

"I'd say you knew him very well, Mr. Scott."

I stared at my cigarette and let that pass.

"You don't believe Eugene Carmichael killed either one of them?" he said.

"I think Symonds killed himself."

He made a sucking noise with his tongue and allowed that Glenn as good as did it over the long run with the kind of life he led. "But wouldn't you say that Carmichael's refusal to come forward and his flight to avoid arrest are pretty incriminating on the face of it?"

"Who shot Owen Keller?" I said.

"Carmichael."

"Is that so? Where did he get the gun?"

"The gun was registered in Glenn Symonds' name. It was found in Carmichael's trailer."

"All right. Why would Eugene have shot Keller?"

"Try jealousy."

I looked away, angry and disgusted. "And who tried to saw Symonds' nuts off, do you think?"

"Carmichael. It all fits in."

"No way!"

"Then who did?"

"Owen Keller. He planned it and he did it. And his mistake was in botching the job."

"Botching it?"

"In not finishing off Symonds after he had him where he wanted him."

This drew no reaction at all from them. They waited, seemingly bland and indulgent, for me to run out of things to say.

"At any rate," I finished up, lamely enough, "Eugene couldn't do anything approaching what you're trying to pin on him."

"What you mean is, you don't believe he could do it."

"I *know*. And anyway, if he wanted to kill them both, then why shoot one and then go through all that complicated procedure to do the other one in?"

"It's the kind of games you people play, isn't it?"

You people . . .

I said, "Has anybody checked Owen Keller's credentials? He may have been crazy, you know."

"We're working on it."

"Fine. Now meanwhile you've assumed Carmichael went up to L.A., cut up a man he cared a great deal about and had immense respect for, then, not content with that, out of jealousy shot a kid he may or may not have known was this man's nephew, and after all this somehow managed to get his now-crippled victim into an automobile, left him in it unattended while he went into a post office to make a phone call to tell somebody—me—what he was up to, then called a half-dozen other people for a get-together, then drove all the way down here, hauled his bleeding captive into his trailer, shackled him and hung him upside down, and shoved a cyanide ampule in his mouth and made him bite it."

I took a deep breath and let it out.

"You'd have to be a fool to believe all that."

"All right, Mr. Scott. You can't accept the possibility of Carmichael killing Keller and castrating his . . . what would you call him?"

"Friend," I said. "Call him his friend."

"All right," he laughed, joined by the others in the room, "his friend. You won't buy that, or Carmichael's stringing him up and forcing cyanide on him. And yet you have his own word for it that he brought Symonds down here. And we know from the condition of that car seat that Symonds was bleeding before he took that ride with *your* friend—no offense."

I said that I would not necessarily object to the idea of Eugene's hanging Symonds up the way he did, or even of putting the ampule in his mouth if Symonds had told him to do it. "Symonds may have used him."

"Used him for what, Mr. Scott?"

"To bring him down here and string him up and then to call everybody in a certain selected circle to come and see what he had arranged for himself."

Don appeared suddenly very interested in this. "Now why would he want to do that?"

"Maybe as an exemplum."

The three of them exchanged glances.

"As a moral to others," I explained.

This made no impression on them. After a moment Don suddenly asked me if I had a tattoo. I smiled, resigned to what I knew they were thinking, and unbuttoned my sleeve. They all crowded close to look at it, the whip and the rose.

But was it really a whip, or an S? And was it a rose? I told them I did not know; Symonds had never explained it. My reading of it (which I did not tell them) saw it as a flair, the soaring path of a signal rocket's climb and sharp descent, its exploding plummet where, in a final, brilliant expansion of light, it burns out, leaving nothing but an enigmatic message seared on the retina of the blinking eyes that witness it.

I cannot remember, if I ever knew, what caused me to have Glenn Symonds' impressa scratched in the flesh of my wrist (a careful place where a watch would cover it) after I left him and came down here. A reminder? A caution? A safeguard? Something. He, at least, never knew I had it done.

"Is that the only one?" Don asked as I put my watch back on. I nodded.

"I don't have the motto, if that's what you want to know, But . . ." I smiled, putting a hand to my fly. ". . . if you'd like to check. . . ."

"That won't be necessary," he said. "But maybe you'd like to tell us what that other one means."

I shook my head. "Sorry, but I don't know any more about it than you do. It was another one of those things I never asked about, and he never volunteered any information."

"It doesn't make any sense."

"Maybe it's just his comment on what was an important aspect of his life."

"'I leaven lees'?" Don snorted. "What kind of comment is that?"

It was my turn to shrug. "Well, Glenn was a doctor, remember. He might've considered this other part of his activities as therapeutic too."

I didn't mean it to be funny, but they got quite a kick out of it anyway.

Oh, and there were other things I remember now from that last interrogation, many of which I would prefer to forget, but the inflamed wounds to self-esteem have long ago healed with the change that came about in the climate of opinion during the intervening years separating that time from this. Don and his kind now lean over backwards to accommodate what they once delighted to hook with the barb of their safe scorn. It galled them, and perhaps properly so, that I had withheld information, but they would have been hard put to derive any criminal intent from what I did although, in the last analysis, what I did was felonious enough had they wished to push it into court. My luck was that they did not want to have me any more on their hands contesting their simple nigger theory of what some crazy perverts had done to each other. Disprobative evidence more than balanced out the factors pointing to Eugene's probable guilt, but I was not going to press upon them the cloudy issue of Symonds' peculiar attitude toward the institution of victimhood or argue with them that he was fully capable of dispreading violence over as large an area as he did and arranging the resulting carnage in such a way as to baffle those who would come seeking to piece it together afterwards. Eugene's skills fell far short of accomplishing Glenn's extinction all by himself. What was more likely was that he went a fair way to being unhinged by trying to fulfill the turnabout's role assigned him by his demented, but still beloved master. It was ironic that, but for the yapping of Eugene's little terrier, the instructive display Glenn had arranged for Saturday night might have come off as planned. I can still almost smile, picturing the Gadarine panic that would have sent those visitors flying in all directions after they had answered the summons to that trailer park.

Well, in neither county was the law all that sure of the case they made of what happened; the untidiness of their wrapup was not lost on Don or his superiors, but they were inclined to accept it in their distaste and hurry to be done with it. No one was hurt by Eugene's death and the uncontested assignment of guilt to him if one discounted his mother—and they did discount her. She, by her own ignorance of the life he led, pointed up the accusation the revealed elements of his life attracted. And so little was known of Owen Keller (no one came forward to speak about or for him—his own mother said not a word) that nothing could be construed from his part in it save that he had taken up with his uncle and been done in because of it by a disgruntled maniac who had, up to then, displayed a quiet and docile manner which all along was no more than a mask covering the pathological insecurity that finally exploded when so valuable a prize as Symonds was taken from him.

This was to be the official view of Eugene's role and this the personality they assigned him to back up their case.

There can be little doubt that both Keller and Symonds were more or less deranged; no more so than many who lead successful lives in psychopathic, yet legally unaccountable, disregard of those around them, but no less so than others whose luck it is to be certified and locked up as public menaces and so denied the full flowering of their discrepant view of things. Perhaps it was a family characteristic, sex-linked and transmitted by the females who remained unaffected by it themselves, although I have often since wondered about Louise Silver, once Keller, née Symonds, out of a Rose. I thought of her up there and what she might be going through at that moment, shrieking or passed out under sedation. With my limited knowledge of the woman, I found it hard to picture her in either posture. I asked Don if she knew about Owen yet. He looked at his watch and said, "They'll probably wait 'til morning to tell her."

She, for certain, would find their summing-up less than convincing, whatever she might say on the subject when she would be able or inclined to talk about it—and to my knowledge she never did. But neither the complexity of the principals involved, nor the wildness of their imaginings impressed these solid, orderly men. As Mrs. Carmichael had said to me in sibylline despair, they did not like her boy, and they did not care who did it, "and they'll say whatever's needed to bring folks around to their way of thinkin'."

Don must really have believed the theory they were ending up with, that Eugene got cold feet after he finished off Symonds and tried to hide.

"Does his going out to Stallion Oaks sound like someone who's trying to hide?" I said.

"Does his holing up in your house sound like someone who isn't?"

It sounded to me like the actions of one who did not understand what was going on and was uncertain which way to turn. Don accepted that with a shrug that indicated he saw no discrepancy between his interpretation and mine. Either one would fit his needs.

"Where would Eugene get hold of cyanide?" I said.

"From Symonds." He smiled. "There were ampules of it in that place he kept in Highland Park, and he had them at his clinic. They haven't looked yet, but some more will probably turn up at his home. People like him keep stuff like that around for when the going gets rough. Anyway, cyanide's not all that hard to get in one form or another."

"And the gun?" I said. "Would he carry a gun down here with him—Eugene, I mean? And would he leave it in his own trailer if he had just shot somebody with it? Whose prints are on it?"

He admitted there were no prints, but according to him Carmichael would have thought of that and wiped it clean.

They were visibly amused at my head-shaking, and I decided to protest no more. I was free to go, and I was ready to pass out from lack of sleep. Don accompanied me to the street where, before I went out the door, he offered one further comment. It was that old one I had heard so many times before.

"It seems a little strange to me, Mr. Scott, that someone like yourself, who says he doesn't like the kind of thing that's involved here, should hang around with people who don't seem to be interested in much else."

I had no answer for him, at least not one that would satisfy him, so I asked him instead what they were going to do with Glenn's body. He said his secretary, Mrs. Fenton, had claimed it since his sister wanted nothing to do with it.

25

WHEN I returned to Encanto the sky was beginning to lighten. I fell straightaway into bed and a deep sleep from which I woke in the middle of the afternoon with a chill. The sun was in the west, the room in shadow, the blanket on the floor, and in a corner by the outside door Ray sat reading a book. I watched him for a long time before he finally glanced up and smiled.

"How long have you been here?" I said, reaching down and retrieving the bedclothes.

"An hour, maybe. How do you feel?"

"I don't know. Thirsty."

I propped myself up in a sitting position and he went to get me some orange juice out of the refrigerator. I longed for a cigarette and wondered if I would ever again break the habit.

Returning from the kitchen, Ray handed me the glass and tossed the day's mail on the bed, then sat down to watch me drink it like a nurse with a recalcitrant convalescent.

"Linda's going to fix some supper for us over at my house tonight," he said. "You need solid food."

"I need a cigarette."

"No you don't. You mustn't go back to that filthy habit just because of what's happened. Everything's going to be all right now."

"What makes you so sure?" The flood of acidic sweetness tightened the masseter muscles in my jaw and I shivered with the shift in metabolism. "Have you got some good news?"

"Not especially. But I did find out that they know Eugene and Symonds were at Owen Keller's apartment before they drove down here. What do you think that means?"

"I have no idea. Who have you been talking to?" As if I did not know.

"Pinkerton. Who else? He called me when he couldn't reach you at the museum. He didn't want to bother you in case you weren't feeling well."

"Jesus!" I put my hand over my head. "The museum!"

"Well, you certainly didn't expect to go to work today, Scotty, after all you've been through!"

"I should've called in."

"They'll understand. Now why don't you get up and come on over to the house?"

Good friend Ray just wanted to fuss over me, but I had to refuse. "I don't think I could put up with listening to Linda right now. I've got to pull myself together and do a few things."

Ray gave me a look of crybaby disappointment. At the same time he reluctantly shook out cigarettes for us both.

"Linda's very low key today," he said. "She's worried about you too."

"What do you mean, 'too'?"

"Well, we're *all* worried, dummy!—after what you've experienced. And naturally we want to know all about what happened up there, and here, and . . ."

Less than ever did I want to have dinner with anyone. I stared at him as his voice trailed off.

"Well." He stood up and said with a plaintive look on his face, "Call me when you feel better. And come over anyway if you change your mind. We'll be there."

After he had gone, leaving the rest of the pack of cigarettes with me, I lay still for a moment, listening to the far-off, stringent snarl of a lonely bike skidding up the rally tracks of the hill to the north of the house. It reminded me that I had a wrecked Harley I might have to dispose of. I looked down at the pieces of mail Ray had brought in. A fat manila envelope of the kind I kept in my desk in the living room caught my eye. It looked familiar. I picked it up. The handwriting was Eugene's. Tearing it open, I almost laughed as his safety valve spilled out.

I found and read through the last entries quickly, then re-read them carefully. It was all there, set down point by carefully noted point, complete with misspellings and his own inimitable style. It was a very sober, stilted report of how he had picked Symonds up outside the Mount Washington place "in a greatly weakened state," and had been "ordered to assist the Master up a long flite of steps to the waiting vehicle"; how he had been made to stop at the Silver Lake Post Office and mail a big envelope of these papers (and how he had been slapped and verbally abused by the Master for his clumsiness in stuffing them into the envelope); how he noted that the "said envelop was adressed to a Mrs. Silver in Cristal Cove, this state [zip code disremembered]"; of how he had then, always under orders, driven down to "my own loggings in N.C."; how he had secured him and hoisted him, "and endevoured to make him comfortable," and apparently was cursed for his solicitude. Then, having accomplished all this, he left the trailer, according to the journal, taking care to leave the door to it unlocked, again as his Master had ordered him to do, but not before placing "a pill" in the commanding mouth and (surreptitiously, one would expect) picking up his journal on the way out in order to record these untoward events.

There followed a wordy account of what was done to him at

Stallion Oaks, and his complaints and comments on how badly it was done. I got the impression while reading this part that he had lost himself in the pleasant and familiar occupation of describing his adventures. No doubt it served to take his mind off the fears that must have been pressing in on him, for it was my guess that he did all the writing right there in my house while waiting for me to return. At any rate, the narrative broke off suddenly while he was still going over the disappointing session at Stallion Oaks, and it was not resumed.

The important thing was that he never knew Owen Keller was lying dead on the floor of that bottom unit, for Glenn had been waiting in the side patio for Gene to help him up that long flight of steps and into the car to begin their terminal exercises.

I picked up the envelope again and examined it. It was one of mine and it was postmarked that very morning, which meant that he had dropped it in a post office—Lemon Grove—on his way to the border Sunday noon. It had far too many stamps affixed to it and they came, as I later confirmed, from a roll I kept in my desk. He was taking no chances on its not reaching me.

So there it was, if the law would believe it. And if the hunch I had about the packet of letters and papers mailed to Louise Silver was right, then they would have no other recourse but to do so. As to whether or not they could be made to understand the implications and the rationale behind Symonds' unreasonable acts, that was outside the scope of what I cared about, since it would not alter their findings one way or another. They probably settled it in their own minds with something to the effect that Owen Keller overreacted to the shameful advances of a demented relative and paid for his embarrassment and bad judgment with his life. Only, in saying as much they would be begging the question rather than answering it. Glenn's undoubted suicide was an insanely virtuous act carried out in consequence of his own judgment against himself. It was meant as a dreadful cautionary punishment self-inflicted for a fault too heinous to bear, a repellent personal flaw that had come too prominently to light, and for which even his killing of the one who was responsible for his predicament could not atone. In his view of things the shame could only be expunged with his own hand and under circumstances as outrageous and demeaning as he could contrive on such short notice and with the kind of help he had at hand. (And, all things considered, one would have to say that Eugene did pretty well by him.) So, in accomplishing this, Glenn was crystallizing the legend he expected to become at the same time that he was declaring what he was. His vanity had no limits save pride, and his pride knew no limits at all.

I picked up the phone and dialed the police. Don was there and expressed no surprise when I told him what the mailman had brought me. He only grumped about my taking so long to let them know.

"Has anyone questioned Louise Silver today?" I said.

"Why?"

"She may have the key to the whole thing."

"Just sit tight, Mr. Scott. We'll be right over and take a look at what you've got."

"What if she burns it?"

"Burns what?"

"Whatever it is she got in the mail today."

"Take it easy. We'll be there in a few minutes."

While waiting I decided to call work and see what the picture there was. The secretary said Dr. Silver had come in shortly before this and was leaving again in a few minutes. I asked her to put me through. There was a long wait on hold, and when he clicked on his voice was an ambiguous blend of courtesy and coolness. It had only been within the past six months that he had warmed up to me, but hearing him drop back into his old way of speaking felt like a doubling of the distance that had separated us before. I knew immediately what was coming. I started to apologize for not showing up for work.

"That's all right, Scott. You'd better get some rest and stay away for a while until things quiet down."

"That's what I wanted to ask you about, Dr. Silver. I wasn't sure how this thing ought to be treated."

"I don't think anyone does right now," he said. "I'm going up to the Cove in a few minutes. I'll be away for several days, perhaps a week."

"Would you express my sympathy to Louise?"

"Certainly! Certainly! And I think it would be best if you didn't call her. She isn't seeing or talking to anyone for a while."

I asked him if it would be all right for me to come in on the evenings when the museum was closed. "There's some work I'm in the middle of."

"That can wait, I'm sure, Scott. You ought to stay away at least until we can get an idea of how this thing is going to affect the board. Why don't you take a leave of absence, get away somewhere? The secretary will fix it up. Take your time, look around, call me in a few weeks, maybe, after this is settled and we've got a better picture. I'm sure you understand."

I understood I was out of a job. I thanked him and said I appreciated all he had done for me, and that I wished this had not happened.

"Worry about yourself, son," he said. "What it comes down to where we're concerned is a private, family matter, and I think it really has nothing to do with you. It will quiet down very soon, I hope."

He was right, even as were those two National City detectives, although they never said as much. Nothing in this had really anything to do with me. Self-absorbed, I had intruded myself on these happenings, and not until Mount Washington and the sight of Owen Keller lying dead in Symonds' workshop, did I realize how inconsequential and meddling a part I had played seeking to leaven the stale lees of an old association in the rising action that came in the wake of Symonds' death. Now here was my boss telling me I had better look for something else to do.

I put the telephone aside, got into some clothes, and fed Fagin. Don appeared shortly afterwards and took the journal off my hands. I accompanied him back outside to his car with the dog keeping close behind. He started the car, then leaned out the window.

"Say," he said, "you wouldn't like another one, would you?"

"Another what?"

"Another dog. We've got Eugene Carmichael's mutt in the pound."

I said no thanks. "Two's more than I can handle."

He shrugged. "Then I guess they'll destroy him."

"It's a bitch," I said.

He nodded. "Yeah, it's rough all over."

"Won't Eugene's mother claim it?"

"No, she doesn't want anything to do with it. She says Carmichael trained it to snap at her. They tell me it's a smart little thing."

"That figures," I said.

26

A VERY BUSINESSLIKE undertaker called me the next day after I came home from cleaning out my office. He wanted to discuss arrangements for Eugene's funeral. The body had been released to his mother who, from a hospital bed in Chula Vista, stipulated that her son's obsequies should be inexpensive. The reason he was telling me this was that she, unable to see to it herself, had fingered me to carry out her wishes. Ray and Linda and Pinkerton and I were the only mourners at the crematorium the next day. Ray and I split the cost of this ultimate and very modest pomp. No one wanted the ashes, so we haggled with the undertaker, who finally consented to dispose of them without additional charge.

I did not learn the details of Mrs. Carmichael's prostration until the latter part of the week when, while crating some books against the day when I should be leaving, I heard a car coming up the hill and got up from my work on the living room floor to quiet Thief, who was back at the house with me.

Outside, I watched a green Mercury sedan turn into the driveway and lurch to a stop, the fright wig outline of Snuffy Wiggins in the driver's seat, and beside her, a tall shadow: Bobbie-Jo. Her daughter was in the back seat from which Pinkerton was extracting himself. While he walked over to me, Snuffy stuck her head out the window.

"How you doin'?" she yowled.

"Fine, thanks." I waved at Bobbie-Jo, who responded listlessly from behind the windshield.

"You don't look too good," Snuffy said.

"Thanks." I nodded to her and gave my attention to Dale.

"What's this I hear about you leavin' here, chum?"

"That's right," I said. "What're you up to?"

"Oh, we're goin' to pick up Linda and then head out to Chula Vista to visit Mrs. Carmichael. You wanna come too?"

I saw no reason to, and told him so.

"Well, I don't either, to tell you the truth. These women like that kinda thing, though, and I told 'em I'd ask you. Now I've done it."

"What kind of shape is she in? Nobody's told me anything except that she's in the hospital."

"She had some kinda stroke when she heard Gene got smashed up."

"She looked to me as if she were made of stronger stuff than that."

"Shit!" he grinned. "That ol' woman'll outlive you and me both unless she dies of cussedness. She's just havin' a personal go-around with worryin' about what people're gonna say. She's been blabbin' a lot of nonsense, blamin' you for what happened to her boy."

"If she's blaming me for it, then why in hell did she commission me to look after his funeral?"

"Oh, I didn't say she was makin' sense, or anything like that. She'll blame you and use you all at the same time if you let 'er. That's the way she treated Gene, ain't it? An' that's the reason her oldest won't give 'er the time of day. Hell! If she hadn't scared the shit outta Eugene yesterday, he'd still be alive today, an' she knows it."

"What do you mean, 'scared' him?"

"Why, didn't they tell you about that? Eugene called 'er while he was holed up here—it must've been close to the time you was headin' back from seein' her—an' she told him you thought he was the one that done in Symonds and that you was gonna turn 'im in if you found 'im, and that he better clear out."

I could hear that hectoring voice loosening his resolve to wait for me, channeling its way back through a generation of conditioning to annul his resolve never to seek the aid of his family in anything, and most particularly never to cross the border to confront his brother.

"How do you know that?" I asked, knowing the answer before he gave it.

"That's what she told the police. She said you come roarin' around to her place and like to've scared her half to death. She knew Eugene put a lotta stock in you, and here you was, tryin' to turn 'im in. She told 'im you was comin' back to take 'im down to the police, and that he had to get away from here and go down to stay with his brother."

"What did the old fool expect to accomplish by that? I didn't think she even knew where Larry was."

"Oh, she knew all right. She was gettin' money from him as well as Gene. She kept it to herself."

"Does she realize she got her son killed by meddling?"

"She'll come to terms with that, I reckon. As a matter of fact, the police are more pissed off at you for goin' out there to talk to her than anything else you done."

"I was only trying to comfort the old bat."

"I don't doubt that. An' I don't think the police does either. But it's the way these things happen when you're dealin' with people that won't listen to reason."

"But what did she do it for?" We were in the house now. "She knew what happened the last time she sent him down to see his brother."

"Beats me. But he didn't get nowhere near to him this time. As soon as he hung up, she got on the blower an' kept at it 'til she traced Larry Carmichael, put in a call to 'im in Mexicali, an' told 'im his brother was comin'. Then it looks like that son of a bitch alerted the border people on both sides of the fence, but Eugene was already across it by that time an' headin' east. When he got to Mexicali, though, there was a carload of Baja police on the lookout for 'im. From what I hear, he made a run for it clean through that dump an' out the crossin' to Calexico before he piled up."

"Who shot him, the Mexicans?"

"Shit! He wasn't shot none. Them goddam greasers shot *at* 'im, but they couldn't hit the side of a barn if they was inside it with the doors closed. They just scared 'im enough so he lost control of that machine of yours. I guess you've seen it."

I said I had not.

"You got it insured?"

"Only liability."

"Well, I guess you can get somethin' for parts. It's in worse shape than Gene, and he wouldn't be in very good workin' order if he was alive, I can tell you."

"Saw him, did you?"

"Somebody had to identify the dumb bastard. His brother wouldn't come over an' do it, and his maw was already bein' carted to the hospital when they wheeled him in. As it was, she had hysterics when they told her what happened. You'd have thought he was her pride and joy the way she carried on before they shut her up with a shot of dope. Imagine a woman her size carryin' on like that!"

"What's size got to do with it?"

"Oh, I don't know. She looked like she had enough grit in 'er to stand up to somethin' like that, wouldn't you say?"

I had already said it, so I said nothing.

Dale sighed. "Well, I guess the bigger they come the harder the fall. Did they fire you at the museum?"

"They're not calling it that, but that's what it amounts to."

"What're you gonna do now?"

"Whatever I can. But the first thing will be to get out of this place." I looked down at the books and abstracts stacked and lying about on the floor. "Tell me, Dale, did you follow the lead of your buddies on the force in thinking Eugene was guilty?"

"They ain't my buddies. How many times I got to tell you? And guilty of what? He sure as hell was guilty of flight to avoid arrest, and withholdin' evidence, and a whole lot more besides."

"That's not what I mean."

"I know it, chum." He put a hand on my shoulder and offered me a cigarette. After lighting it, he said, "What's this Silver babe like?"

"Louise Silver? I thought she was all right. At least I did until after I saw her the last time. I don't know what to think now. And God knows what she thinks either, now that she's got another dead son on her hands."

"And a dead brother."

"I don't think she cared much about the brother."

"Have you talked to her?"

"No, Dale. And you can tell *them* I haven't. I've got no reason to talk to her. She probably wouldn't want to talk to me if I did, and her husband wouldn't let me near her if she were so inclined. Why do you ask?—as if I didn't know."

"I hear she's in somethin' like the same kinda state Mrs. Carmichael's in."

"In a hospital?"

"Naw. She's still up the coast at that place of hers, but she's under a doctor's care."

My hunch had proved correct. The police were being very close-mouthed about what was in that packet Eugene mailed from Silver Lake, but Don had unbent enough to admit to me that Eugene was now cleared of killing anybody but himself. The coroner had established the time of Owen's death at somewhere around twenty-four hours before Gene arrived at Symonds' place to pick him up. Eugene never knew there was a body in the house.

"Ain't it somethin'?" Pinkerton went on. "There was Gene with the evidence in his hand and what did he do but go and send it to that woman fourth class so's he'd save a few cents! Now you *know* Symonds woulda wanted that thing sent first class."

I agreed. "He probably did it as a way of screwing up so as to get punished for it—conditioned reflex."

"Maybe so, but he'd be alive today if he had sent it like he was supposed to do."

"How do you figure that?"

"Well, for crissake, it would've got to the Silvers the next morning insteada takin' 'til Monday to git there. Then they'd have known what had happened and he wouldn't have gone high-tailin' it over the border thinkin' somebody was after pinnin' it on him."

I conceded that he had a point.

"An' you know somethin' else? She tried to make out to the

police that she never got no package when they come to ask her about it."

"I told them she might well burn it."

"Naw, she just hid it 'til her husband talked some sense into her."

He looked at me with his head cocked to one side, seeing how I took this piece of information. I said nothing, so he pursued it.

"Why d'you suppose she wouldn't want people to know she got a letter from her brother for? Like you said, it might clear everything up if it was a suicide note."

"It wouldn't bring her son back," I said. "And if she did get something from Symonds, then she's probably got a very solid interest in *not* having it cleared up."

"Blood's thicker'n water, is that it?"

It was not exactly it, but I didn't argue the point. I began instead to look for some way of getting rid of him; he was getting on my nerves. As if unconsciously wanting to exacerbate my state of mind, he said, by way of changing the subject, "Boy, you and Faulkner really lifted the lid off it up yonder in L.A.!"

I took him by the arm and steered him to the door.

"There's a question or two I'd like you to answer for me now, Dale."

"Shoot!"

We stopped in the patio, and I shot.

"Why did you put something in my drink Saturday night?"

He looked at me in an openmouthed facsimile of amazement, then his hand settled on my shoulder again.

"I heard that you'd got an idea like that, chum." His voice was hollow with regret and disappointment. "An' I can't for the life of me understand why you think I'd do a thing like that. I got no reason."

I was sorry I had asked.

"Forget it," I said, and brought out my second query, expecting to be served no better than on the first.

"When I talked to you outside the theater Saturday night, you asked me where I'd *really* been the night before. What did you mean?"

He frowned, as if collecting his thoughts.

"Did the *po*-lice ask you that?"

"No, I'm asking you. Should they have?"

"Can't think why they would. To tell you the truth, I don't recall sayin' anything like that. Look-a-here!" He squeezed my shoulder, and the smile he gave me tucked the lines in deeper between his nose and the corners of his mouth, exposing those yellowed horse teeth. "You been through a lot, but it's all over now. You gotta put this behind you, chum. You gotta pick up the pieces and carry on. You got your whole life ahead of you."

I put him behind me right then by conducting him out the door, still spouting, and back to the girls waiting for him in the Merc. I said a word of farewell to Bobbie-Jo and Snuffy, but the latter decided she wanted a turn at questioning me too.

"What d'you plan on doin' now?" she burped.

"Leaving."

They all looked at each other as if I had told them I intended to paddle down to Patagonia in a canoe.

"You oughta think twice about it." she said. "It's your life, naturally, but you won't find no better place in the world to get started again than you will right here where you're at. My advice to you is to stick it out."

"Stick it where?" I said.

Dale doubled up laughing and Bobbie-Jo permitted herself a smirk, but Snuffy's eyes widened like an indignant parrot's.

"There's a child in this car!" she said. And, turning to Dale, she told him to shut up and get in, as if he were another. She started the engine, still glaring at me while she wrestled with the brake.

"Good luck's all I can wish you.'

And with this apopemptic grump, she exhausted her slim supply of interest. I waved them off down the hill, and that was the last I saw of any of them.

I went back into the house and to the packing of books. Ray came over to keep me company and help me sort things out, bringing me something to eat, as he had taken to doing, as if I were an invalid or an indigent. We talked dogs and parting.

Two dogs were too much, espeially since they would gladly have torn each other to pieces in their rivalry. I was not about to give up

Thief, but I felt responsible for Fagin too. By rights he was the burden of Symonds' estate, but I doubted that anyone was going to come forward and claim him, any more than they offered to claim his master's body.

Jessica Fenton, in the first bloom of sentiment, asked me if she could have him when I called her and, in passing, mentioned the problem. What she wanted was for me to bring him up so that we could talk. She needed someone who knew what had been going on to listen to her version of what was then taking place.

The following Saturday in Santa Monica, after I had briefed her on the handling and feeding of the animal, we had dinner together in her apartment and she unburdened herself of the final tangled ends of Symonds' leaving her (this was how she put it).

She had first arranged a private funeral, but the swarms of Glenn's admirers and friends would not have it that way, and so the bills were going to have to be paid by the estate. The estate soon contested it, of course. And that was not all. She had no more than returned home with the urn under her arm, she explained in whiskey-voiced indignation, when Louise Silver's attorneys served papers on her for the return of some unassigned stock they claimed she had filched from the safe at the clinic on learning of his death. With unexpected candor, she told them that those securities were hers as a gift from Glenn, a compensation set aside against just such an eventuality as had transpired—*if* they got her drift.

With the smile of a crocodile, she assured me they had. And lest they have any second thought on the matter, she instructed them to return to Crystal Cove and let "those people down there" know that any attempt to get the stocks away from her would result in some very ugly revelations—all of them true and provable. The struggle went on quietly for several months without attracting any attention until the claim was finally dropped by the Silvers.

Glenn, so much the master of order and good closure, had neglected ever to make a will, so that Louise Silver got everything: the house and the other properties, the clinic, all the securities and other paper except the shares Jessica laid hold of, the Rolls, the Harley, and the money salted away in Geneva (to settle which she and Morris left the country shortly afterwards for an extended vacation). I never saw Jessica again, but I found out from her a few months after I turned Fagin over to her that she had tolerated him less than a fortnight, then, unable to contact me, had had him done away with at the county pound.

Before Ray went back home on that Monday night, we decided between us that I would put the house on the market and he would have the power of attorney to sell it for me while I went north to look for a job. The furniture I gave to him and Linda. In less than two weeks after Symonds' death, I was in Berkeley, and then in San Francisco, where I lived until the following summer when a teaching job came through up on the north coast at a state university campus located, ironically, in the county where Eugene grew up. Ray's sale of the house staked me through those months before I worked again.

27

AFTER RAY left and the night had settled in, and there was no one else come to bother me, I picked up the phone and dialed San Francisco. I have said that it was only the players, and not the games, that I cared for, and a familiar tug was pulling at me as I talked to Nelson over the wires. but at the same time that I gave with the pull of his voice, I resisted with all my soul another commitment such as I had only now been through. When I went up to Berkeley, I went to see him in San Francisco, and during the months I stayed there, waiting to take the job further north, we saw a lot of each other. But in that mutual schooling, we worked ourselves by careful stages into a different, gentler route along the paths of friendship. We did this rather than follow what would have been at first an easier way, but a way that leads unalterably in such mis-matchings into a state of siege, into that civil war that was the issue of my love for Symonds and his, I think, for me.

I had been self-righteous with the pride of bearing no man's yoke (although, of my own doing, I bore the indelible mark of one man's rose and whip), and Faulkner had about him the scorn of one who disdained to ever find his own mark, if he had one, personally borne, or given, to another. At first together, and then gradually in our separate ways, we were able to refine these sublime pretensions into humbler strengths without, however, falling into the other ambush that awaits the soft of heart and infirm of purpose. For the heart commands great reverence and service from those it would use because it leads them to believe that it offers the greatest beauty and adventure this life can bestow. And in the lives of most there is little that is rare, or remarkable, or poetic, or profound, or truly terrible with that reflective terror which gives substance and bite to any experience. And what little of these there is passes like an angel in the night, unapprehended and unmissed. We, Nelson and I, all unyielding, felt the wind of those wings close by, but we missed the soaring flight no doubt prepared for us. And we missed it by choice. Once is enough.

After I moved to the north coast, I came down to the city rarely, and rarely saw Nelson when I did. We moved, as we always had, in different circles. But because of the long intervals between our chance meetings, he seemed to change in sudden turnings back towards what he might have been when younger, even as he seemed to have aged perceptibly each time I saw him again. That trim body wore good suits as beautifully as it once donned leather in a manner unmistakably his own, and the unstinting probe of those startling eyes remained the same, carrying always something of the hunter in them. We never talked of the events I have been describing except once. It was six or seven years afterwards, and I forget how the subject came up, but, as if in response to a subliminal clue discovered in his manner, I asked Nelson if he too had not gone down to that trailer in National City on the night that Symonds got it. And, "Sure," he said, a hint of mockery faintly turning up the corners of his mouth. "You always did suspect it was me you saw holding him, didn't you?"

"Why are you telling me this now?" I said.

"Because you asked."

"What if I had asked you in L.A. that night at Owen Keller's place?'

"Who knows?" He shrugged. "What if the L.A. police had asked me about you being down there?"

"Why would they have done that?"

"I might have told them, since you told me. Or Dale Pinkerton might have spilled it to his San Diego protectors that he saw both of us go in and out of there."

"He saw you too?"

He nodded. "He got there about the time you came along on your bike, and waited until you were gone. Then he opened the door to come in just as I was reaching for it to go out."

"I wonder why he never told the law that?"

"Why should he? It would've involved him in it too. Besides, after he got a look at Symonds he knew neither you nor I had done it. He was already cold and getting stiff."

"Weren't you worried I might say something under the pressure of those two cops?'

"It was a gamble, but we figured you to be pretty close-mouthed."

"So the two of you had it all figured out."

"No, friend. We had less to go on than you did. And the thing that had us worried was that you wouldn't keep still, running all over the place trying to find Eugene. It was Dale's idea to put you out of commission for a while. He was sure the police would uncover Eugene in a matter of hours."

After that, I cannot remember anything of importance that we found to discuss in what were no more than extended exchanges of commonplaces whenever we ran into each other. The last time I encountered him was on a downtown street at noontime a year or so after this. I asked him then if he ever made the scene anymore, and he almost smiled, shaking his head in reply.

"It's not the same, you know." There was neither sadness nor scorn in his voice, still scarred, as if from a rasp. "Leather is in, and everybody's into it. It's not what it used to be."

"It probably never was," I offered.

"No," he said, "you know better than that."

His eyes, which through glasses lent the severity of his face a scholarly cast, still held that old, secret flame of rage burning like points of light deep dancing, beckoning with equivocal glints past the pale portals of the iris.

"There's no concentration or quietness in it anymore," he said. "It's all a parody."

And, looking around, I saw that it was so. The passion was gone, and the darkness with it. The guilt, the fear, and the uncertainty are done away with and there is nothing to shiver at any more. But people continue to get their skin peeled, and atrocious killings still take place. Yet this in such times is parody. For what was once a game performed in counterpoint to reality in order to heighten that reality has usurped reality. And everthing is common now to all—or at least open and available to men of common stripe. In these common times boredom has succeeded passion, and there remains now only order, like love, to be reinvented as successors to the deadly, humorless play of disorder that ornaments our lives and crowns our works.

It was a dilemma Symonds recognized and solved in his own way. For Symonds was never one to suffer gladly the common man or the common man's view of things, save as the common man might come to him for instruction. When the choice was forced upon him whether he should become a parody or remain forever an enigma in the minds of those who knew him whom he cared about, he chose what his special wisdom told him he must choose, knowing that he could not have both, and that he was not made for a common end, and that anyway parodied enigma is enigma no more.

And Faulkner was right: I knew better.

Mad River, 1966
San Francisco, 1981